saur

INTERNATIONAL COUNCIL ON ARCHIVES

Guides to the Sources for the History of the Nations

3rd Series: North Africa, Asia and Oceania

Volume 4: Sources of the History of Asia and Oceania in the Netherlands

Part II: Sources 1796–1949

CONSEIL INTERNATIONAL DES ARCHIVES

Guides des Sources de l'Histoire des Nations

3ème Série: Afrique du Nord, Asie et Océanie

Volume 4: Sources de l'Histoire d'Asie et d'Océanie conservées aux Pays-Bas

Part II: Les Sources 1796–1949

Netherlands State Archives Service
Royal Institute of
Linguistics and Anthropology

# Sources of the History of Asia and Oceania in the Netherlands

## Part II: Sources 1796–1949

Compiled by Frits G.P. Jaquet

K·G·Saur
München · New York · London · Paris 1983

Translated from Dutch into English by J. W. Veenendaal-Barth

Edited under the auspices of UNESCO

CIP-Kurztitelaufnahme der Deutschen Bibliothek

**Sources of the history of Asia and Oceania in the Netherlands** / Netherlands State Archives Service ; Royal Inst. of Linguistics and Anthropology. – München ; New York ; London ; Paris : Saur
(Guides to the sources for the history of the nations : Ser. 3, North Africa, Asia and Oceania ; Vol. 4)

NE: Netherlands State Archives Service; Guides to the sources for the history of the nations / 03

Pt. 2. Sources 1796 – 1949 / comp. by Frits G. P. Jaquet. [Transl. from Dutch into Engl. by J. W. Veenendaal-Barth]. – 1983.
ISBN 3-598-21477-4

NE: Jaquet, Frits G. P. [Bearb.]



Phototypesetting by Verlag Politisches Archiv GmbH, Landshut
Printed by Weihert-Druck, Darmstadt
Bound by Verlagsbuchbinderei Kränkl, Heppenheim/Bergstraße

ISBN 3-598-21477-4

# CONTENTS

## CHAPTER IV: STATE ARCHIVES



## CHAPTER V: MUNICIPAL ARCHIVES



## CHAPTER VI: UNIVERSITIES

## CHAPTER VII: OTHER SCIENTIFIC INSTITUTIONS

# INTRODUCTION

Although this guide covers about 28 running kilometres of paper containing documents on Asia and Oceania, distributed over *c.* 750 archives and collections and located in 64 different repositories, it cannot claim to be complete, for the following reasons:

A number of institutions are not accessible to the public, on account of trade secrets or due to the fact that the material kept contains many strictly personal data.

After completion of the research in the various institutions acquisition of course continued as usual. Hence it was impossible to mention the most recent acquisitions. An example of such an institution that is most exasperating in this respect is the ARA in The Hague.

Of a few of the institutions described, certain archives were not allowed to be mentioned because of their secret character.

Some institutions and/or archives may simply have been overlooked, for it is impossible to trace everything connected with 150 years of Asian and Oceanic history.

The present guide is to be regarded as the continuation of *Sources of the history of Asia and Oceania in the Netherlands, part I: sources up to 1796* (München, New York, London, Paris : Saur, 1982) which refers to relevant sources for the VOC period. Naturally there will be some overlap, as archives are not as a rule very amendable to divisions of periods and subjects introduced by compilers of guides.

The geography used of this guide is the UNESCO one. So Asia begins in Persia and comprises all the countries further east from there. Oceania is the Pacific area including Australia and New Zealand.

The text has been translated into English with the exclusion of titles of articles, monographs and periodicals. Specific terms have likewise been left in their original languages, followed where possible by an English translation between brackets. For further comments on the translation see the introduction to the index.

The emphasis of this guide lies with the archive material. Collections of oriental manuscripts, maps, drawings, photographs etc., are of secondary importance. Of institutions which do not as a rule keep archive material, such as, for example, museums, only the addresses have been included. For this see chapter IX.

The project was begun in 1968 with the despatch of circulars to *c.* 200 institutions, selected from all kinds of address books. Only those institutions which gave a positive reply to the question as to whether they kept archives relevant for this guide were examined more closely, in most cases by means of a personal visit to the institution concerned. The results of this examination were published in 8 preliminary surveys up to 1977. This guide is a cumulation of these preliminary surveys, supplemented with the descriptions made since.

This procedure of publishing preliminary surveys was chosen in order to provide information on research possibilities in the Netherlands in advance. I also had some hope that possible shortcomings would be brought to my notice. This turned out to be an illusion. Only Mr. A. van Marle went to the trouble to point out errors, for which I wish to express my sincere thanks to him.

The method of description evolved in these preliminary surveys does appear to have met with appreciation, however, witness the similar arrangement and formulation of the historical information in the descriptions of various archives and collections which have been used in the *Guide to the sources of the history of Africa South of the Sahara in the Netherlands* published in this same series by M.P.H. Roessingh and W. Visser in 1978.

In the arrangement of the institutions described in this guide I have followed the classifications used in that of Roessingh and Visser as much as possible, though deviations were unavoidable, as will be apparent from the table of contents.

For details concerning the compilation of the "list of bibliographical works", the "list of institutions which may be important for research in the Netherlands", and the "index", the reader is referred to the introductions prefacing the parts concerned.

Regarding the descriptions of the institutions the following should be observed. The libraries, publications and documentation of institutions have been mentioned only where these are relevant for this guide and/or where they have some connection with the archive material described. The archives and collections themselves (for purposes of this guide the word "collection" has been used to refer to papers originating from one or more persons) are introduced by a summary historical sketch giving the dates of the principal activities of the institution or person forming the archives. This introduction is meant to provide information on the papers which are likely to be found in the archive or collection concerned. For by no means all inventories provide information on the subjects which may be found in the relevant archives, not to mention the archives and collections which lack inventories. Notably formal descriptions of series, such as "minutes" and "correspondence", give little information other than on the period in which these papers were drawn up and received. With some historical knowledge and the aid of indexes, postbooks etc. drawn up by the administrative departments themselves, as well as other entries (inventories etc.) drawn up later, the researcher will be able to look into the material relevant for him rather more systematically.

The data for the above-mentioned historical introductions have been drawn from the archives themselves, the inventories to these archives, where present, the *Encyclopaedie van Nederlandsch-Indië* or other encyclopaedias, the *Nieuw Nederlandsch biografisch woordenboek* and *Wie is dat?* Where other sources have been used, this has been mentioned in brackets after the introduction.

The description of the archives and collections themselves is based exclusively on the inventories. The actual archive documents were only consulted if the inventory was confusing, or if the archives lacked an inventory. In the latter case samples were taken to see if the archives concerned contained relevant material. In the description of the archives themselves, in principle only the composition and arrangement of the archives were taken into consideration. Names and subjects have only been mentioned if space permitted. Where the enumeration of names and subjects would have taken up too much space, the integration of these data into the index had to suffice. This has been consistently indicated with the sentence: "For the names and subjects mentioned in the inventory see the index".

The description of each archive or collection is concluded with information on its accessibility, size, expressed in running metres (and number of inventory items), conditions for consultations where applicable, and any publications based on the archives described.

This preface should be concluded with my acknowledgments to all the institutions described here; without their co-operation this guide could not have been completed. Not only was I spontaneously given all the help I needed in all these institutions, but those I turned to for help also went to much trouble to check my draft descriptions of their respective institutions.

In conclusion I wish to thank the Koninklijk Instituut voor Taal-, Land- en Volkenkunde (Royal Institute of Linguistics and Anthropology) at Leyden for the opportunity given me to compile this guide. I am particularly grateful to Mr. R. Nieuwenhuys, who lost in me a documentalist for his department, but hopefully promoted a wider interest in the process.

# LIST OF BIBLIOGRAPHICAL WORKS

(Research finished in 1979)

This list has been compiled as an aid for the researcher when he has chosen his subject or is still looking for one, and before he proceeds to consult the fundamental sources.
The library of the Koninklijk Instituut voor Taal-, Land- en Volkenkunde (Royal Institute of Linguistics and Anthropology) has served as a base for this list, with the use of the following method:

Only those titles were included which mention a reasonable number of works on Indonesia, 1796–1949. Indonesia was taken as a guidance because the great majority of the nineteenth and twentieth century "Asian archive material" in the Netherlands are mainly pertaining to Indonesia.

Titles which were found bibliographically, but are not present in the library of the KITLV, were not included, with the exception of no. 3–19, 27, 31 and 35. These titles the researcher will find in the university library of Leyden and/or the Koninklijke Bibliotheek (Royal Library) at The Hague. Other libraries were not consulted for the compilation of this list.

Catalogues and lists of acquisitions of separate libraries were left out, just as publishers' catalogues, and bibliographies and readings lists included in textbooks, monographs, periodicals etc., with the exception of no. 14, 21, 23, 31, 34, 35, 78, 93, 111, 153.

Also bibliographical works with a limited range of circulation (e.g. examination papers of schools for librarians and works in typescript) were not included.

The list is arranged regionally, beginning with the heading "general" which means the non-regional subject bibliographies. Within the headings the titles are arranged chronologically to first edition.

The list consists of the following headings:



**General**

1. Engelmann, W. Bibliotheca geographica; Verzeichniss der seit der Mitte des vorigen Jahrhunderts bis zu Ende des Jahres 1856 in Deutschland erschienen Werke über Geographie und Reisen; mit Einschluss der Landkarten,

Pläne und Ansichte; mit einem ausführlichen Sach-Register. Leipzich, 1858.

2. Friederici, K. Bibliotheca oriëntalis; or a complete list of books, papers, serials and essays published, 1876–1883 in England and the colonies, Germany and France on the history, languages, religions, antiquities, literature and geography of the East. Leipzig etc., 1876–1884.
   Reprint 1967.
3. Orientalische Bibliographie; bd. 1 (1887)–25 (1911, 1926). Berlin, 1877–1922, 1928.
4. Bibliographie géographique faisant suite à la bibliographie géographique annuelle des Annales de Géographie; vol. 1 (1891)– . Paris, 1894– (Union Géographique Internationale).
   From 1931 published under the title: "Bibliographie géographique internationale".
5. Bibliotheca Geographica; Jahresbibliographie der geographischen Literatur, 1891/92–1911/12. Berlin, 1895–1917. (Gesellschaft für Erdkunde).
6. International bibliography of historical sciences; vol. 1 (1926)– . Washington, 1930– . (International Committee of Historical Sciences).
7. Goodland, R. A bibliography of sex, rites and customs; an annotated record of books, articles and illustrations in all languages. London, 1931.
8. Dutcher, G.M. A guide to historical literature, New York, 1931.
   2nd ed. 1936.
   3rd ed. published anonymously 1961.
9. Bibliographie cartographique internationale, 1936– . Paris, 1938– . (Publiée sous les auspices du Comité Nationale Français de Géographie et de l'Union Géographique Internationale).
10. Foreign affairs bibliography; a selected and annotated list of books on international relations, 1932– . New York, 1945– . Vol. 1– . (Council of Foreign Relations).
    Vol. 1. 1932–1942. 1945.
    Vol. 2. 1942–1952. 1955.
    Vol. 3. 1952–1962. 1964.
11. Conover, H.F. Non self-governing areas; with special emphasis on mandates and trusteeships; a selected list of references. Washington, 1947. 2 vol. (Library of Congress; general reference and bibliography division).
12. Bibliographie linguistique des années 1939– . – Linguistic bibliography for the years 1939– ; vol. 1– . Utrecht etc., 1949– .
13. International political science abstracts. – Documentation politique internationale; vol. 1– . Paris, 1951– . (UNESCO and International Political Science Association).
14. International bibliography of sociology. – Bibliographie internationale de sociologie; vol. 1 (1951)– . Paris, 1952– . (UNESCO publication).
    Vol. 1 (1951)–5(1955) in *Current sociology.*
    From vol. 10 (1960) continued in the series: "International bibliography of

the social sciences. – Bibliographie internationale des sciences sociales, 1960– ." Paris, 1963– . (ICSSD.–CIDSS).

15. Sociological abstracts; vol. 1– . New York, 1953–
16. International bibliography of the history of religions. – Bibliographie internationale de l'histoire des religions; for the year 1952– ; compiled by H. Boas; under supervision of C.J. Bleeker. Leiden, 1954– . (International Association for the Study of the History of Religions).
17. International bibliography of political science. – Bibliographie internationale de science politique; vol. 1(1952)– . Paris, 1954– . (UNESCO publication).
    From vol. 9(1960) continued in the series: "International bibliography of the social sciences. – Bibliographie internationale des sciences sociales; 1960– ." Paris, 1963– . (ICSSD.–CIDSS).
18. International bibliography of economics. – Bibliographie internationale de science économique; vol. 1 (1952)– . Paris, 1955– . (ICSSD.–CIDSS).
    From vol. 9 (1960) continued in the series: "International bibliography of the social sciences. – Bibliographie internationale des sciences sociales; 1960– ." Paris, 1963– . (ICSSD.–CIDSS).
19. Historical abstracts, 1775–1945; a quarterly of abstracts of historical articles, appearing currently in periodicals the world over; editor E.H.Boehm. New York, 1955– .
    Subsequent volumes were published without the division into the periods 1775–1945.
20. International bibliography of social and cultural anthropology. – Bibliographie internationale d'anthropologie sociale et culturelle; vol. 1(1955)– . Paris, 1958– . (UNESCO publication).
    From vol. 6(1960) continued in the series: "International bibliography of the social sciences. – Bibliographie internationale des sciences sociales; 1960– ." Paris, 1963– . (ICSSD.–CIDSS).
21. Pearson, J.D. Oriental and Asian bibliography; an introduction with some reference to Africa. London, 1966.
    Supplement in *Progress in library science 1967;* ed. by R.L. Collinson. London, 1967.
22. Nedo-abstracts; vol. 1– . Amsterdam, 1974– .

Currently appearing bibliographical works no. 4, 6, 9, 10, 12, 13, 14, 15, 16, 17, 18, 19, 20, 22.

**Asia**

23. Bibliography of Asian studies, 1957– . New York, 1958– .
    In *The journal of Asian studies;* vol. 5 of each year.
    Continuation of: Far Eastern bibliography.
    See no. 34.
24. Documentation on Asia; editors Girja Kumar and V. Machwe; vol. 1 (1960)– . New Delhi, 1963– . (Issued under the auspices of the Indian Council of World Affairs).

25. Stucki, C.W. American doctoral dissertations on Asia, 1933–1962; including appendix of master theses at Cornell University. Ithaca, 1963. (Cornell University Southeast Asia program; data paper no. 50).
26. Asien und Ozeanien; behandelnde Zeitschriften und ihre Bestände in Bibliotheken der Bundesrepublik Deutschland; zusammengestellt und als Manuskript gedruckt vom Institut für Asienkunde. Hamburg, 1970.
27. Cumulative bibliography of Asian studies, 1941– . Boston, 1970– .
28. Nunn, G.R. Asia; a selected and annotated guide to reference works. Cambridge, 1971.
29. Bishop, E. Australian theses on Asia; a unionlist of higher degree theses accepted by Australian universities to 31 December 1970; occasional paper 12. Canberra, 1972.
30. Shulman, F.J. Doctoral dissertations on Asia; an annotated bibliographical journal of current international researches. Ann Arbor, 1975. 2 vol.

Currently appearing bibliographical works no. 23, 24, 27.

**Southeast Asia/Far East**

31. Bouterwerk, K. Hinterindien und Indonesien, 1913–1925. Gotha, 1927.
    In *Geographisches Jahrbuch;* vol. 42(1927), p. 22–86.
    See also no. 35.
32. Annual bibliography of Indian archaeology for the year 1926– . Leiden, 1928– . (Instituut Kern).
33. Streit, P.R., and P.J. Dindinger. Missionsliteratur Indiens und Indonesiens, 1800–1909. Aix-la-Chapelle, 1934. (Bibliotheca Missionum; vol. 9).
    2nd ed. Vienna, 1966.
34. Bulletin of Far Eastern bibliography; vol. 1(1936)–5(1940). Washington, 1936–1940.
    Continued under the title: "Far Eastern bibliography, 1941–1956". New York, 1941–1957. In *Far Eastern quarterly.*
    Continued under the title: "Bibliography of Asian studies, 1957– ."
    See no. 23.
35. Helbig, K. Hinter- und Insel-Indien. Gotha, 1942.
    In *Geographisches Jahrbuch;* vol. 57(1942), I, p. 138–343, II, p. 547–769.
    II is the Indonesian part of the bibliography.
    See also no. 31.
36. Pelzer, K.J. Selected bibliography on the geography of Southeast Asia. New Haven, 1949–1956. (Behavior science bibliographies; Yale University; Southeast Asia studies; by arrangements with Human Relations Area Files).
    Vol. 1. Southeast Asia; general. 1949.
    Vol. 2. Philippines. 1950.
    Vol. 3. Malaya. 1956.

37. Embree, J.F. Selected bibliography on Southeast Asia. New York, 1950. 2nd ed. revised by B. Lasker. 1952.
3rd ed. revised under the title: "Southeast Asia: a select bibliography". 1955.
4th ed. revised under the title: "Books on Southeast Asia: a select bibliography". 1956.
5th ed. with: "Supplement covering the period June 1956 to May 1957.". 1957.
6th ed. with: [2] supplements covering the period June 1956 to March 1959. 1959.
7th ed. with: [3] supplements covering the period June 1956 to April 1960. 1960.

38. Hobbs, C. [C.] Southeast Asia; an annotated bibliography of selected reference sources. Washington, 1952. (Library of Congress; Orientalia division).
2nd revised ed. published under the title: "Southeast Asia; an annotated bibliography of selected reference sources in Western languages". 1964.
3rd ed. revised and enlarged. 1968.

39. Sharp, J.C. In Japanese hands; a list of books dealing with prisoner of war and internment camps in the Far East, 1941–1945; with a supplementary list of other works written in prison camps. Birmingham, 1952, 1953.

40. Irikura, J.K. Southeast Asia; selected annotated bibliography of Japanese publications. New Haven, 1956. (Behavior science bibliographies; Yale University; Southeast Asia studies; in association with Human Relations Area Files).

41. Hay, S., and M.H. Case. Southeast Asian history; a bibliographic guide. New York, 1962.
Continued as: "Morrison, G. A guide to books on Southeast Asian history (1961–1966); ed. by S. Hay". Santa Barbara, 1969.

42. Union catalogue of documentary materials on Southeast Asia. Tokio, 1964. 5 vol.
1. General and Southeast Asia in general.
2. India (I); the social sciences.
3. India (II); the humanities and natural sciences.
4. Other countries in Asia.
5. Index.

43. Ichikawa, Kenjiro. Southeast Asia viewed from Japan; a bibliography of Japanese works on Southeast Asian societies, 1940–1963. Ithaca, 1965. (Cornell University; Southeast Asia program; data paper no. 56).

44. Anderson, G.H. Christianity in Southeast Asia; a bibliographical guide; an annotated bibliography of selected references in Western languages. New York etc., 1966.

45. Nunn, G.R. South and Southeast Asia; a bibliography of bibliographies. Honolulu etc., 1966. (Occasional paper of East-West Center Library no. 4).

46. Premiers éléments bibliographiques relatifs aux problèmes actuels du Sud-Est de l'Asie; [par . . . Bonew]. Bruxelles, 1966. (Centre d'Étude du Sud-Est Asiatique).
47. Survey, A, of bibliographies in Western languages concerning East and Southeast Asian studies. Tokio, 1966, 1969. 2 vol. (Centre for East Asian Cultural Studies; bibliography no. 4, 5).
48. Berton, P., and A.Z. Rubinstein. Soviet works on Southeast Asia; a bibliography of non-periodical literature, 1946–1965; with a contribution by A. Allot. Los Angeles, 1967.
49. Sternstein, L., and C. Springer. An annotated bibliography of material concerning Southeast Asia; from Petermanns Geographische Mitteilungen, 1855–1966. Bangkok, 1967.
50. Index to Chinese periodical literature on Southeast Asia, 1905–1966. Singapore, 1968.
51. The, L., and P.W. van der Veur. Treasures and trivia; doctoral dissertations on Southeast Asia accepted by universities in the United States. Athens, 1968. (Papers in international studies; Southeast Asia series; no. 1).
52. Tregonning, K.G. Southeast Asia; a critical bibliography. Tucson, 1969.
53. Johnson, D.C. A guide to reference materials on Southeast Asia; based on the collections in the Yale and Cornell university libraries. New Haven etc., 1970.
54. Nevadomsky, J.J., and A. Li. The Chinese in Southeast Asia; a select and annotated bibliography of publications in Western languages, 1960–1970. Berkeley, 1970. (Occasional paper; no. 6).
55. Sardesai, D.R., and Bhanu D. Sardesai. Theses and dissertations on Southeast Asia. Zug, 1970. (Bibliotheca Asiatica; no. 6).
56. Southeast Asia; a bibliography for undergraduate libraries; by D.C. Johnson, E. Ehrman, K. Hale a.o. New York, 1970. (Occasional publication; no. 13).
57. Pedersen, P. Youth in Southeast Asia; a bibliography; modified and expanded by J.P. Tamney a.o. Singapore, 1971. (Institute of Southeast Asian Studies; occasional paper; no. 6).
58. Breese, G. Urban Southeast Asia; a selected bibliography of accessible research reports and related materials on urbanism and urbanization in Hong Kong, Indonesia, Malaysia, the Philippines, Singapore, Thailand and Vietnam; contributions by A.E. Goodman, E.L. Guttierrez, T.G. McGee a.o. New York, 1973.
59. Jenner, Ph. N. Southeast Asian literature in translation; a preliminary bibliography. Honolulu, 1973. (Asian studies at Hawaii; no. 9).
60. Bixler, P. Southeast Asia; bibliographic directions in a complex area. Middletown, 1974. (Choice bibliographical series; no. 2).
61. Johnson, D.C. Index to Southeast Asian journals, 1960–1974; a guide to articles, book reviews and composite works. Boston, 1977.
62. Nunn, G.R. Southeast Asian periodicals; an international union list; with contributions from D. Wyatt, Ch. Bryat, Do Van Anh a.o. London, 1977.

63. Partaningrat, Winarti. Masterlist of Southeast Asian microforms. Singapore, 1978.
64. Willer, Th.F. Southeast Asian references in the British parliamentary papers, 1801–1972/73; an index. Athens, 1978.
65. Moon, B.E. Periodicals for South-East Asian studies; a union catalogue of holdings in British and selected European libraries. London, 1979.

Currently appearing bibliographical work no. 32. See also no. 34.

**Pacific/Oceania**

See also no. 26

66. Lewin, E. The Pacific region; a bibliography of the Pacific and East Indian islands, exclusive of Japan. London, 1944. (Royal Empire Society bibliographies; no. 11).
67. Taylor, C.R.H. A Pacific bibliography; printed matter relating to the native peoples of Polynesia, Melanesia and Micronesia. Wellington, 1951. (Memoirs of the Polynesian Society; no. 24).
    2nd ed. Oxford, 1965.
68. Capell, A. A linguistic survey of the South-Western Pacific. Noumea, 1954. (South Pacific Commission; technical paper; no. 70).
    2nd revised ed. 1962. (Technical paper; no. 136).
69. Leeson, I. A bibliography of bibliographies of the South Pacific. London etc., 1954.
70. Streit, P.R., and P.J. Dindinger. Missionsliteratur von Australien und Ozeanien, 1525–1950. Freiburg, 1955. (Bibliotheca Missionum; bd. 21).
71. Kliencberger, H.R. Bibliography of Oceanic linguistics. London, 1957. (London Oriental bibliographies; vol. 1).
72. Hollyman, K.J. A checklist of Oceanic languages (Melanesia, Micronesia, New Guinea, Polynesia). Auckland, 1960. (Linguistic Society of New Zealand; Te Reo monographs).
73. Australian national bibliography, 1961– . Canberra, 1961– .
74. Cammack, F.M., and Shiro Saito. Pacific Island bibliography. New York, 1962.
75. Dickson, D., and C. Dossor. World catalogue of theses on the Pacific islands. Canberra, 1970.
76. Zantkuijl, M. Tentative bibliography of cargo cults and other manifestations of cargo ideology in the South Pacific. Utrecht, 1976. (ICAU mededelingen; no. 9).
77. McLean, M. An annotated bibliography of Oceanic music and dance. Wellington, 1977.

Currently appearing bibliographical work no. 73.

**Malaysia/Indonesia**

78. Sherborn, C. Davies. A bibliography of Malaya, 1888–1894. Singapore, 1890–1894.
    In *Journal Straits Branch RAS;* vol. 22, 24, 26, 27, 29.
79. Cheeseman, H.R. Bibliography of Malaya: being a classified list of books, wholly or partly in English, relating to the Federation of Malaya and Singapore. London, 1959.
80. Teeuw, A., and H.W. Emanuels. A critical survey of studies on Malay and bahasa Indonesia. The Hague, 1961. (Bibliographical series KITLV; no. 5).
81. Bibliografi Negara Malaysia. – Malaysian national bibliography, 1967– . Kuala Lumpur, 1969– .
82. Deakin, Chr., R. Jones and N. Phillips. Indonesian reading list; a provisional reading list for courses in Indonesian and Malay studies. London, 1974.
    2nd ed. 1975.
    3rd ed.; revised and enlarged by E.U. Kratz. 1978.

Currently appearing bibliographical work no. 81.

**Indonesia**

See also no. 31, 32, 33, 35, 58, 66, 78–82

83. Chijs, J.A. van der. Proeve eener Nederlandsch Indische bibliographie (1659–1870). Batavia, 1875. (VBGKW; vol. 37).
    Amended and enlarged revised ed. for the years 1659–1720; with supplement and amendments for the years 1721–1870. (VBGKW; vol. 39).
    2nd supplement. 1903. (VBGKW; vol. 55).
    See also no. 95, 96.
84. Proeve van een overzicht der in Nederlandsch Indië gepubliceerde geneeskundige boeken en der verhandelingen over geneeskundige onderwerpen, voorkomende in de tijdschriften in Nederlandsch Indië; uitgegeven tot 1 januari 1876. Batavia, 1877.
85. Repertorium op de literatuur betreffende de Nederlandsche koloniën; voor zoover zij verspreid is in tijdschriften en mengelwerken (1595–1932). 's-Gravenhage, 1877–1935. 11 vol.
    Vol. 1, 2. 1595–1865; by C.J. Hooykaas. 1877, 1880. 2 vol. Forthcoming.
    Alphabetic register to names of persons; compiled by D. [B.M.] Buur.
    Vol. 3–5. 1866–1893, 1894–1900, 1901–1905; by A. Hartmann. 1895, 1901, 1906. 3 vol.
    Vol. 3. With: Alfabetisch persoonsnamenregister; compiled by D. [B.M.] Buur. Leiden, 1974.

Vol. 6–9. 1906–1910, 1911–1915, 1916–1920, 1921–1925; by W.J.P.J. Schalker and W.C. Muller. 1912, 1917, 1923, 1928. 4 vol.
Vol. 10, 11. 1926–1930, 1931–1932; by D. Sepp. 1935, 1934. 2 vol.
See also no. 133.

86. Kan, C.M. Proeve eener geographische bibliographie van Nederlandsch Oost-Indië voor de jaren 1865–1880. Utrecht, 1881.

87. Tiele, P.A. Nederlandsche bibliographie van land- en volkenkunde. Amsterdam, 1884. (Frederik Muller-Fonds; bijdragen tot eene Nederlandsche bibliographie; no. 1).

88. Abendanon, J.H. De Nederlandsch-Indische rechtspraak en rechtsliteratuur van 1849–1930. 3 vol.
Vol. 1. 1849–1907. 2nd revised ed. Leiden, 1908. 2 vol.
Vol. 2. 1908–1917. 's-Gravenhage etc., 1918.
Vol. 3. 1917–1930; ed. by O.W. van Ewijk. 's-Gravenhage etc., 1930.

89. Verzameling van Nederlandsch Indische rechtspraak en rechtsliteratuur. 1898–1929. 's-Gravenhage, 1909–1931. 5 vol.
Vol. 1, 2. 1889–1907, 1908–1909; by J. Duparc. 1909, 1910.
Vol. 3. 1910–1919; by A.S. Hirsch. 1920.
Vol. 4, 5. 1920–1924, 1925–1929; by J. Paulus. 1925, 1931.

90. Adatrechtbundels; bezorgd door de Commissie voor het Adatrecht en uitgegeven door het Koninklijk Instituut voor Taal-, Land- en Volkenkunde van Nederlandsch-Indië. 's-Gravenhage, 1910–1955. 45 vol.
Vol. 30. Index.
Vol. 40. Index; with list of literature.

91. Muller, W.J. Literatuurlijst voor het adatrecht van Indonesië; uitgegeven door de Adatrechtstichting te Leiden. 's-Gravenhage, 1920.
2nd ed. revised and brought up to date by J.A. Quist. 1927.
Supplementary list of literature (1927–1937). 1938.

92. Kleiweg de Zwaan, J.P. Anthropologische bibliographie van den Indischen archipel en van Nederlandsch West-Indië. Weltevreden, 1923. (Mededeelingen van het Bureau voor de Bestuurszaken der Buitengewesten; bewerkt door het Encyclopaedisch Bureau; no. XXX).
Supplement. Leiden, 1940. (Mededeelingen van de Afdeeling Volkenkunde van het Koloniaal Instituut; special series; no. 3).

93. Systematische katalogus van tijdschrift en dagbladartikelen; bijdrage van Koloniale Studiën; no. 3–6; vol. 10; met een lijst der tijdschriften die gedocumenteerd zullen worden; samengesteld door het Bureau van Ekonomische en Sociale Documentatie. Weltevreden, 1926.

94. Literatuur-overzicht voor taal-, land- en volkenkunde en geschiedenis van Nederlandsch-Indië voor het jaar 1936–1939. 's-Gravenhage, 1937–1940. 4 vol. (Koninklijk Instituut voor Taal-, Land- en Volkenkunde van Nederlandsch-Indië).
Vol. 1, 2. 1926, 1937; compiled by F.H. van Naerssen. 1937, 1938.
Vol. 3, 4. 1938, 1939; compiled by H. van Meurs. 1939, 1940.

95. Ockeloen, G. Catalogus van boeken en tijdschriften; uitgegeven in Nederlandsch Oost-Indië van 1870–1941. Batavia etc., 1940, 1942. 2 vol.
    Vol. 1. 1870–1937.
    Vol. 2. 1938–1941.
    See also no. 83, 96.
96. Ockeloen, G. Catalogus dari boekoe-boekoe dan madjalah-madjalah jang diterbitkan di Hindia Belanda dari tahoen 1870–1954. 5 vol.
    Vol. 1. 1870–1937. Batavia etc., 1940.
    Vol. 2. 1945–1949/1937–1941. Bandung, 1950.
    Vol. 3–5. 1950–1951, 1952–1953, 1954. 1952, 1954, 1955.
    From vol. 2 published under the title: "Catalogus dari buku-buku jang diterbitkan di Indonesia".
    See also no. 83, 95, 105.
97. Conover, H.F. The Netherlands East Indies; a selected list of references; compiled under the direction of F.G. Hellman. Washington, 1942. (Library of Congress; division of bibliography).
98. Netherlands East Indies; a bibliography of books published after 1930, and periodicals after 1932 available in US libraries. Washington, 1945. (Library of Congress; Netherlands studies unit of the general reference and bibliography division).
99. Kennedy, R. Bibliography of Indonesian peoples and cultures. New Haven, 1945. 2 vol. (Behavior science bibliographies; Yale University; Southeast Asian studies; by arrangement with Human Relations Area Files).
    2nd ed. revised by Th.W. Maretzki and H.Th. Fischer. 1955. 2 vol.
    3rd ed. 1962.
100. Boschma, H. Het systematisch zoölogisch onderzoek in Nederlandsch Oost-Indië en naburige landen. Leiden, 1946. (Maatschappij ter Bevordering van het Natuurkundig Onderzoek der Nederlandsche Koloniën; bulletin 100).
101. Schrieke, B.J.O. Report of the scientific work, done in the Netherlands on behalf of the Dutch overseas territories during the period between approximately 1918 and 1943; published by the Werkgemeenschap van Wetenschappelijke Organisaties in Nederland (Association of Scientific Organizations in the Netherlands). Amsterdam, 1948.
102. Postwar scientific bibliography for Indonesia; survey 15th August 1945–31st December 1949; [compiled by Kho Han Tiong, Husni Muscal, Triman Prasadio a.o.] Djakarta, 1950, 1951. (Organization for Scientific Research; bulletin 5, 7, 9).
    Bulletin 5. Postwar scientific bibliography for Indonesia. 1950.
    Bulletin 7. Indexes. 1950.
    Bulletin 9. Medical bibliography for the period 15th August 1945–31st December 1950. 1951.
103. Schleiffer, H., and D.S. Paauw. Bibliography on the economic and political development of Indonesia. Cambridge, 1953–1955. 6 vol. (Massachusetts Institute of Technology; Center for International Studies).

104. Berita bulanan dari Kantor Bibliografi Nasional. Bandung, 1953–1955. Vol. 1–7.
    See also no. 116.
105. Berita bibliografi, Djanuari 1955– . Djakarta, 1955– .
106. Postmus, S., R. Luyken and P.J. van der Rijst. Nutrition bibliography of Indonesia; compiled and annotated. Honolulu, 1955.
107. [Pamuntjak, R.] A regional bibliography of social science publications; Indonesia. Djakarta, 1957.
108. List of Japanese documents for a study of the impact of the Japanese military government to the Indonesian independence. Tokio, 1958.
109. Indonesian abstracts; abstracts on current scientific Indonesian literature; vol. 1–[17]. Djakarta, 1958–[1975].
110. Index of Indonesian learned periodicals. – Indeks madjalah ilmiah, [1958]-bulletin no. 1– . Djakarta, [1959]– . (Lembaga Ilmu Pengetahuan Indonesia).
111. Jaspan, M.A. Social stratification and social mobility in Indonesia; a trend report and annotated bibliography. Djakarta, 1959. (Science and society series; no. 1).
    2nd enlarged ed. 1961.
    Also published under the title: "Rural sociology in Indonesia". In *Current sociology. – La sociologie contemporaine;* vol. VIII; no. 1. London, 1959.
112. Coolhaas, W.Ph. A critical survey of studies on Dutch colonial history. The Hague, 1960. (Bibliographical series KITLV; no. 4).
    2nd ed. revised by G.J. Schutte. 1980.
113. Checklist of serials in Indonesian libraries. – Katalogus induk sementara madjalah$^2$ pada perpustakaan$^2$ di Indonesia; vol. III (Indonesian serials A–Z and supplements). Djakarta, 1962. (Biro Perpustakaan Dep. P.D. dan K.).
114. Echols, J.M. Preliminary checklist of Indonesian imprints during the Japanese period, March 1942–August 1945. Ithaca, 1963. (Cornell University; modern Indonesia project; bibliography series; no. 1).
115. Bhatta, J.N. A social science bibliography on Indonesia; with special emphasis on demographic and geographic materials. Djakarta, 1965. (Dinas Geografi; Direktorat Topografi; Departemen Angkatan Darat; publication no. 13).
116. Bibliografi nasional Indonesia; kumulasi 1945– ; diusahakan oleh projek Perpustakaan Nasional dan Biro Perpustakaan Dep. P.D. dan K. Djakarta, 1965– .
    Kumulasi 1964– . S.a.– .
    See also no. 104.
117. Penerbitan Indonesia, 1945– ; diterbitkan oleh Ikatan Penerbit Indonesia (IKAPI). Djakarta, 1965– .
    First number published under the title: "Daftar buku; 20 tahun penerbitan Indonesia".

118. Research di Indonesia, 1945–1965; editor koordinator M. Makagiansar. Djakarta, 1965. 4 vol.
    Vol. 1. Soedarmo, Poorwo. Bidang kesehatan.
    Vol. 2. Soemantri, R.M. Bidang teknologi dan industri.
    Vol. 3. Sumintawikarta, Sadikin. Bidang pertanian.
    Vol. 4. Nitisastro, Widjojo. Bidang ekonomi sosial dan budaja.
119. Hicks, G.L., and G. McNicoll. The Indonesian economy, 1950–1965; a bibliography. New Haven, 1967. (Yale University; Southeast Asia Studies; Bibliography series; no. 9).
    With bibliographical supplement, 1965–1967. 1967. (Bibliography series; no. 10).
120. The Ling Gie. Bibliografi ilmu administrasi dalam bahasa Indonesia. Jogjakarta, 1968.
121. Singarimbun, Masri. The population of Indonesia, 1930–1968; a bibliography. London, 1969.
    2nd revised and enlarged ed.; 1930–1972. Yogyakarta, 1974.
122. Excerpta Indonesica; no. 1– . Leiden, 1970– . (KITLV, afdeling Documentatie Modern Indonesië).
123. Nunn, G.R. Indonesian newspapers; an international union list. Honolulu, 1971. (Occasional series; no. 14.)
124. Damian, E., and R.N. Hornick. Bibliografi hukum Indonesia, 1945–1970; dengan bantuan A. Sobardy, Memet Harlan dan Moestopo. Bandung, 1971. (Lembaga Penelitian Hukum dan Kriminologi).
125. Veur, P.W. van der. The Eurasians of Indonesia; a political-historical bibliography. Ithaca, 1971. (Bibliography series, Cornell Modern Indonesia project).
126. Ramlie, P. Rice bibliography of Indonesia, 1842–1971; underwritten by the Ford Foundation. Bogor etc., *c.* 1972. (Bibliotheca Bogoriensis).
127. Supit, Wanny W.W. A bibliography on Indonesian material for the humanities and social sciences, 1960–1970; associate editors Rusdi Muchtar, Istinganah Sukarno and Sulandjani Sukardjo. Jakarta, 1972. (Seri bibliografi; no. 1. Lembaga Research Kebudajaan Nasional (Bibliography series; no. 1. Research Institute for National Culture)).
128. Buur, D. [B.M.] Persoonlijke documenten, Nederlands-Indië/Indonesië; keuzebibliografie. Leiden, 1973. (KITLV, afdeling Documentatie Geschiedenis Indonesië).
129. Hoenderkamp, M. De Verenigde Naties en de Indonesische kwestie, 1947–1963; een literatuurlijst. Amsterdam, 1973. (Koninklijk Instituut voor de Tropen).
130. Tairas, J.N.B. Daftar karya bibliografi Indonesia. Jakarta, 1973.
    2nd enlarged ed. published under the title: "Indonesia; a bibliography of bibliographies". 1974.
131. Thomas, M., Sutan Zanti Arbi and Soedyarto. Indonesian education; an annotated bibliography. Santa Barbara, 1973. (Indonesian bibliography project; Graduate School of Education; vol. 1).

132. Müller, W. Bibliographie deutscher Literatur über Indonesia. Hamburg, 1974.
133. Nagelkerke, G.A. Bibliografisch overzicht uit periodieken over Indonesië. – Bibliographical survey based on periodicals on Indonesia, 1930–1945. Leiden, 1974. (KITLV, afdeling Bibliotheek).
See also no. 85.
134. Bibliografi perkamusan dan ensiklopedi. Jakarta, 1974. (Perpustakaan Lembaga Bahasa Nasional).
135. Sukanda-Tessier, V. Bibliographie d'une documentation indonésienne contemporaine, 1950–1970; avec la collaboration de H. Sukanda Natasasmita. Paris, 1974. (Publications de l'École Française d'Extrème Orient; vol. XCVII).
136. Jongeneel, J.A.B. Bibliografi ilmu agama dan theologia Kristen dalam bahasa Indonesia. Jakarta, 1975, 1976. 2 vol.
137. Koentjaraningrat. Anthropology in Indonesia; a bibliographical review. The Hague, 1975. (Bibliographical series KITLV; no. 8).
138. Nas, P.J.M., M. Pols-Haighton and Satyawan Sunito. Bibliografie urbanisatie in Indonesië. Leiden, 1975.
139. Purawidjaja, I[pon] Sukarsih. Bibliografi bahasa Indonesia. Jakarta, 1975.
140. Bibliografi perkamusan Indonesia. Jakarta, 1976. (Dep. Pendidikan dan Kebudayaan).
141. Bibliografi sastra Indonesia; edisi kesatu; disusun oleh Ipon S[ukarsih] Purawidjaja, Gina Ginanta, S. Slabiyah dll. Jakarta, 1976.
142. Soekanto, Soerjono. Bibliografi hukum adat Indonesia; akhir abad 19 sampai 1975. Bandung, 1976.
143. Bibliografi tentang bibliografi. Jakarta, 1977. (Team teknis sub-proyek bibliografi).
144. Bastin, J., and B. Brommer. Nineteenth century prints and illustrated books of Indonesia; with particular reference to the Tropenmuseum, Amsterdam; a descriptive bibliography. Utrecht, 1979.

Currently appearing bibliographical works no. 105, 110, 116, 117, 122.

### Separate Indonesian islands and districts

For Nieuw Guinea/Irian Jaya see also no. 66–79

145. Joustra, M. Litteratuuroverzicht der Bataklanden. Leiden, 1907.
146. Literatuurstudie van Halmaheira en Morotai. Batavia, 1918. (Mededeelingen van de Vereeniging tot Bevordering van het Bibliotheekwezen in Nederlandsch-Indië; no. 1).
147. Lekkerkerker, C. Bali en Lombok; overzicht der litteratuur omtrent deze eilanden tot einde 1919. Rijswijk, 1920. (Uitgave van het Bali-Instituut).
See also no. 152, 153.
148. Wellan, J.W.J., en O.L. Helfrich. Zuid-Sumatra; overzicht van de literatuur der gewesten Benkoelen, Djambi, de Lampongsche districten en Palem-

bang; loopende tot en met 1925. 's-Gravenhage, 1923, 1925. 2 vol. (Uitgave van het Zuid-Sumatra-Instituut).
Vol. 1. Up to 1915.
Vol. 2. 1916–1925.

149. Joustra, M. Overzicht der litteratuur betreffende Minangkabau. Amsterdam, 1924. (Uitgaaf van het Minangkabau-Instituut; no. 2).
See also no. 151.

150. Overzicht van de literatuur betreffende de Molukken. Amsterdam, 1928, 1935. 2 vol. (Uitgave van het Molukken-Instituut).
Vol. 1. 1550–1921; by W. Ruinen. 1928.
Vol. 2. 1922–1933 (Nieuw Guinea not included); by A.B. Tutein Nolthenius. 1935.

151. [Tideman, J.] Overzicht der litteratuur betreffende Minangkabau, 1923–1936. Amsterdam, 1936. (Uitgaaf van het Minangkabau-Instituut; no. 3).
See also no. 149.

152. [Goris, R.] Overzicht over de belangrijkste litteratuur betreffende de cultuur van Bali over het tijdvak 1920–1935. Leiden, 1937. (Mededeelingen Kirtya Liefrinck-Van der Tuuk; no. 5).
See also no. 147, 153.

153. Hooykaas, C., en . . . van Baal. Aanvullende bibliografie aangaande de Sasaks van Lombok, 1920–1941. S.l., 1942.
In *Djawa;* vol. 21; no. 6.
See also no. 147, 152.

154. Galis, K.W. Bibliographie van Nederlands-Nieuw-Guinea. 's-Gravenhage, 1952.
2nd revised and supplemented ed. Hollandia, 1956.
3rd revised and enlarged ed. 's-Gravenhage, 1962.

155. Hoogerwerf, A. An ornithological bibliography; having particular reference to the study of the birds of Java. Djakarta, 1953. In *Organization for scientific research in Indonesia;* bulletin 13–16.

156. Voorhoeve, P. Critical survey of studies on the languages of Sumatra. The Hague, 1955. (Bibliographical series KITLV; no. 2).

157. Cense, A.A., and E.M. Uhlenbeck. Critical survey of studies on the languages of Borneo. The Hague, 1958. (Bibliographical series KITLV; no. 7).

158. Suzuki, P. Critical survey of studies on the anthropology of Nias, Mentawei and Enggano. The Hague, 1958. (Bibliographical series KITLV; no. 3).

159. Uhlenbeck, E.M. Critical survey of studies on the languages of Java and Madura. The Hague, 1964. (Bibliographical series KITLV; no. 7).

160. McGrath, W.A. A select annotated bibliography on land tenure in the Territory of Papua and New Guinea. Port Moresby, 1964.

161. Ethnographic bibliography, An, of New Guinea. Canberra, 1968. 3 vol. (Department of Anthropology and Sociology [of the] Australian National University).

162. Danandjaja, J. An annotated bibliography of Javanese folklore. Berkeley, 1972. (Center for South and Southeast Asia Studies; occasional papers; no. 9).
163. Muttalib, Jang Aisjah. The history and society of South Sumatra, 1800–1920; publications in New York libraries; compiled and annotated. New York, 1972. (Southern Asian Institute, Columbia University; occasional bibliographical papers; no. 2).
164. Bibliografi Irian Jaya. Jakarta, 1973.
165. Naim, Asma M., and Mochtar Naim. Bibliografi Minangkabau. Singapore, [1975].
166. Srikayadi, R. Bibliografi daerah Jawa-Tengah; daftar buku dan berkala terbit di daerah Jawa Tengah, no. 1. Semarang, 1978.

## Some Dutch reference books

167. Jong, J. de. Alphabetische naamlijst van boeken, welke sedert het jaar 1790 tot en met 1831 in Noord-Nederland zijn uitgekomen. 's-Gravenhage etc., 1832.
    Continued by C.J. Brinkman under several titles.
    From 1937 under the title: "Brinkman's cumulatieve catalogus van boeken en verder in den boekhandel voorkomende artikelen die in het jaar 1930– in Nederland zijn uitgegeven of herdrukt". Leiden, 1937– .
168. Encyclopaedie van Nederlandsch-Indië. 's-Gravenhage etc., 1896–1905. 4 vol.
    2nd [revised] ed. 's-Gravenhage, 1917–1921. 4 vol.
    With 4 supplements; vol. 5–8. 1927–1939.
    With supplements and amendments. 1939, 1940.
170. Repertorium der verhandelingen en bijdragen betreffende de geschiedenis des vaderlands in tijdschriften en mengelwerken tot op 1956 verschenen. Leiden, 1907–1963. 12 vol.
171. Enzyklopaedie des Islam; geographisches, ethnographisches und biographisches Wörterbuch der Muhammedanischen Völker. Leiden, 1908–1934. 4 vol.
    New ed. published under the title: "The encyclopaedia of Islam". Leiden, 1960– .
172. Nieuw Nederlandsch biografisch woordenboek. Leiden, 1911–1937. 10 vol.
    In course of preparation: "Biografisch woordenboek van Nederland".
173. Nederlandse overheidsuitgaven; lijst van officieele uitgaven verschenen in Nederland (en Nederlandsch Oost- en West-Indië); bewerkt door de Koninklijke Bibliotheek. 's-Gravenhage, 1930– .
    Vol. 1(1929)– .
    From vol. 25(1953) published under the title: "Bibliografie van in Nederland verschenen officiële en semi-officiële uitgaven".

174. Catalogus van academische geschriften in Nederland (en Nederlandsch-Indië) verschenen, 1924– ; met naam- en zaakregister; vol. 1– . Utrecht, 1931– .
175. Koeman, C. Collections of maps and atlasses in the Netherlands; their history and present state. Leiden, 1961.
176. Landwehr, J. Studies in Dutch books with coloured plates, published 1662–1875; natural history, topography and travel costumes and uniforms. The Hague, 1976.
177. Centrale catalogus van periodieken en seriewerken in Nederlandse bibliotheken; 2e uitgave. 's-Gravenhage, 1978. 7 vol.

Currently appearing bibliographical works no. 167, 173, 174.

## Some Dutch surveys of archives and publications of sources

178. Matthes, B.F. Kort verslag aangaande alle mij in Europa bekende Makasaarsche en Boeginesche handschriften, vooral die van het Nederlandsch Bijbelgenootschap te Amsterdam. Amsterdam, 1875.
    . Supplement. 1881.
179. Kern, R.A. I La Galico; catalogus der Boegineesche tot den I La Galico-cyclus behorende handschriften, bewaard in het Legatum Warnerianum te Leiden, alsmede in andere Europeesche bibliotheken. Leiden, 1939. Supplement published under the title: "Catalogus van de Boeginese tot de I La Galico-cyclus behorende handschriften van Jajasan Matthes te Makassar". Makasar, 1954.
180. Poerbatjaraka, Ng., P. Voorhoeve en C. Hooykaas. Indonesische handschriften. Bandung, 1950.
181. Rijksarchieven, De, in Nederland; overzicht van de inhoud van de rijksarchiefbewaarplaatsen. 's-Gravenhage, 1953.
    2nd ed. revised to date 1973. 1973. 2 vol.
182. Voorhoeve, P. Handlist of Arabic manuscripts in the library of the university of Leiden and other collections in the Netherlands. 's-Gravenhage, 1957. (Bibliotheca Universitas Leidensis; codices manuscripti; vol. 7).
183. Wal, S.L. van der. Het onderwijsbeleid in Nederlands-Indië, 1900–1940. – Educational policy in the Netherlands-Indies, 1900–1940. Groningen, 1963. (Uitgaven van de Commissie voor Bronnenpublicatie betreffende de Geschiedenis van Nederlands-Indië, 1900–1942, van het Historisch Genootschap gevestigd te Utrecht; no. 1).
184. Wal, S.L. van der. De Volksraad en de staatkundige ontwikkelingen van Nederlands-Indië. – The Peoples Council and the political development of the Netherlands-Indies. Groningen, 1964, 1965. 2 vol. (Uitgaven van de Commissie voor Bronnenpublicatie betreffende de Geschiedenis van Nederlands-Indië, 1900–1942, van het Historisch Genootschap gevestigd te Utrecht; no. 2, 3).
    Vol. 1. 1891–1926. 1964.
    Vol. 2. 1927–1942. 1965.

185. Wal, S.L. van der. De opkomst van de nationalistische beweging in Nederlands-Indië. – The emergence of the nationalistic movement in the Netherlands-Indies. Groningen, 1967. (Uitgaven van de Commissie voor Bronnenpublicatie betreffende de Geschiedenis van Nederlands-Indië, 1900–1942, van het Historisch Genootschap gevestigd te Utrecht; no. 4).

186. Pigeaud, Th.G.Th. Literature of Java; catalogue raisonné of Javanese manuscripts in the library of Leiden and other public collections in the Netherlands. The Hague, 1967–1970. 3 vol.
Vol. 1. Synopsis of Javanese literature, 900–1900. 1967.
Vol. 2. Descriptive list of Javanese manuscripts. 1968.
Vol. 3. Illustrations and facsimiles of manuscripts, maps, addenda, and a general index of names and subjects. 1970.

187. Wal, S.L. van der. Officiële bescheiden betreffende de Nederlands-Indische betrekkingen, 1945–1950. 's-Gravenhage, 1971– . (Rijks geschiedkundige publicatiën; kleine serie; no. 36).

188. Creutzberg, P. Het ekonomisch beleid in Nederlands-Indië. – Economic policy in the Netherlands-Indies; capita selecta; een bronnenpublikatie; 1e–3e stuk. Groningen, 1972–1975. 4 vol. (Uitgaven van de Commissie voor Bronnenpublicatie betreffende de Geschiedenis van Nederlands-Indië, 1900–1942, van het Historisch Genootschap gevestigd te Utrecht; no. 5–7).

189. Kwantes, R.C. De ontwikkeling van de nationalistische beweging in Nederlands-Indië. – Development of the nationalistic movement in the Netherlands-Indies; 1e– stuk. Groningen, 1975– . (Uitgaven van de Commissie voor Bronnenpublicatie betreffende de Geschiedenis van Nederlands-Indië, 1900–1942, van het Historisch Genootschap gevestigd te Utrecht; no. 8– ).
Part 1; 1917–1923. 1975.
Part 2; 1923–1928. 1978.
See also no. 185.

## Index

The index is based principally on the catchwords mentioned in the titles in this list of bibliographical works. In preparation of this list I have followed in principle the same rules as for the index to the archive material.

# LIST OF ABBREVIATIONS

This list consists of abbreviations found in the various inventories. Only those abbreviations were included which had been solved in the course of the research.

| | |
|---|---|
| AAF | Allied Air Force |
| AAG | Assistant Adjutant General |
| AAT | Aan- en Afvoertroepen |
| ABC | Australian Broadcasting Company |
| ABDACOM | American British Dutch Australian Command |
| ABIA | Annual Bibliography of Indian Archaeology |
| ABIZ | Amsterdamsch Bureau voor Indische Zaken |
| ACOS | Assistant Chief of Staff |
| ACSEA | Air Command South East Asia |
| ADB | American Dutch British Conversations Singapore |
| ADCOS | Assistant Deputy Chief of Staff |
| ADCS | Anglo-Dutch Country Section |
| ADRI | Angkatan Republik Indonesia |
| ADS | Anglo Dutch Section |
| Advoland | Adviesbureau voor Landbouwzaken |
| AFNEI | Allied Forces Netherlands East Indies |
| AFPAC | Allied Forces Pacific |
| AFPEL | Anti-Fascist People's Freedom League |
| AFSEA | Allied Forces South East Asia |
| AG | Adjudant-Generaal |
| AHK | Algemeen Hoofdkwartier |
| AHQ | Allied Headquarters |
| AIA | Australia-Indonesia Association |
| AIB | Allied Intelligence Bureau |
| ALFSEA | Allied Land Forces South East Asia |
| ALHQ | Allied Land Headquarters |
| ALO | Allied Officer Commanding |
| ALPF | Algemeen Landbouw Pensioenfonds |
| ALRI | Angkatan Laut Republik Indonesia |
| ALS | Algemeen Landbouw Syndicaat |
| AMACAB | Allied Military Administration Civil Affairs Branch |
| AMF | Australian Military Forces |
| AMSI | Admirality Merchant Shipping Instructions |
| ANEP | Algemeen Nederlands Persbureau |
| ANETA | Algemeen Nieuws en Telegraaf Agentschap |
| ANIEM | Algemeene Nederlandsch-Indische Electriciteits Maatschappij |
| ANIP | Algemeen Nederlandsch-Indisch Persbureau |
| ANKOMA | Angkatan Komunis Muda |
| ANLO | Australian Navy Liaison Officer |

| | |
|---|---|
| ANP | Algemeen Nederlands Persbureau |
| ANWB | Algemene Nederlandse Wielrijdersbond |
| ANZAK | Australia and New Zealand Army Corps |
| ANZC | Algemeene Nederlandsche Zendingsconferentie |
| AOC | Allied Officer Commander |
| AOK | Ministerie van Algemene Oorlogvoering van het Koninkrijk |
| AP | Associated Press |
| API | Angkatan Pemuda Indonesia |
| APRA | Angkatan Perang Ratu Adil |
| APRIS | Angkatan Perang Republik Indonesia Serikat |
| APWI | Allied Prisoners of War and Internees |
| ARA | Algemeen Rijksarchief |
| ASB | Algemeen Schoolbegeleider |
| ASGC | Algemeen Soerabajaasch Gezondheids Comité |
| ASSI | Algemeen Syndicaat voor Suikerondernemingen in Indonesië |
| ATC | Air Traffic Control |
| AURI | Angkatan Udara Republik Indonesia |
| AVC | Afvoer en Vernielingscommissie |
| AVROS | Algemeene Vereeniging van Rubberplanters ter Oostkust van Sumatra |

| | |
|---|---|
| BB | Binnenlands Bestuur |
| BdZ | Bevelhebber der Zeestrijdkrachten |
| BENISO | Bond van Eigenaren van Nederlandsch-Indische Suikeronderne-mingen |
| BFO | Bijeenkomst Federaal Overleg |
| BGS | British General Staff |
| BiZa | Ministerie van Binnenlandse Zaken |
| BKI | Bijdragen tot de Taal-, Land- en Volkenkunde (van Neder-landsch-Indië) |
| BKR | Barisan Keamanan Rakjat |
| BMGH | Bijdragen en Mededelingen van het Historisch Genootschap te Utrecht |
| BNI | Bank Negara Indonesia |
| BNS | Bevelhebber der Nederlandse Strijdkrachten |
| BO | Boedi Oetomo |
| Borsumij | Borneo Sumatra Handel Maatschappij |
| BOW | Departement van Burgerlijke Openbare Werken |
| BPL | Bibliotheca Publica Latina |
| BPM | Bataafsche Petroleum Maatschappij |
| BPRI | Barisan Pemberontakan Rakjat Republik Indonesia |
| BS | Binnenlandse Strijdkrachten |
| BSO | Bevelhebber Strijdkrachten Oosten |
| BuZa | Ministerie van Buitenlandse Zaken |

| | |
|---|---|
| CAD | Centrale Accountantsdienst |
| CAS | Carpentier Alting Stichting |
| CBG | Centraal Bureau voor Genealogie |
| CBN | Centraal Bureau NIWIN |
| CBOAR | Commandant Bewakingsdienst Oost Archipel |
| CCO | Chief Commanding Officer |
| CCS | Combined Chiefs of Staff |
| Cdt | Commandant |
| CGD | Commissie van Goede Diensten |
| CGO | Committee of Good Offices |
| CGS | Chef Generale Staf |
| CHH | Chung Hwa Hue |
| CHU | Christelijk Historische Unie |
| CICEF | Commander in Charge Eastern Forces |
| CICEI | Commander in Charge East Indies |
| CICSWP | Commander in Charge South West Pacific |
| CICUS | Commander in Charge United States |
| CID | Centrale Inlichtingen Dienst |
| CID | Colonial Infantry Division |
| CIDSS | Comité International pour la Documentation des Sciences Sociales |
| Cie | Compagnie |
| CIGS | Chief of the Imperial General Staff |
| CK Ins | Commandant Korps Insulinde |
| CLg | Legercommandant |
| CLG | Compagnie Leger Genietroepen |
| CMI | Centrale Militaire Inlichtingendienst |
| CMLDO | Commandant Marine Luchtvaart Dienst Oosten |
| COG | Commissie Onderzoek Gedragingen |
| CONICA | Commanding Officer NICA |
| COP | Commissie Overname Personeel |
| COS | Chiefs of Staff |
| CPA | Chief Political Adviser |
| CPN | Communistische Partij Nederland |
| CS | Chief of Staff |
| CUM | Comité Uitzending Missionarissen |
| CUSNSWP | Commander US Navy South West Pacific |
| CZM | Commandant Zeemacht |
| CZMO | Commandant Zeemacht Oosten |
| | |
| D of I | Director of Intelligence |
| D of P | Director of Plans |
| DCO | Dienst Centrale Opleidingen |
| DD of I | Deputy Director of Intelligence |
| DEI | Dutch East Indies |

| | |
|---|---|
| DEPPEN | Departemen Penerangan |
| DIRVO | Directie Verre Oosten |
| DIVI | Deviezen Instituut voor Indonesië |
| DNLO | Dutch Naval Liaison Officer |
| Dosta | Dagelijkse Overzichten Staf A-divisie |
| DPAO | Deputy Principal Administration Officer |
| DSM | Deli Spoorweg Maatschappij |
| DVG | Dienst Volksgezondheid |
| EC | Eskader Commandant |
| ECAFE | Economic Commission for Asia and the Far East |
| ECK | Evangelisch Christelijke Kerk |
| ECOSOC | Economic and Social Council |
| EIS | Economic Intelligence Summary |
| EM | Expeditionaire Macht |
| EVD | Economische Voorlichtingsdienst |
| EZ | Ministerie van Economische Zaken |
| FAI | Federatie van Ambtenarenorganisaties in Indonesië |
| FAO | Food and Agriculture Organization |
| FCHJG | Federatie van Christelijk-Historische Jongeren Groepen |
| FDR | Front Demokrasi Rakjat |
| FELO | Far Eastern Liaison Officer |
| FIWI | Federatie van ex-Illegale Werkers in Indië |
| FNI | Front Nasional Indonesia |
| GAPKI | Gabungan Pendjagaan Kemerdekaan Indonesia |
| GB | Gouvernementsbesluit |
| GEA | Greater East Asia |
| GEBEO | Gemeenschappelijk Electriciteits-Bedrijf Bandoeng en Omstreken |
| GG | Gouverneur-Generaal |
| GGG | Nationaal Verbond Godsdienst Gezin Gezag |
| GGM | Gezaghebber Gouvernements-Marine |
| GHQ | General Headquarters |
| GIA | Gouvernements Indisch Arts |
| GM | Gouvernements-Marine |
| GMC | Gedelegeerd Missie Comité |
| GNA | Groter Nederland Actie |
| GPC | Gemengde Plaatselijke Commissie |
| GPP | Gabungan Pembela Proklamasi |
| GRR | Gerakan Revolusi Rakjat |
| GS | Generale Staf |
| GSO | General Staff Officer |
| GU | Gemeenteuniversiteit Amsterdam |
| GZB | Gereformeerde Zendingsbond |

| | |
|---|---|
| HBM | Handelsbeschermingsmededelingen |
| HBO | Handelsbeschermingsofficier |
| HIS | Hollandsch-Inlandsche School |
| HK | Hoofdkwartier |
| HKAG | Hoofdkantoor Adjudant Generaal |
| HKMG | Hoofdkantoor Kwartiermeester Generaal |
| HKML | Hoofdkwartier Militaire Luchtvaart |
| HKS | Hogere Krijgsschool |
| HMG | Hoog Militair Gerechtshof |
| HPD | Head P-division |
| HQ | Headquarters |
| HTP | Handleiding Tropisch Pioniersmateriaal |
| HVK | Hoge Vertegenwoordiger van de Kroon |
| ICAO | International Civil Aviation Organization |
| ICSSD | International Committee for Social Sciences Documentaton |
| IDI | Infanterie Depôt Indië |
| IEFC | International Emergency Food Council |
| IEV | Indo-Europeesch Verbond |
| II | Indische Instelling |
| IISG | Internationaal Instituut voor Sociale Geschiedenis |
| ILO | International Labour Organization |
| IMKI | Ikatan Mahasiswa Katolik Indonesia |
| IN | "Indonesië-Nederland" |
| Internatio | Internationale Crediet- en Handelsvereniging Rotterdam |
| IRAA | Inventarissen van Rijks- en Andere Archieven |
| IRRA | International Routine and Reporting Authority |
| ISC | Indologisch Studenten Corps |
| ISD | Interallied Service Department |
| ISDV | Indische Sociaal-Democratische Vereeniging |
| IVM | Indische Missie-Vereeniging |
| JC | Java-Comité |
| JCJL | Japan-China-Japan-Lijn |
| JCS | Joint Chiefs of Staff |
| JIB | Jong Islamieten Bond |
| JPS | Joint Planning Staff |
| KB | Koninklijk Besluit |
| KB | Koninklijke Bibliotheek |
| KCM | Klattensche Cultuurmaatschappij |
| KDC | Katholiek Documentatie Centrum |
| KDP | Kantoor Displaced Persons |
| KIT | Koninklijk Instituut voor de Tropen |
| KITLV | Koninklijk Instituut voor Taal-, Land- en Volkenkunde |

| | |
|---|---|
| KJB | Katholieke Jeugdbeweging |
| KKM | Kerdja di Kalangan Musuh |
| KL | Koninklijke Luchtmacht |
| KLM | Koninklijke Luchtvaart-Maatschappij |
| KM | Koninklijke Marine |
| KMA | Koninklijke Militaire Academie |
| KMG | Kwartiermeester Generaal |
| KMT | Kuo Min Tang |
| KNAG | Koninklijk Nederlands Aardrijkskundig Genootschap |
| KNID | Koninklijk Nederlands Infanterie Depôt |
| KNIL | Koninklijk Nederlandsch-Indisch Leger |
| KNILM | Koninklijke Nederlandsch-Indische Luchtvaartmaatschappij |
| KNIP | Komite Nasional Indonesia Pusat |
| KNM | Koninklijke Nederlandse Marine |
| KNMI | Koninklijk Nederlands Metereologisch Instituut |
| KOONI | Kort Overzicht van het Officieel Nieuwsblad Indonesië |
| KOSI | Kort Overzicht Staatsbladen Indonesië |
| KOWANI | Kongres Wanita Indonesia |
| Kpg | Kampung |
| Kpl | Korporaal |
| KPM | Koninklijke Paketvaart Maatschappij |
| Kpt | Kapitein |
| KRIS | Kebaktian Rakjat Indonesia Sulawesi |
| KRO | Katholieke Radio Omroep |
| KSM | Kediri Stoomtrammaatschappij |
| Ktz | Kapitein ter Zee |
| KVP | Katholieke Volkspartij |
| | |
| L of C | Line of Communications |
| LAAPLN | Lembaga Alat-Alat Pembajaran Luar Negeri |
| Landops | Land operaties |
| LCI | Landing Craft Infantery |
| LCT | Landing Craft Tank |
| LEIDWEST | Cultuurmaatschappij Leidong West |
| LHQ | Land Headquarters |
| LIB | Lichte Infanterie Bataljon |
| LO | Leger Order |
| LO | Liaison Officer |
| LOC | Leger Organisatie Centrum |
| LOGHQ | Liaison Officer General Headquarters |
| LSI | Landing Ship Infantery |
| LSK | Luchtstrijdkrachten |
| LST | Landing Ship Tank |
| Lt | Luitenant |
| LTD | Leger Technische Dienst |

| | |
|---|---|
| LtGG | Luitenant Gouverneur-Generaal |
| Ltk | Letterkunde |
| LVD | Leger Voorlichtingsdienst |
| | |
| Marva | Marine Vrouwen Afdeling |
| Marvo | Marine Voorlichting |
| Masjumi | Madjelis Sjuro Muslimin Indonesia |
| MAW | Marine Attaché Washington |
| MB | Mariniers Brigade |
| MC | Marine Commandant |
| MCA | Marine Commandant Australië |
| MD | Muhamadiah |
| ME | Marine Etablissement |
| MEWINI | Maatschappij tot Exploitatie van Waterleidingen in Nederlandsch-Indië |
| MEWLO | Ministry of Economic Warfare Liaison Officer |
| MG | Militair Gezag |
| MGD | Militaire Geneeskundige Dienst |
| MIAI | Madjelis Islam A'laa Indonesia/Madjilsul Islamil A'laa Indonesia |
| Micharnold | Maatschappij tot Exploitatie van Rijstlanden op Java Michiels-Arnold |
| MID | Militaire Inlichtingendienst |
| Milobs | Military Observers Teams |
| Minog | Ministerie van Overzeese Gebiedsdelen |
| MIT | Mahkamah Islam Tinggi |
| MLD | Marine Luchtvaartdienst |
| MONI | Maatschappij voor Ondernemingen in Nederlandsch-Indië |
| MP | Militaire Politie |
| MS | Malang Stoomtrammaatschappij |
| MSC | Missionarii Sacri Cordis |
| MSF | Missionarii Sanctae Familiae |
| MSM | Madoera Stoomtram Maatschappij |
| MT | Military Transport |
| MTB | Motor Torpedo Boats |
| MWO | Militaire Willemsorde |
| | |
| NASSI | Comité Nationale Actie Steunt Spijtoptanten Indonesië |
| NBG | Nederlandsch Bijbelgenootschap |
| Nedinreg | Nederlands-Indische Regering |
| NEDINRUB | Nederlandsch-Indische Rubber en Koffie Cultuur Maatschappij |
| NEFIS | Netherlands Eastern Forces Intelligence Service |
| NEHA | Nederlands Economisch-Historisch Archief |
| NEITS | Netherlands East Indies Transport Service |

| | |
|---|---|
| NEM | NEFIS Echolong "M" |
| NEMAT | Nederlandse Militair Attaché |
| NEMILOD | Nederlandse Militair Liaison Officier New Delhi |
| NG | Nieuw Guinea |
| NGZV | Nederlandsche Gereformeerde Zendingsvereeniging |
| NHM | Nederlandsche Handel-Maatschappij |
| NI | Nederlands-Indië |
| NIBEG | Nederlands-Indische Bond van Ex-Krijgsgevangenen en Geïnterneerden |
| NICA | Netherlands Indies Civil Administration |
| NICC | Netherlands Indies Cooperative Committee |
| NIGEO | Netherlands Indies Governmental Export Organization |
| NIGIS | Netherlands Indies Government Informations Service |
| NIHB | Nederlandsch-Indische Handelsbank |
| NILS | Nederlandsch-Indische Land Syndicaat |
| NIMVA | Nederlandsch-Indische Maatschappij voor Affinage |
| NIPO | Nederlands Instituut voor Publieke Opinie |
| NIROM | Nederlandsch-Indische Radio Omroep Maatschappij |
| NIRUB | Nederlands-Indisch Rubberfonds |
| NISM | Nederlandsch-Indische Spoorweg Maatschappij |
| NIT | Negara Indonesia Timur |
| NIVAS | Nederlands-Indische Vereniging voor Afzet van Suiker |
| NIVB | Nederlandsch-Indische Vrijzinnigen Bond |
| NIWEM | Nederlandsch-Indische Waterkracht Exploitatie Maatschappij |
| NIWIN | Nationale Inspanning Welzijnsverzorging in Indië |
| NIWOE | Netherlands Indies Welfare Organization for Evacuees |
| NNG | Nederlands Nieuw-Guinea |
| NOT | Nederlandsche Overzee Trustmaatschappij |
| NPC | Netherlands Purchasing Commission |
| NPPM | Nederlandse Pacific Petroleum Maatschappij |
| NRC | Nieuwe Rotterdamsche Courant |
| NRK | Nederlandsche Roode Kruis |
| NRLO | Netherland Routing Liaison Officer |
| NSASEAC | Nederlandse Staf Afdeling SEAC |
| NSB | Nationaal Socialistische Beweging |
| NSHS | Nederlandse Zendingshogeschool |
| NSO | Netherlands Special Operation/Organization |
| NSSSEAC | Netherlands Staff Section SEAC |
| NSTC | Netherlands Shipping and Trading Committee |
| NTM | Nederlandse Turfgasgenerator Maatschappij |
| NUFFIC | Netherlands Universities Foundation for International Co-operation |
| NZG | Nederlandsch Zendelinggenootschap |
| NZS | Nederlandsche Zendingsschool |
| NZV | Nederlandsche Zendingsvereeniging |

| | |
|---|---|
| O en E | Onderwijs en Eeredienst |
| OARS | Overseas Approach Routes |
| OAZ | Oost-Aziatische Zaken |
| OBI | Ondernemersbond Indonesië |
| OBSO | Onderbevelhebber der Strijdkrachten in het Oosten |
| ODO | Opsporingsdienst Overledenen |
| OFM | Ordo Fratres Minores |
| OHB | Officier Handelsbescherming |
| OIC | Oost-Indische Compagnie |
| OJC | Oost-Java Stoomtram-Maatschappij |
| ONI | Office of Naval Intelligence |
| ONRANI | Ondernemersraad voor Nederlands-Indië |
| OOT | Overzichten Ontwikkelingen van de Toestand |
| OSS | Office Strategic Services |
| OSVIA | Opleidingsschool voor Inlandsche Ambtenaren |
| OVT | Oostelijk Verkennings- en Transportsquadron |
| OVW | Oorlogsvrijwilliger |
| OWI | Office of War Information |

| | |
|---|---|
| P & T-Landen | Pamanukan- en Tjiasem-Landen |
| Pardafi | Partai Demokrasi Asosiatif Federal Indonesia |
| PARI | Partai Republik Indonesia |
| PARINDRA | Partai Indonesia Raja |
| PARKI | Partai Kebangsaan Indonesia |
| PARKINDO | Partai Kristen Indonesia |
| PAT | Pao An Tui |
| PbSM | Probolinggo Stoomtram-Maatschappij |
| PBT | Persatuan Buruh Tionhoa |
| PBTB | Perserikatan Borneo Timur Besar |
| PEB | Politiek-Economische Bond |
| PERMAI | Persatuan Rakjat Marhaen Indonesia |
| PERPIN | Persatuan Pengatjara Indonesia |
| PESINDO | Pemuda Sosialis Indonesia |
| Peta | Pembela Tanah Air |
| PHOHI | Philips Omroep Holland-Indië |
| PI | Perhimpunan Indonesia |
| PID | Politieke Inlichtingendienst |
| PIR | Persatuan Indonesia Raja |
| PKB | Protestantsch Kerkbestuur |
| PKI | Partai Kebangsaan Indonesia |
| PKI | Partai Komunis Indonesia |
| PKI | Perserikatan Komunis di India |
| PKM | Persatuan Kebangsaan Madura |
| PKO | Pendjagaan Keamanan Oemoem |
| PKRI | Partai Katolik Republik Indonesia |

| | |
|---|---|
| Plenaps | Plans for employment of naval and air forces of the associated powers in the events of war with Japan |
| PMI | Palang Merah Indonesia |
| PMKRI | Perhimpunan Mahasiswa Katolik Republik Indonesia |
| PNI | Partai Nasional Indonesia |
| POAS | Pacific Overseas Airways of Siam |
| POW | Prisoner of War |
| PPK | Kementerian Pendidikan, Pengetahuan dan Kebudajaan |
| PPMI | Perserikatan Perhimpunan Perhimpunan Mahasiswa Indonesia |
| PPN | Panitia Persiapan Negara |
| PPTR | Putera Pusat Tenaga Rakjat |
| PPU | Pan Pacific Union |
| PRD | Partai Rakjat Djelata |
| PRDT | Persatuan Rakjat Djawa Timur |
| PRI | Partai Rakjat Indonesia |
| PRI | Pemuda Republik Indonesia |
| PRIS | Partai Rakjat Indonesia Serikat |
| PRP | Partai Rakjat Pasundan |
| PRRI | Pemerintah Revolusioner Republik Indonesia |
| PSI | Partai Sosialis Indonesia |
| PSII | Partai Sjarikat Islam Indonesia |
| PSS | Partai Selebes Selatan |
| PsSM | Pasoeroean Stoomtram-Maatschappij |
| PTT | Post-, Telegraaf- en Telefoondienst |
| PvdA | Partij van de Arbeid |
| PW | Pyrotechnische Werkplaats |
| PWD | Persatuan Wanita Deppen |
| PWI | Prisoners of War and Internees |
| | |
| RAAF | Royal Australian Air Force |
| RAF | Royal Air Force |
| RAN | Royal Australian Navy |
| RAPWI | Rehabilitation (Recovery) Allied Prisoners of War Internees |
| RAS | Royal Asiatic Society |
| RAvNI | Regeringsalmanak voor Nederlandsch-Indië |
| RECOMBA | Regeringscommissaris voor Bestuursaangelegenheden |
| RGP | Rijks Geschiedkundige Publicatiën |
| RH van Boreel | Regiment Huzaren van Boreel |
| RI | Regiment Infanterie |
| RI | Republik Indonesia |
| RIOD | Rijksinstituut voor Oorlogsdocumentatie |
| RIS | Republik Indonesia Serikat |
| RKSP | Roomsch Katholieke Staatspartij |
| RN | Royal Navy |
| RNIA | Royal Netherlands Indies Army |

| | |
|---|---|
| RNIG | Royal Netherlands Indies Government |
| RNMFS | Royal Netherlands Military Flying School |
| RNN | Royal Netherlands Navy |
| RPI | Republik Persatuan Indonesia |
| RRI | Radio Republik Indonesia |
| RSAP | Revolutionair-Socialistische Arbeiders-Partij |
| RSOB | Regeling van de Staat van Oorlog en Beleg |
| RSP | Revolutionair-Socialistische Partij |
| RTC | Ronde Tafelconferentie |
| RVA | Regiment Veldartillerie |
| RVD | Regerings Voorlichtingsdienst |
| RVCCS | Regeringsvertegenwoordigers bij de Combined Chiefs of Staff |
| S en TC | Sangi- en Talaud-Comité |
| SAC | South Asia Command |
| SAC | Supreme Allied Commander |
| SACSEA | Supreme Allied Command(er) South East Asia |
| Sarbupri | Sarekat Buruh Perkebunan Republik Indonesia |
| SbN | Schout bij Nacht |
| SCAP | Supreme Command Allied Powers |
| SCS | Semarang-Cheribon Stoomtram-Maatschappij |
| SCJ | Societas Cordis Jesus |
| SDAP | Sociaal-Democratische Arbeiders-Partij |
| SDS | Serajoedal Stoomtram-Maatschappij |
| SEAC | South East Asia Command |
| SEATIC | South-East Asia Translation and Interrogation Centre |
| SEM | Solosche Electriciteit Maatschappij |
| SHAEF | Supreme Headquarters of the Allied Expeditionary Forces |
| SI | Sarekat Islam |
| Sitreps | Situation reports |
| SJ | Societas Jesu |
| SJS | Semarang-Joana Stoomtram-Maatschappij |
| SM | Sergeant-majoor |
| SMM | Societas Mariae Montfortanis |
| SMN | Stoomvaart Maatschappij Nederland |
| SMW | Sint Melania-Werk |
| SO | Special Operation |
| SO | Staff Officer |
| SOB | Staat van Oorlog en Beleg |
| SOBSI | Sentral Organisasi Buruh Seluruh Indonesia |
| SOEIF | Senior Officer East Indies Fleet |
| SONICA | Staff Officer NICA |
| SOP | School Opleiding Parachutisten |
| STICUSA | Stichting Culturele Samenwerking |
| STOVIA | School tot Opleiding van Inlandsche Artsen |

| | |
|---|---|
| SUMCAMA | Sumatra Caoutchouc Maatschappij |
| SWPA | South-West Pacific Area |
| SZC | Samenwerkende Zendings-Corporaties |
| | |
| TAAS | Trans Asiatic Airlines Siam |
| TBA | Territoriaal Bestuursadviseur |
| TBG | Tijdschrift voor de Indische Taal-, Land- en Volkenkunde |
| TEP | Team Penjelesaian |
| TKNAG | Tijdschrift van het Koninklijk Nederlandsch Aardrijkskundig Genootschap |
| TM-boats | Torpedo Motorboats |
| TNI | Tentara Nasional Indonesia |
| TNI | Tijdschrift voor Nederlandsch-Indië |
| TRI | Tentara Republik Indonesia |
| TWAPRO | Twaalfde Provincie |
| | |
| UBL | Universiteitsbibliotheek Leiden |
| UDC | Universele Decimale Classificatie |
| UKSN | Unie van Katholieke Studenten in Nederland |
| UMA | United Maritime Authority |
| UNCI | United Nations Commission for Indonesia |
| UNESCO | United Nations Educational, Scientific and Cultural Organization |
| UNO | United Nations Organization |
| UNRAPWI | United Nations RAPWI |
| UNRRA | United Nations Relief and Rehabilitation Administration |
| UNWCC | United Nations Warcrimes Commission |
| USAAF | United States of America Air Forces |
| USIS | United States Information Service |
| USN | United States Navy |
| UZV | Utrechtsche Zendingsvereeniging |
| | |
| V en W | Verkeer en Waterstaat |
| VBGKW | Verhandelingen van het Bataviaasch Genootschap van Kunsten en Wetenschappen |
| VC | Vaderlandsche Club |
| VENINDOS | Vereniging van Nederlands-Indische Oud-Strijders en Oorlogsslachtoffers |
| VFR | Voorlopige Federale Regering |
| VHK | Vrouwen Hulpkorps |
| VIRO | Vereniging voor Internationale Rechtsorde |
| VJ-day | Victory of Japan-day |
| VJSP | Vereenigde Java Suiker-Producenten |
| VKCM | Vereenigde Klattensche Cultuur-Maatschappijen |
| VKNG | Vereniging "Kolonisatie Nieuw-Guinea" |

| | |
|---|---|
| VML | Maatschappij tot Exploitatie der Vereenigde Majang-landen |
| VN | Verenigde Naties |
| VNISTM | Vereeniging van Nederlandsch-Indische Spoor- en Tramweg-maatschappijen |
| VOC | Vereenigde Oost-Indische Compagnie |
| VPJS | Vereeniging het Proefstation voor de Java Suikerindustrie |
| VROA | Verslagen omtrent 's Rijks Oude Archieven |
| VSI | Verenigde Staten van Indonesië |
| VSTP | Vereeniging van Spoor- en Tramwegpersoneel |
| VU | Vrije Universiteit Amsterdam |
| VZB | Vrouwen Zendingsbond |
| WET | War Equipment Table |
| WHG | Werken van het Historisch Genootschap |
| Wmr | Wachtmeester |
| WSA | War Shipping Administration |
| YMCA | Young Men's Christian Association |
| ZB | Zendingsbureau |

# CHAPTER I

# ALGEMEEN RIJKSARCHIEF (GENERAL STATE ARCHIVES)

## ALGEMEEN RIJKSARCHIEF (GENERAL STATE ARCHIVES), ARA 1802

20 Prins Willem Alexanderhof, 2595 BE The Hague.
Admission free. Lending to institutions.
Photocopying equipment available.
Research finished in 1976.

### Introductory data [1–3]

**Library** [1]

45,000 books; 400 currently received periodicals and other serial publications.

Reference library, primarily for the use of the staff of the ARA.

In addition to reference books, handbooks, monographs etc., there are many archival publications and inventories of archives and collections.

Directed to the archives in general and in especial to the archives of the Netherlands and the former colonies; history and its auxiliary sciences, including those of the law, economics, sociology (also church history) and biography.

For literature concerning the ARA see also
[R. C. Bakhuizen van den Brink]. *Overzicht van het Nederlandsche Rijks-Archief,* 's-Gravenhage, 1854.

*The General State Archives and their contents*. 's-Gravenhage, 1932.

*De Rijksarchieven in Nederland;* overzicht van de inhoud van de rijksarchiefbewaarplaatsen. 's-Gravenhage, 1973.

**Publications** [2]

*Verslagen omtrent 's Rijks oude archieven;* [eerste serie]; vol. 1(1878) – vol. 50 (1927). 's-Gravenhage, 1879–1928 (VROA). Included as annexes are inventories of archives and collections. The VROA 1865–1877 were published in 1914.

*Verslagen omtrent 's Rijks oude archieven;* tweede serie; vol. 1(1928) – .'s-Gravenhage, 1929– . Included are lists of acquisitions.

*Inventarissen van Rijks- en andere archieven van rijkswege uitgegeven, voor zoover zij niet afzonderlijk zijn afgedrukt;* vol. 1(1928) – vol. 4(1931). 's-Gravenhage, 1929–1933.

Inventories, printed, in offset or cyclostyled. See sub "entries" to the seperate archives and collections.

**Introduction to the archives and collections of the ARA** [3]

*Eerste Afdeling (First Section).* This section is in charge of the archives of the central administrative bodies in the Netherlands and of collections and miscellaneous acquisitions up to 1795. Some collections and miscellaneous acquisitions are mentioned because they contain papers concerning Asia after 1795.

*Tweede Afdeling (Second Section).* This section is in charge of the archives, collections and miscellaneous acquisitions from 1795 up to the present.

*Entries to the whole property of archives. De Rijksarchieven in Nederland.* 's-Gravenhage, 1973.

Alphabetic and systematic index for the first hundred collections deposited at the Tweede Afdeling (Second Section).

*Regulations for the use of the archives and collections.* The archives deposited at the ARA are in principle public. If permission for consultation is required, this is stated with the separate archives. For consultation of non public archives one should apply in writing to the General State Archivist, stating the subject of study. In case of a publication from non public archives, the manuscript has to be approved first. One copy of a publication from archives deposited at the ARA should be presented to the library of the ARA.

## Highest government bodies [4–17]

The data concerning the history of these bodies were drawn from the inventories and from G.W. Bannier. *Grondwetten van Nederland;* teksten der achtereenvolgende staatsregelingen en grondwetten sedert 1975; met verschillende andere staatsstukken, historische toelichtingen en eenige tabellen; met een voorwoord van B.C. Savornin Lohman. Zwolle, 1936.

**Wetgevende colleges (Legislative bodies), 1796–1801** [4]

After the downfall of the Republic in 1796, national elections were held in the Netherlands for the Eerste Nationaale Vergadering (First National Assembly), meeting for the first time March 1st 1796. It was composed of 128 members and was the highest authority in the new republic. Its tasks included the drawing up of a new constitution. Because the differences between federalists and unitarians appeared to be unsolvable, a Tweede Nationaale Vergadering (Second National Assembly) was elected, meeting September 1st 1797. January 22nd 1798 a coup took place, bringing the unitarians to power. An Uitvoerend Bewind (Executive Government; see no.7) was appointed. The Constitueerende Vergadering (Constituent Assembly) became the legislative body. By plebiscite the constitution of this assembly was approved and a Vertegenwoordigend Lichaam (Representative Body) elected. This Body, meeting for the first time May 4th 1798, divided into a Eerste en Tweede Kamer (First and Second Chamber). However, June 12th 1798 a federalistic coup took place. An Intermediair Wetgevend Lichaam (Intermediary Legislative Body) carried on the government, till on July 31st of that year a newly elected Vertegenwoordigend Lichaam (Representative Body) met. That Body appointed a new Uitvoerend Bewind (Executive Government), but was dissolved by the coup of September 18th 1801 (see no. 5).

The archives consist of the decrees (resolutions) 1796–1801 with an index of respectively the Eerste-, Tweede Nationaale Vergadering (First, Second National Assembly), Constitueerende Vergadering (Constituent Assembly), Vertegenwoordigend Lichaam (Representative Body) composed of Eerste- en Tweede Kamer (First and Second Chamber), Intermediair Wetgevend Lichaam en Vertegenwoordigend Lichaam (Intermediary Legislative Body and Representative Body) composed of the First and Second Chamber; secret decrees 1796–1801 with index; letters, memorandums, advice etc. received and despatched 1796–1801; petitions received 1796–1801 with index; papers of several committees, including the Comité tot de Zaken van de Oost-Indische Handel en Bezittingen (Committee for the Affairs of East Indies Trade and Possessions), viz. papers received by the Committee 1796–1800, replies 1797 of the Committee to questions of the committee charged with examination of the activities of the several committees.

Entries: In addition to the indexes mentioned above there is also the "groene index" (green index; see no. 10); inventory.

Size: *C.* 79 running metres (554 inventory items).

See also

*Decreten der Nationaale Vergadering, representeerende het volk van Nederland;* vol. 1 (Maart 1796) – vol. 23 (January 1798). 's-Gravenhage, 1796–1798.

*Decreeten der Constitueerende Vergadering, representeerende het Bataafsche volk;* vol. 1 (January 1798) – vol. 4 (4 Mey 1798). 's-Gravenhage, 1798.

*[Register der] decreeten [van de Eerste Kamer van het Vertegenwoordigend Lichaam des Bataafschen volks] van 4 Mey tot 12 Juny 1798.* 's-Gravenhage, 1798.

*[Register der] decreeten [van de Tweede Kamer van het Vertegenwoordigend Lichaam des Bataafschen volks] van 4 Mey tot 12 Juny 1798.* 's-Gravenhage, 1798.

*Decreeten van het Intermediair Wetgevend Lichaam des Bataafschen volks;* vol. 1 (Juny 1798) – vol. 2 (July 1798). 's-Gravenhage, 1798.

*Besluiten der Eerste Kamer van het Vertegenwoordigend Lichaam des Bataafschen volks;* vol. 1 (Augustus 1798) – vol. 38 (September 1801). 's-Gravenhage, 1798–1801.

*Besluiten van de Tweede Kamer en decreeten van het Vertegenwoordigend Lichaam des Bataafschen volks;* vol. 1 (Augustus 1798) – vol. 38 (September 1801). 's-Gravenhage, 1798–1801.

*Dagverhaal der handelingen van de Nationaale Vergadering, representeerende het volk van Nederland;* vol. 1 (Maart 1796) – vol. 9 (4 Mey 1798). 's-Gravenhage, 1796–1798. Vol. 8 and 9 published under the title: "Dagverhaal der handelingen van de Constitueerende Vergadering".

*Dagverhaal der handelingen van het Vertegenwoordigend Lichaam des Bataafschen volks;* vol. 1 (4 Mey 1798) – vol. 13 (18 September 1801). 's-Gravenhage, 1798–1801. Vol. 1 published under the title: "Dagverhaal der handelingen van het Intermediair Wetgevend Lichaam".

Tables of contents of the printed series mentioned above.

**Wetgevend Ligchaam van het Bataafsch Gemeenebest (Legislative Body of the Batavian Commonwealth), 1801–1810** [5]

After the coup of September 18th 1801 the Staats-Bewind (State Government; see no. 8) became the most important body in the Netherlands. This Body appointed the 35 members of the Wetgevend Ligchaam van het Bataafsch Gemeenebest (Legislative Body of the Batavian Commonwealth), which possessed much less power than the Vertegenwoordigend Lichaam (Representative Body; see no. 4). Its only task was to approve or disapprove laws. After R.J. Schimmelpenninck had come to power as Raadpensionaris (Grand Pensionary) in 1805 (see no. 9), the number of the members was reduced to 19, but under the government of Louis Napoleon (see no. 10) in 1806 this number went up to 39. From 1805 the Body was titled Vergadering van Hun Hoog Mogenden, vertegenwoordigende het Bataafsch Gemeenebest (Assembly of Their High Mightinesses, representing the Batavian Commonwealth). After the incorporation of the Netherlands into France in 1810, the Vergadering (Assembly) was dissolved.

The archives consist of non-secret and secret minutes with papers received 1801–1810.

Entries: "Groene index" (green index; see no. 10); inventory.

Size: *C.* 5 running metres (52 inventory items).

See also

*Notulen van het Wetgevend Ligchaam van het Bataafsch Gemeenebest.* 's-Gravenhage, 1801–1805. 3 vol. With index.

*Register der resolutiën van Hun Hoog Mogenden, vertegenwoordigende het Bataafsch Gemeenebest, 3 January 1806–1807.* 's-Gravenhage, 1807.

**Tweede Kamer (Second Chamber), 1815–** [6]

In the Constitution of 1815 two chambers were created (see also no. 275). The Second Chamber was composed of 110 members, elected by the Provincial States. They sat for 3 years, every year one third retired. The Chamber possessed initiative power and voted the decennial budgets. The meetings were public. After the secession of Belgium the number of the members was reduced to 58 in the

Constitution of 1840. The budgets became biennial, the Chambers acquired control of the expenditure of the colonial surplus. The Constitution of 1848 provided that the King can do no wrong and that the ministers are answerable to the States General. The Second Chamber acquired the right to introduce amendments, to make inquiries and to put questions to ministers. The budget was presented annually, the conduct of colonial affairs was subject to the sanction of the chambers. The members were elected directly by constituencies and according to the system of right of voting on the ground of a property qualification. One member on a population of 45,000, session 4 years, one half retired every two years. The Constitution of 1887 provided the number of the members to be fixed on 100. They still sat for 4 years, but all the members retired at the same time. The Constitution of 1917 brought the introduction of universal suffrage (universal women's suffrage in 1919) and proportional representation. From 1940–1945 the States General did not meet. In 1956 the number of the members was fixed on 150.

The archives consist of a series of "notulen" (minutes) with annexes from 1815. The papers are filed in a folder for each meeting. They include the attendance-book, signed by the members of the Chamber during the meeting, and the "punten" (points), taken down by the clerk during the meeting, from which the progress of the subjects concerned may be concluded and from which the decisions of the Chamber during the meeting may be known. These "points" are of guidance for the publication of the *Verslag der handelingen van de Tweede Kamer van de Staten-Generaal.* Furthermore the report of such a meeting contains all the papers, printed or not, received by the Chamber, such as Royal messages and decrees, letters from the government, the First Chamber and third parties, petitions, declaratory resolutions and the (final) reports produced by the several committees. The papers sent to the Chamber by the government are printed and distributed among the members, and added to the proceedings as annexes at the end of a session. The "notulen" (minutes) and annexes 1815–1830 are in Dutch and French. Besides this series there are "overgelegde stukken" (papers produced) from 1910 forward and the "officiële verslag" (official report) (this is the concise report of a meeting approved by the Chamber) from 1920. There is also a series of "notulen van de Comité's-Generaal" (minutes of the Committees General); the report of the proceedings of the (non public) committees general, represented by the clerk or by the stenographic service, respectively in short or in full. The series runs from 1817–*c.*1920 (2 running metres). Finally there are some non published reports of the committee of inquiry concerning Indonesia during the second world war (no consultation for the moment).

Entry: Register to the printed proceedings.

Size archives up to 1950: *C.* 800 running metres, of which 240 metres up to 1919.

N.B. The archives of the period 1919–1950 (*c.* 560 running metres) are deposited with the Second Chamber. Address: Binnenhof 1a, The Hague. Admission by appointment. For the archives of the First Chamber see no. 275.

See also *Verslag der handelingen van de Staten-Generaal, 1814–* . (Up to 1848 only of the Second Chamber, with a concise report of the proceedings in the First Chamber 1827/28–1845/46; 1940–1945 not published).

With annexes, including the budget of the Dutch East Indies (1871–1932 separately published as annex B, after 1932 as annex no. 4) and the colonial report from 1849 forward (1849–1877 also published as a separate publication, 1877–1932 as annex C, 1932–1939 as a separate publication titled *Indische verslag).*

With annual and from 1847 forward decennial indexes to personal names and subjects. From *c.* 1825 the catchword "overzeesche bezittingen" (overseas possessions) refers to proceedings and annexes concerning loans for colonial purposes, from 1830 the most important catchwords are "koloniën" (colonies) and "Indië" (Indies), but also towns and countries (outside Indonesia) are inserted as catchwords if one of these territories had been discussed in one of the Chambers.

**Uitvoerend Bewind (Executive Government), 1798–1801** [7]

The 5 directors of the Executive Government were appointed by the Constituent Assembly (see no. 4). From May 5th 1798 it was subordinate to the Representative Body (see no. 4). The directors were assisted by 8 agents (ministers). The Executive Government was principally responsible for the enforcement of the laws and resolutions of the Representative Body. After the coup of June 12th 1798 an Intermediair Bestuur (Intermediary Administration) was established, getting its definite staffing on August 17th 1798 under its old name of Executive Government. On the nomination of the Representative Body the 5 directors appointed 8 agents, viz. of Foreign Affairs (see no. 38), Naval Affairs (see no. 69), War (see no. 71), Finance, Justice, Interior Police, National Education and National Economy. They also appointed the members of the Committee for the Affairs of East Indies Trade and Possessions, later the Council for the Asian Possessions and Establishments (see no. 18 and 19). The government continued in office till the coup of September 18th 1801 (see no. 8).

The archives consist of the secret minutes with annexes 1798–1801 of respectively the Executive Government, the Intermediary Administration and the Executive Government; non secret minutes 1798–1801 with annexes up to March 31st 1798; papers received 1798–1801, including papers of the Committee for the Affairs of East Indies Trade and Possessions 1798–1800 and the Council for the Asian Possessions and Establishments 1800–1801; papers despatched 1798–1801; registers of publications 1798–1801 (the alphabetic register is to be found in the archives of the Grand Pensionary, see no. 9); bundles of miscellaneous papers.

Entries: "Groene index" (green index, see no. 10); inventory.

Size: *C.* 76 running metres (715 inventory items).

See also *Publicatiën gearresteerd by de Nationaale Vergadering, representeerende het volk van Nederland;* vol. 1 (17 January 1798) – vol. 10 (29 December 1809). 's-Gravenhage, 1798–1809. With alphabetic indexes. From January 22nd 1798 published under the title: "Publicatiën gearresteerd by de Constitueerende Vergadering".

From January 30th 1798 published under the title: "Publicatiën gearresteerd by het Uitvoerend Bewind".

From June 12th 1798 published under the title: "Publicatiën gearresteerd by het Intermediair Bestuur".

From August 17th 1798 published under the title: "Publicatiën gearresteerd by het Uitvoerend Bewind".

From October 17th 1801 published under the title: "Publicatiën gearresteerd by het Staatsbewind".

From May 31st 1805 published under the title: "Publicatiën van Hun Hoog Mogenden, vertegenwoordigende het Bataafsch Gemeenebest".

From June 9th 1806 published under the title: "Publicatiën van Zijne Majesteit den Koning van Holland".

**Staats-Bewind (State Government), 1801–1805** [8]

After the coup of September 18th 1801 the State Government became the most important body in the Netherlands. As executive power it took over the responsibilities of the Executive Government (see no. 7), but it also appointed the members of the Legislative Body (see no. 5). It was composed of 12 members, of which 7 were elected by the Executive Government. The other 5 were elected by the 7 appointed members. The Body was assisted by i.a. the Secretaries of State for Foreign Affairs (see no. 38), Naval Affairs (see no. 69), War (see no. 71), Interior Affairs and a Council of Finance. The Council for the Asian Possessions and Establishments (see no. 19) also became subordinate to the State Government. April 29th 1805 the body was dissolved and the Grand Pensionary (see no. 9) took over its responsibilities.

The archives consist of non secret minutes with annexes October 17th 1801 – April 29th 1805; secret minutes 1801–1805; papers received 1801–1805 of several Secretaries of State and councils, including those of War and Naval Affairs, Council for the Asian Possessions and Establishments, Committee for Naval and Colonial Affairs, War Committee; papers despatched 1801–1805; registers of publications 1801–1805 (the alphabetic register to this is to be found in the archives of the Grand Pensionary, see no. 9); deposited archives of various regional departments; secret minutes of the meeting 1805 of the Commissie betreffende de Oost-Indische Handel (Committee for the East Indies Trade).

Entries: "Groene index" (green index, see no. 10); inventory.

Size: *C.* 59 running metres (538 inventory items).

See also

*Notulen van het Staats-Bewind der Bataafsche Republiek, October 1801 – 9 Juny*

*1806*. 's-Gravenhage, 1801–1806. 17 vol. With index. From April 29th 1805 published under the title: "Staatsbesluiten der Bataafsche Republiek". With index.

*Publicatiën gearresteerd by de Nationaale Vergadering, representeerende het volk van Nederland;* vol. 1 (17 January 1798) – vol. 10 (29 December 1809). 's-Gravenhage, 1798–1809. From October 17th 1801 – May 31th 1805 published under the title: "Publicatiën gearresteerd by het Staats-Bewind". With alphabetic index.

**Raadpensionaris (Grand Pensionary), 1805–1806** [9]

On April 29th 1805, at the insistence of Napoleon, R.J. Schimmelpenninck took over the responsibilities of the State Government (see no. 8) as Grand Pensionary. His tasks included appointing the members of the Staatsraad (State Council; see no. 13) and the Secretaries of State for Foreign Affairs (see no. 38), Naval Affairs (see no. 69), War (see no. 71), Interior Affairs, Finance and the Council of Finance. The Council for the Asian Possessions and Establishments (see no. 19) was also subordinate to him. June 4th 1806 he was compelled to resign and Louis Napoleon became King of Holland.

The archives consist of secret and non secret resolutions 1805–1806; papers received of several Secretaries of State and councils, including those of Naval Affairs and War and the Council for the Asian Possessions and Establishments; papers despatched 1805–1806; petitions 1805–1806; registers of publications 1805–1806 with alphabetic indexes 1798–1809.

Entries: "Groene index" (green index, see no. 10); inventory.

Size: *C.* 11.50 running metres (98 inventory items).

See also

*Notulen van het Staats-Bewind der Bataafsche Republiek, October 1801–9 Juny 1806*. 's-Gravenhage, 1801–1806. 17 vol. With index. From April 29th 1805 published under the title: "Staatsbesluiten der Bataafsche Republiek". With index.

*Publicatiën gearresteerd by de Nationaale Vergadering, representeerende het volk van Nederland;* vol. 1 (17 January 1798) – vol. 10 (29 December 1809). 's-Gravenhage, 1798–1809. From May 31st 1805 – June 9th 1806 published under the title: "Publicatiën van Hun Hoog Mogenden, vertegenwoordigende het Bataafsch Gemeenebest".

**Staatssecretarie Koning Lodewijk Napoleon (State Secretariat King Louis Napoleon), 1806–1810** [10]

June 5th 1806 Louis Napoleon became King of Holland. He took over the responsibilities of the Grand Pensionary (see no. 9), which included appointing the ministers of Foreign Affairs, Naval and Land Forces, Finance, Interior Affairs. The supreme government of the colonies was also vested in him. After the incorporation of the Netherlands into France the King abdicated July 12th 1810.

The French representative in the Netherlands then became the Prins-Stedehouder (Prince-Viceregent). These archives include papers of him up to 1811.

The archives consist of secret and non secret royal decrees June 9th 1806 – July 12th 1810, including secret resolutions concerning J. van Kal and the VOC; papers received 1806–1810 of nearly all higher public bodies, including of the Ministry of Trade and Colonial Affairs 1806–1807 and of the Ministry of Naval and Colonial Affairs 1808–1810; minutes 1806–1810, including those of the Minister-Secretary of State and the Minister-Vice-President of the Staatsraad (State Council); miscellaneous papers, including papers received 1810 of the Ministry of Naval and Colonial Affairs and accounts of the colonies; papers 1810–1811 of the period of the Empire, originating from the Prince-Viceregent, including minutes October 1810–December 1811.

Entries: "Groene index" (green index); this is a repertory for all the proceedings and resolutions of the central governing bodies from 1798 (see no. 4, 5, 7–10). The indicator to this shows which archives contain papers mentioned in the "groene index". The most important headings concerning Asia in the "groene index" are "Oost-Indische handel" (East Indies trade), 1798–1811; "Aziatische bezittingen" (Asian possessions), 1802–1806; "Koloniale troepen" (Colonial forces), 1803–1810 and "Coloniën van de staat" (Colonies of the State), 1808. In addition to the "groene index" with indicator there is also an inventory.

Size: *C.* 73 running metres (668 inventory items).

See also

*Decreten en besluiten van Zijne Majesteit den Koning van Holland, 9 Juny 1806–31 December 1808.* 's-Gravenhage, 1808, [1809]. With index.

*Publicatiën gearresteerd by de Nationaale Vergadering, representeerende het volk van Nederland;* vol. 1 (17 January 1798) – vol. 10 (29 December 1809). 's-Gravenhage, 1798–1809. From June 9th 1806 published under the title: "Publicatiën van Zijne Majesteit den Koning van Holland". With alphabetic index.

**Prins-Stedehouder van de Keizer van Frankrijk in de Hollandsche Departementen (Prince-Viceregent of the Emperor of France in the Dutch Departments), 1810–1813** [11]

After the incorporation of the Netherlands into France, July 12th 1810, Louis Napoleon (see no. 10) was discharged from his function and a Prince-Viceregent was appointed to execute the policy of the French government in the Netherlands. In the middle of November 1813 French troops left the Netherlands and Willem I took over the sovereignty. Naturally the Prince-Viceregent became unnecessary then.

The archives, incomplete and rather inaccessible, in large part consist of papers concerning accountancy and clerical matters, viz. resolutions January 12th 1811 – November 6th 1813; papers of functionaries in the Dutch departments, i.a. concerning the navy and the colonies; papers originating from functionaries in

France. The minutes 1810–1811 of the Prince-Viceregent are to be found in the archives of Louis Napoleon (see no. 10).

Entry: Inventory.

Size: *C.* 7 running metres (59 inventory items).

**Algemeene Staatssecretarie/Kabinet des Konings (General State Secretariat/The King's Cabinet), 1813–1840, 1840–1897** [12]

The State Secretariat started to work November 1813. It was meant as a link between the King and his ministers and its responsibilities included making copies of royal decrees and despatching them. Next to the State Secretariat the King's Cabinet was active. The Cabinet was intended more for the King personally and its tasks included dealing with reports, petitions, etc. directed to him. In 1840 both administrations were abolished. A new institution, named the King's Cabinet, came in their place. The principal task of this body was to copy, despatch and preserve the royal decrees.

The archives consist of two parts, viz. the archives of the General State Secretariat and the King's Cabinet 1813–1840 and the archives of the King's Cabinet 1840–1897. The nucleus of these archives is a series of royal decrees 1813–1897 with systematic and alphabetic indexes, and in addition, especially in the King's Cabinet for the period 1813–1840, papers concerning Asia: under the heading "algemeen" (general) petitions 1813–1814 of J.A. Kelfken for a post in Batavia, under the heading "buitenlandse zaken" (foreign affairs) consular reports 1830 from Canton, under the heading "koloniën" (colonies) reports 1808–1815 by P. van IJzendoorn, D. van Hogendorp, G.A.G.Ph. van der Capellen, D.F. van Alphen about finance, military matters, civil servants, Java, Banka, Billiton, correspondence 1815–1824 between Willem I and G.A.G.Ph. van der Capellen, lists of consignments of money 1815, reports 1815–1818 about colonial trade and troops, papers 1827 of L.P.J. du Bus de Gisignies with a letter from J. Hunt about the coffee-culture, papers 1763–1816 about Chr. Schreuder who died on the way to Banda, papers 1787–1815 of W.C.H. van Lynden van Blitterswijk about slave trade, in the papers not on the post-book 1840–1897 i.a. papers 1840–1848 concerning J.D. Kruseman, H.G. Nahuys van Burgst and military matters.

Entries: Systematic and alphabetic indexes. Papers concerning Asia are to be found under the headings "koloniale troepen" (colonial troops), "Oost-Indische militaire zaken" (East Indies military matters), "Oost-Indische zaken" (East Indies affairs), occasionally also under other headings like "handel" (trade), Koloniaal Werfdepôt (Colonial Recruiting Depot), China. In addition to the above mentioned entries there are also inventories 1813–1840 and 1840–1897. The inventory 1813–1840 was published in cyclostyle: H. Bonder. *De archieven van de Algemeene Staatssecretarie en van het Kabinet des Konings met de daarbij gedeponeerde archieven over 1813–1840.* 's-Gravenhage, 1938.

Size: *C.* 1525 running metres (respectively 6809 and 4534 inventory items).

**Staatsraad (State Council), 1805–1810** [13]

The Grand Pensionary (see no. 9) appointed the 5 to 9 members of the State Council. When in 1806 Louis Napoleon (see no. 10) came to the throne the Council was enlarged to 13 members, including the 4 ministers of state. The Council was an advisory body and was presided by the Grand Pensionary and later by the King. It was divided into the following sections: Legislation; Finance; Navy; War; Trade and the Colonies; War, Navy, Trade and the Colonies (1807). After the incorporation of the Netherlands into France in 1810 the Council was abolished.

The archives consist of the minutes May 6th 1805 – January 1808 with systematic and alphabetic indexes; minutes of the several sections, including Trade and the Colonies July 17th 1806 – March 8th 1807, July 8th 1807 – May 20th 1809 and War, Navy, Trade and the Colonies March 9th – July 7th 1807 with systematic and alphabetic indexes; papers about various subjects, including minutes 1807 of a meeting with the Council for the Asian Possessions and Establishments concerning P.H. Breton.

Entries: Besides the above said indexes also the "groene index" (green index, see no. 10); inventory.

Size: *C.* 11 running metres (148 inventory items).

See also

*Register der besluiten van de Staats-raad der Bataafsche Republiek, 6 Mey 1805–25 November 1805.* 's-Gravenhage, 1805.

**Raad van State (Council of State), 1814–1914, 1914–1957** [14]

The Council was established by Royal Decree of April 6th 1814 as an advisory body to the King, up to 1848 to the King personally, later to the King and ministers. In 1843 the Council was also charged with attuning of the governments in the home country and the colonies. In 1862 five specialized sections were established besides the section Geschillen van Bestuur (Differences of Government), viz. of Interior Affairs, Finance and Foreign Affairs, Justice, Colonies, War and Navy. From 1881 forward there were besides the section Geschillen van Bestuur as many sections as there were ministries. After *c.* 1930 few colonial matters were dealt by the Council.

The archives consist of three parts, viz. the archives 1814–1914, the archives of the section Geschillen van Bestuur 1914–1944 and the archives of the section Dienstplichtzaken (Matters of Military Service) 1946–1957. The nucleus of the archives 1814–1914 is a series of minutes 1814–1914 with postbooks, systematic and alphabetic indexes, and in addition secret minutes 1818–1912, minutes concerning East and West Indies matters 1859–1861 and papers of the several sections including the section of the Colonies 1862–1923.

Entries: Postbooks, systematic and alphabetic indexes. Papers concerning the colonies are mainly to be found under the following headings of the index:

"Koophandel en koloniën" (Trade and the colonies) from 1814; "Publiek onderwijs, nationale nijverheid en koloniën" (Public education, national industry and the colonies) from 1819; "Nationale nijverheid en koloniën" (National industry and the colonies) from 1824; "Marine en koloniën" (Naval and colonial affairs) from 1826; "Koloniën" (Colonies) from 1829. In addition to the above mentioned entries also inventories 1814–1914 and 1914–1957.

Size: *C.* 250 running metres (2476 inventory items).

Consultation of papers more recent than 50 years only with permission.

**Raad van Ministers (Council of Ministers), 1823–1945** [15]

The Raad van Ministers was established by Royal Decree of September 19th 1823. It was a meeting of all the ministers together about common affairs. The Raad was established to promote the uniformity of government policy.

The archives mainly consist of a series of minutes 1823–1939 with annexes from 1840 and postbooks and systematic and alphabetic indexes.

Entries: Postbooks and systematic and alphabetic indexes, and in addition an inventory 1823–1925. The part 1926–1945 has no inventory.

Size: *C.* 30 running metres.

Consultation of papers more recent than 50 years only with permission.

**Rekenkamer (Audit Department), 1796–1810** [16]

The responsibility of the Audit Department was the financial check of the public revenue. They were successively called Generaliteitskamer (Chamber of the Generalty), 1796–1799; Commissarissen voor de Nationale Rekening (Commissioners for the National Account), 1799–1802; Nationale Rekenkamer (National Audit Department), 1802–1809 and Hof van Rekeningen (Court of Accounts), 1809–1810. After the incorporation into France the Court was abolished. The Algemene Rekenkamer (General Audit Department), see no. 17, was re-established in 1813.

The archives consist of minutes with annexes with postbooks, systematic and alphabetic indexes, and in addition series and papers kept separately.

Entry: Besides the above mentioned indexes there is also a list..

Size: *C.* 200 running metres.

**Algemene Rekenkamer (General Audit Department), 1813–1900** [17]

The Algemene Rekenkamer is a continuation of the Hof van Rekeningen (Court of Account), see no. 16. Its responsibility was to check the financial management of the public revenue.

The archives consist of secret and non secret minutes with annexes with postbooks, systematic and alphabetic indexes, and in addition divisional archives and series and papers kept separately, including registers of Dutch-East Indies pensions, accounts and statements Dutch East Indies 1874–1900 and papers Nederlandsche Handel-Maatschappij (Dutch Trading Company).

Entry: Besides the above mentioned indexes an inventory on index cards.

Size: *C.* 1040 running metres.

The part of the archives after 1900 is deposited with the Algemene Rekenkamer itself.

## Ministries [18–81]

### Colonial Affairs [18–37]

For the other archives see no. 276–277.

#### Comité tot de Zaken van de Oost-Indische Handel en Bezittingen (Committee for the Affairs of East Indies Trade and Possessions), 1796–1800 [18]

December 24th 1795 the States General replaced the Administration of the VOC by a new administration called Committee for the Affairs of East Indies Trade and Possessions. It was composed of 4 departments, viz. 1. Department for Defence, 2. Department for Interior Administration of the Indies and the Cape, 3. Department for Trade and Equipage, 4. Department for Finance. March 1st 1796 the Committee started its activities. It was completely subordinate to the National Assembly (see no. 4), which had set up a visiting committee for the colonies. In 1798 the Executive Government (see no. 7) took over the power of the National Assembly concerning the Committee. By decree of the Executive Government of May 6th 1800 the Committee was dissolved dating from May 15th 1800.

The archives consist of secret and non secret resolutions and papers despatched 1796–1800 with alphabetic subject indexes; papers received 1791–1799 from i.a. commissioners general, governor general, Raden van Indië (Councils of the Indies) and Council of Justice at Batavia, functionaries in Bengal, Coromandel, Ceylon, Galle, Malabar and Surat; archives 1796–1800 of the above mentioned departments of the Committee and of special committees including the Commissie tot de Retourschepen in Portugal (Committee for the Returnships in Portugal), the Commissie tot de Indische Expeditiën (Committee for the Indies Expeditions) and the Commissie tot de Directe Vaart en Handel op China (Committee for the Direct Trade to China); papers received later, i.a. from the Departement tot het Huishoudelijk Bestuur van Indië en de Kaap (Department for the Interior Administration of the Indies and the Cape), 1796; the Departement tot de Commercie en Equipage (Department for the Trade and Equipage), 1800; miscellaneous papers, 1793–1799. For the names and subjects mentioned in the inventory see the index.

Entry: Inventory of the archives belonging to the Ministry of Colonial Affairs, 1796–1813.

Size: *C.* 21 running metres (244 inventory items).

**Raad der Aziatische Bezittingen en Etablissementen (Council for the Asian Possessions and Establishments), 1800–1806** [19]

By decree of the Executive Government of May 6th 1800 the Committee for the Affairs of East Indies Trade and Possessions (see no. 18) was dissolved and replaced by the Council for the Asian Possessions and Establishments. The Council was composed of 3 departments, viz. 1. Department for Indies Affairs, 2. Department for Trade and Maritime Matters, 3. Department for Finance. From 1801 (January 1st of that year the final organization became operative) the Ministries of War and Naval Affairs (see no. 69, 71) took care of matters concerning colonial defence. The Council was subordinate to the Executive Government (see no. 7), from 1801 to the State Government (see no. 8) and from 1805 to the Grand Pensionary (see no. 9). By Royal Decree of July 29th 1806 the Council was abolished dating from August 1st.

The archives consist of resolutions 1800–1806 with alphabetic subject index, secret resolutions 1800–1806 with annexes and alphabetic subject index and papers despatched with alphabetic subject index; papers received 1796–1805 from the governor general and councillors, Council of Justice at Batavia and supercargo's and Commercieraad (Council of Commerce) at Canton; archives 1800–1806 of the above mentioned departments of the Council and of special committees including the Commissie tot de Indische Expeditiën (Committee for the Indies Expeditions), the Commissie tot de Directe Vaart en Handel op China (Committee for the Direct Trade to China) and the Commissie tot Onderzoek naar het Tekort van wijlen de Commissaris over de Inlanders Mr. C.H.C. Wegener (Committee for Inquiry into the Defecit of the late Commissioner for the Natives Mr. C.H.C. Wegener); archives from the Indies and the Cape including those of the commissioners general 1793–1799; papers sent over of i.a. the commissioners general 1795–1801, papers of S.C. Nederburg 1798–1800 and of A.H. Wiese 1802; miscellaneous papers 1781–1806. For the names and subjects mentioned in the inventory see the index.

Entry: Inventory of the archives belonging to the Ministry of Colonial Affairs, 1796–1813.

Size: *C.* 34 running metres (407 inventory items).

**Ministerie van Koophandel en Koloniën/Marine en Koloniën (Ministry of Trade and Colonial Affairs/Naval and Colonial Affairs), 1806–1807, 1808–1810** [20]

By Royal Decree of July 6th 1806 P. van der Heim was appointed director general of Trade and Colonial Affairs. July 29th 1806 the Asia Council (see no. 19) and the America Council (to which from 1801 the West Indies Committee was added) were abolished and dating from August 1st 1806 united into the Ministry of Trade and Colonial Affairs, subordinate to King Louis Napoleon (see no. 10). The

Ministry was composed of a First Division for the East Indian Affairs, a Second Division for the West Indian Affairs and a Third Division for the Affairs of Trade. The reorganization, mainly for the purpose of bringing the colonial affairs under a one-head government, was only meant to be temporary. January 14th 1808 the Ministry of Naval and Colonial Affairs was established (see also no. 69). It was composed of a General Secretariat, a First Division for the accountancy of the Ministry, a Second, Third and Fourth Division for naval affairs, a Fifth Division for supervision of the military forces in the colonies, a Sixth Division for the government of the colonies and a Seventh Division for the sale of colonial products. In November 1808 the organization was simplified. Besides the General Secretariat for i.a. accountancy of Naval and Colonial Affairs, a First and Second Division for naval affairs were instituted, a Third Division for the civil, military and political affairs of the colonies (the former Sixth Division) and a Fourth Division for the sale of colonial products and supply of the colonies (the former Seventh Division). After the incorporation of the Netherlands into France the Ministry became by Imperial Decree of June 27th 1810 and dating from January 1st 1811 part of the French Ministry of Naval and Colonial Affairs.

The archives consist for the greater part of papers concerning Asia. The papers relating to the West Indies, America, Trade and the Navy have been made into separate archives. The "Aziatische archief" (Asian archives) consist of 3 parts, viz. 1. the archives of the Ministry of Trade and Colonial Affairs consisting of minutes, reports etc. 1806–1807 of the minister with annexes, alphabetical and chronological registers and index, "verbalen" (minutes), reports etc. 1806–1807 of the several bureaux of the First Division; 2. the archives of the Ministry of Naval and Colonial Affairs consisting of "verbalen" (minutes), reports etc. 1808–1810 of the minister with alphabetic index and chronological register, "verbalen" (minutes), reports etc. 1808–1812 of the several bureaux of the Sixth Division (later Third Division), "verbalen" (minutes), reports etc. 1808–1813 of the Seventh Division (later Fourth Division); 3. papers received later consisting of reports and memorandums 1806–1813, papers sent over from the Indies 1805–1810, papers of the First Division of the Ministry of Trade and Colonial Affairs 1807, papers of the Sixth Division (later Third Division) of the Ministry of Naval and Colonial Affairs 1806–1810, papers of the Commissie tot de Kerkelijke Zaken (Committee for the Church Affairs) 1801–1810 and miscellaneous papers 1807–1812. For the names and subjects mentioned in the inventory see the index.

Entry: Inventory of the archives belonging to the Ministry of Colonial Affairs, 1796–1813.

Size: *C.* 12 running metres (225 inventory items).

**Hollandse Divisie bij het Ministerie van Marine en Koloniën te Parijs (Dutch Division of the Ministry of Naval and Colonial Affairs at Paris), 1810–1814** [21]

After the incorporation of the Netherlands into France the policy concerning the colonies was made in Paris. The Dutch Division was only concerned with clerical duties.

These archives, containing few papers relating to the policy concerning the colonies, consist of papers received and despatched 1810–1814 of the head of the Dutch Division; papers 1806–1816 received from the Indies and China by the Ministry in Holland and at Paris; "verbalen" (minutes) 1808–1814 of the commissioner for the affairs of the colonies at Amsterdam; papers 1810–1813 concerning the Coast of Guinea. For names and subjects mentioned in the inventory see the index.

Entry: Inventory of the archives belonging to the Ministry of Colonial Affairs, 1796–1813.

Size: *C.* 1.50 running metres (53 inventory items).

**Comptabiliteit betreffende de Oost-Indische bezittingen (Accountancy concerning the East Indies possessions), 1795–1813** [22]

After the liquidation of the VOC a new Commerciekantoor (Commercial Office) was established, mainly concerned with the accountancy and the trade of the colonies and with the liquidation of the former Company.

The archives consist of journals, ledgers, cashbooks and other accounting series and papers 1785–1814 of the collector general and chief bookkeeper; accounts, lists etc. 1736–1825, mainly concerning pay and pensions of the servants of the Company, coming from the pay-office; muster-rolls etc. 1788–1807 of various ships coming from the muster-office; journals, ledgers and other accounting papers 1787–1808 coming from the office of Amsterdam; accounting papers 1765–1810 coming from the office of Delft; journal, ledger etc. 1730–1813 coming from the office of Enkhuizen; journal, ledger etc. 1729–1813 coming from the office of Hoorn; accounting papers 1713–1813 coming from the office of Rotterdam; accounting papers 1747–1813 coming from the office of Zeeland. For the names and subjects mentioned in the inventory see the index.

Entry: Inventory of the archives belonging to the Ministry of Colonial Affairs, 1796–1813.

Size: *C.* 14.50 running metres (275 inventory items).

**Oost-Indische troepen (East Indian troops), 1796–1808** [23]

This collection contains papers concerning the garrison at Leeuwarden; minutes, papers received and despatched, weekly and daily reports 1796–1806 coming from the troop-depot for the Indies and Cape of Good Hope at Hoorn; papers 1802–1807 concerning the 22nd infantry battalion at Cape of Good Hope.

Entry: Inventory of the archives belonging to the Ministry of Colonial Affairs, 1796–1813.

Size: *C.* 2 running metres (53 inventory items).

**Uit Engeland overgezonden stukken betreffende Azië en Kaap de Goede Hoop (Papers concerning Asia and Cape of Good Hope sent over from England), 1791–1797** [24]

This collection consists for the greater part of papers before 1796. The following papers (25 inventory items) are relating to Asia from 1796 forward: secret letters 1792–1797 from and to Bima and Bulekomba; secret daily record with annexes, secret resolutions and secret letters 1792–1797 from Makasar; secret letters 1792–1797 from and to Malabar (Maros); secret letters 1792–1797 from and to Saleyer.

Entry: Inventory of the archives belonging to the Ministry of Colonial Affairs, 1796–1813.

Size: *C.* 3 running metres (118 inventory items).

**Ministerie van Koloniën (Ministry of Colonial Affairs), 1814–1900, 1932** [25]

By Royal Decree of April 6th 1814 the Department of Trade and Colonial Affairs was established, with a secretary of state at its head. From 1818 the following ministries were established: in 1818 Education, National Industry and Colonial Affairs, in 1824 National Industry and Colonial Affairs, in 1825 Naval and Colonial Affairs, in 1830 Water Management, National Industry and Colonial Affairs, in 1834 Colonial Affairs, in 1840 Naval and Colonial Affairs, from 1842 Colonial Affairs continued as an independent ministry.

Although Colonial Affairs were part of greater ministries up to 1842 (with the exception of the period 1834–1840), colonial matters were treated separately from the beginning and separate archives were made with their own entries such as postbooks, systematic and alphabetic indexes. The policy concerning the Dutch colonies was made at the Ministry. The archives of Colonial Affairs are its paper residue. Besides political papers drawn up in the Netherlands, these archives contain also reports, memorandums, letters etc. from the colonies, which have been important for the making of the policy.

The archives consist of 3 parts, viz. the ministerial archives, deposited archives and documentation.

A. *Ministerial archives*
These archives consist of series and miscellaneous papers concerning the East and West Indies and the Coast of Guinea. They are divisible into:
   a. The non secret archives consisting of the non secret "verbaal" (minutes, also called resolutions) 1814–1900, with postbooks, systematic and alphabetic indexes; smaller series and bundles of miscellaneous papers of a general nature and concerning special subjects.
   b. The secret/cabinet archives consisting of the secret/cabinet "verbaal" (minutes) 1814–1900 with postbooks, systematic and alphabetic indexes; smaller series and bundles of miscellaneous papers.
   c. Series miscellaneous papers coming from the Section East Indies Affairs, including secret and non secret registers of the "handelingen en besluiten"

(proceedings and resolutions) of the governor general 1816–1932 and "stamboeken" (rolls) of civil servants from 1814 forward.

d. Series and miscellaneous papers coming from the Section Military Affairs, including "stamboeken" (rolls) from 1815 forward.
e. Series and miscellaneous papers coming from the Section West Indies and the Coast of Guinea.
f. Mail reports 1869–1900 consisting of those reports from the East Indies colonies that have not been inserted in the non secret or secret/cabinet "verbaal" (minutes).

For the names and subjects mentioned in the inventories see the index.

B. *Deposited archives*

a. Raad van Koophandel en Koloniën (Council of Trade and the Colonies), including minutes 1814–1817.
b. Commissaris voor de Koloniën, etablissementen Amsterdam en Middelburg (commissioner for the Colonies, establishments of Amsterdam and Middelburg), including papers received and despatched 1813–1829 with indexes.
c. Nederlandsche Geoctroyeerde Maatschappij voor den Chinaschen Theehandel (Dutch Chartered Company for the Chinese Tea Trade), consisting of papers 1791–1818.
d. Miscellaneous papers 1814–1850 concerning i.a. China and British India.

For the names and subjects mentioned in the inventory see the index.

C. *Documentation*

Included among the documentation are: indexes to East Indies newspapers 1810–1923 and to Dutch newspapers concerning East Indies matters 1843–1947; a collection of maps (see no. 274). The library of the Ministry has been transferred to the University Library of Leyden (see no. 473).

*Classification.* The archival documents are filed chronologically in sequence of the date of attention. This means: a paper received by the Ministry was entered in the postbook on the date of receipt and numbered. Then it was submitted to the minister, who passed it to the appropriate section. After the matter concerned had been dealt with the draft direction was submitted to the minister for his sanction. This sanction given, papers were drawn up for despatch and sent to the interested party. The draft direction itself remained at the Ministry and was filed with the annexes (the paper received and possible preceding papers which had been used in connection with the matter in hand) on the date and number of the despatched direction, after having been indexed (see "entries"). In this way the series of "verbalen" (minutes) came into being, of which the non secret "verbaal" (minutes) became by far the most voluminous.

In principle all ministerial papers should have been entered up either in the non secret "verbaal" (minutes) or in the secret/cabinet "verbaal". Up to 1850, however, there was a tendency to keep several smaller series and bundles of miscellaneous papers at the section where they had been attended to.

The series of registers of the proceedings and resolutions of the governor general (see A.c) have always been kept separately, being an isolated case, sent over from the Indies. They were copies of registers drawn up in the Indies, which also served the purpose of being evidence of the proceedings of the government there. In addition, after 1869 the separate series of mail reports (see A.f) came into existence. These reports were sent to the minister for information about the policy of the East Indies government. At first they were drawn up as one consecutive account. Later this was changed into lists stating the numbers and summaries of the most important papers drawn up in the Indies, with copies of these papers included as annexes. The reports are filed partly in the non secret "verbaal", partly in the secret/cabinet "verbaal". The reports not filed in these "verbalen" have been made into separate series.

*Entries.* Apart from the aforesaid postbooks the principal entries to the "verbalen" are the systematic indexes (with alphabetic indexes to the systematic ones), written up annually. In the systematic indexes the papers were entered according to heading, with statement of the contents and reference to the place where they can be found (that is the date and number on which the direction is filed in the "verbaal"). To the systematic indexes "klappers" (alphabetic indexes of catchwords and names) have been compiled, with references to the headings and to pages in those headings of the systematic indexes. The alphabetic and systematic indexes are attuned to the need of the clerical department of the Ministry. Headings (they were called "hoofden") and catchwords were drawn up or left out whenever the clerical department thought it expedient for its own work, or when they were forced by practice, e.g. when matters came up which could not be entered in the existing headings. The entry of papers under certain headings naturally depends of the interpretation of the contents of those papers by the civil servants of that time. Frequently the researcher, whose questioning being different from the questioning of the clerical department, will find the papers under other headings or catchwords than he would have expected. So he will have to imagine himself in the place of the clerk of the time to find the papers he is looking for. In many cases it will be easier to find the papers by means of "objective" data such as exact dates, place and personal names, than by means of the "subjective" headings and catchwords.

Besides the number of headings also their contents sometimes changed with time. During the first half of the nineteenth century matters concerning non Europeans: "Inlanders, Chinezen en Vreemde Oosterlingen" (Natives, Chinese and Alien Orientals) were mainly entered under the headings "Inlanders" (Natives), "Justitie en politie" (Justice and police), "Onlusten" (Disturbances) and "Zeeroverij" (Piracy). So officials like regents, wedono's and assistant-wedono's were not entered under the heading "Ambtenaren en civiele personen" (Officials and civil persons), which was set apart for Europeans, but under "Inlanders" (Natives), unless it was a matter in which the European government was also mixed up. From 1861 there was a growing tendency to enter all matters relating to the government, which means also relating to the "Inheems bestuur" (Native government), under the new made heading "Bestuur van Nederlandsch-Indië" (Government of the Dutch East Indies). From 1898 only papers concerning the governor general and the Raad van Indië (Council of the Indies) were entered under this heading. In addition to this new heading "Binnenlandsch Bestuur" (East Indies Civil Service) was created. In 1909 the last mentioned heading was divided into "Binnenlandsch Bestuur; algemeen" (East Indies Civil Service; general) and "Binnenlandsch Bestuur; decentralisatie" (East Indies Civil Service; decentralization). In 1918 the "broader" headings were superseded by "narrower" catchwords. After 1919 there is no index to the non secret "Register der handelingen en besluiten "buiten rade" van de gouverneur-generaal" (Register of the proceedings and resolutions "without interference of the Council of the Indies" of the governor general), from which the above mentioned examples have been drawn. Indexes and registers of the proceedings and resolutions and of the "verbalen" are in principle alike. The difference is that also papers concerning e.g. the West Indies, were entered under the headings of the indexes to the "verbalen". In this connection the headings of the last mentioned indexes were often divided according to colony.

Finally the "klapper" (alphabetic index of catchwords and names to the systematic index) naturally only mentions those names and subjects which are mentioned in the systematic index. The result of this is that there are very few Indonesian personal names mentioned compared with the number of European names. For, as a rule only the function or quality of Indonesian persons was stated. The name of the "hoofdmuiter in Oost-Indië, Diepo Negoro" (chief mutineer in the East Indies Diponegoro) was mentioned in the "klapper" of the non secret "verbaal" in connection with attempts at negotiations as recently as 1829 and only in 1830 he was mentioned in the "Register der handelingen en besluiten "buiten rade" van de gouverneur-generaal", viz. in connection with his arrest. Only the secret "klapper" mentions him as early as 1827. Under the heading of the systematic index to the secret "verbaal": "Indische vorsten (betrekkingen met de)" (East Indian princes, relations with) a letter from the King to the commissioner general has been entered by which he "verlangt de gedachte te vernemen omtrent het geven van een gedeelte van het Mataramsche rijk aan den opstandeling Dipo Negoro" (desired to know his opinion on the idea to give a part of the Mataram empire to the rebel Diponegoro). Papers concerning this matter are also to be found by means of the catchwords "Java (onkosten van den oorlog op)" (Java, expenses of the war on) or "Djocjocarta" in the "klapper". In the systematic index they have principally been entered under the heading "Onlusten" (Disturbances).

*Inventories.* Besides the aforesaid postbooks, and systematic and alphabetic ("klappers") indexes to the non secret and secret "verbaal", there are also two inventories, giving a systematically arranged description of the archives. The first inventory covers the period 1814–1849. The second covers the period 1850–1900, 1932.

*Size.* The archives 1814–1849 *c.* 500 running metres (4713 inventory items), 1850–1900 *c.* 1000 running metres (9151 inventory items). For the archives from 1900 and the "stamboeken" (rolls) from *c.* 1850 see no. 276.

**Koloniën, supplement (Colonial Affairs, supplement), 1664–1952** [26]

This collection contains papers originating from the Ministry of Colonial Affairs, the Regeringsvoorlichtingsdienst (Government Information Service) and military intelligence services (mainly concerning the period 1945–1949) and of papers of origin unknown.

The collection has the following subdivisions: Dutch East Indies before 1942 consisting of the headings "staatkundig" (political), including treaties with Indonesian princes from 1664, military affairs, economics and finance, geography, sundries; Dutch East Indies from 1942 consisting of the headings "Japans militair bestuur" (Japanese military government), period 1945–1952. For the names and subjects mentioned in the inventory see the index.

Entry: Inventory.

Size: *C.* 3.50 running metres (145 inventory items).

**Directie van de Indische Pensioenfondsen (Direction of the East Indies Pension Funds), 1815–1961** [27]

These archival materials are principally important because of the personal notes of civil servants and soldiers in the East and West Indies and of their widows and orphans. They consist of a series of pension lists, registers 1815–1940, notes, correspondence 1900–1927, annual reports of the Stichting Administratie Indonesische Pensioenen (Foundation Administration of Indonesian Pensions). The remaining part of these archives are filed with the archives of the Colonies from 1900 (see no. 277).

Entry: In addition to the registers above, also alphabetic indexes to personal names and an inventory of part of the archives.

Size: *C.* 80 running metres.

**Commissie tot Ontwikkeling van de Fabrieksnijverheid in Nederlandsch-Indië (Committee for the Development of the Manufacturing Industry in the Dutch East Indies), 1915–1940** [28]

The Committee was instituted by decree of the governor general of September 14th 1915. Its object was to investigate the possibilities of new branches of manufacturing industry, mainly for processing raw produce from the Dutch East Indies, and to advise the government on that subject.

The archives include: minutes up to *c.* 1921; statistical records; reports; documents concerning i.a. blast furnaces in South-East Borneo, coke works at Sawah Lunto, steel industry in Central Celebes, oil fields near Djambi, Chinese and Javanese labour force, social issues, technical education, people's credit system; correspondence.

Size: *C.* 1.50 running metres.

**Dienst Legercontacten (Army Contacts Service), 1939–1950** [29]

The duty of this service was to give information about the army activities in Indonesia. The papers are relating to the period 1945–1949. The archives consist of correspondence, reports etc., and of propagandistic photographs.

Size: *C.* 2.50 running metres.

See also no. 370.

**Gouverneur-Generaal ter Kuste van Guinea (Governor General on the Coast of Guinea), 1816–1872** [30]

The Dutch possessions on the Coast of Guinea were transferred to Great Britain in 1871.

The archives contain papers of the Afrikaansche Werfdepôt (African Recruiting Depot) about Africans destined for the army of the Dutch East Indies.

Entry: Inventory. In *VROA*, 1921, p. 353–382.

Size: *C.* 33 running metres.

Deposited at the First Section.

**Archival materials sent over from Indonesia** [31–37]

The numbers 31–34 were sent from Batavia to the Netherlands in 1863. The numbers 35–37 were sent from Jakarta in 1953.

**Hooge Regeering te Batavia (Supreme Government at Batavia), 1602–1827** [31]

This collection contains papers originating from the Supreme Government, mainly concerning the administration of the outpost factories in Asia, over which it exercised no longer any authority after the "Londens tractaat" (London treaty) of 1824. There are papers concerning the Dutch East Indies.

The collection, to a great extent consisting of papers of before 1796, has the following subdivision: general, Borneo, Timor, China, Japan, Malaya, Sumatra, Bengal, Coromandel, Ceylon, Malabar, Surat, Persia, Cape of Good Hope, North East Coast of Java, Cheribon, Bantam, Batavia and district. Of these *c.* 250 inventory items are relating to Asia from 1796. For the names and subjects mentioned in the inventory see the index.

Entry: Inventory on index cards.

Size: *C.* 20 running metres (1101 inventory items).

Deposited at the First Section.

**Nederlandse bezittingen in Voor-Indië (Dutch possessions in India), 1703–1826** [32]

This collection consists of papers drawn up by the Dutch authorities in India. The papers were sent along to Batavia by the English after they had taken over the Dutch possessions in India in 1825.

The collection has the following subdivision: Bengal 1703–1795, 1819–1822 viz. the resolutions 1819–1822 of the resident and council; Coromandel and Madura 1818–1825 consisting of the "verbaal" (minutes) of the chief factor, papers received and despatched by the governor general and the residencies Palicol, Paleacatta and Tuticorin, instructions of the chief factor for the civil servants; Surat 1722–1803 including resolutions 1794–1797 of the director and council, daily records 1787–1803 by H. Sissingh and papers received 1795–1797, Surat 1818–1826 consisting of papers received and despatched by i.a. the governor general.

Entry: Inventory.

Size: *C.* 3 running metres (162 inventory items).

Deposited at the First Section.

**Nederlandse Factorij in Japan (Dutch Factory in Japan), 1609–1860** [33]

In 1609 the Netherlands hired an establishment on Hirado (Firando) for the purpose of trade to Japan. In 1641 the Factory was compelled to move to Deshima and remained there till its closure in 1860. The personnel of the Factory (from 1639–1854 the only Europeans in Japan) consisted of a merchant with the title 'opperhoofd' (chief factor), from 1858 called government-commissioner, a second merchant, three under merchants and from 1678: 14–15 assistants. Every year at the end of June 2–3 ships sailed from Batavia to Japan and returned at Batavia at the end of October. Imported in Japan were i.a. silk, skins, sapan wood, mercury and in the 19th century textile, sugar, lead and tin. Exported were silver, gold, copper, camphor, lacquered ware and porcelain. After 1840 the Factory changed its nature. In stead of a trading station it became much more a diplomatic post.

The archives are divided into two periods, viz.:

A. *Period 1609–1842,* consisting of:

a. Archives of the chief factor and the council including papers about the establishment of the Factory, resolutions, instructions, daily records, correspondence, papers concerning the relations with the Japanese authorities, trade and internal affairs, ships, personnel, justice;

b. Archives of the warehousekeeper-bookkeeper-secretary, including secretarial and notarial deeds and papers concerning deceased's estates;

c. Archives of the Societeit van Particuliere Handel op Japan (Society of Private Trade to Japan), founded in 1826 at Deshima on the initiative of G.F. Meylan and discontinued in 1831, consisting of resolutions with annexes, papers received, daybooks, accounting papers;

d. Deposited papers and miscellanies.

B. *Period 1843–1860,* consisting of:

a. Archives of the chief factor including the same series as in the archives mentioned under A.a. and also secret archives 1843–1854 relating to the opening of Japan;

b. Archives of the warehousekeeper-bookkeeper-secretary, consisting of papers received and despatched, deeds, papers concerning auction sales;

c. Archives of J.H. Donker Curtius as envoy to the court at Edo 1859–1860.

For the names and subjects mentioned in the inventory see the index.

Entry: M.P.H. Roessingh. *Het archief van de Nederlandse Factorij in Japan. – The archives of the Dutch Factory in Japan, 1609–1860*. 's-Gravenhage, [1968].

Size: *C*. 30 running metres (1952 inventory items).

Deposited at the First Section.

**Nederlandse Factorij in Canton (Dutch Factory at Canton), 1739–1828** [34]

From 1734 the trade to China (i.a. tea, silk and porcelain) was carried on from Batavia. The merchants stayed in a rented house in a suburb of Canton during the trading season. As recently as 1776 a permanent settlement was permitted. From 1756 the trade was conducted from the Netherlands. In Amsterdam the "Chinasche Commissie" (China Committee) was set up. The management in China was entrusted to a director together with merchants (supercarga's) constituting the Commercieraad (Commercial Council). Also assistants, bookkeepers, surgeons and others were employed at the Factory. The responsibility of the Council consisted of the trade, jurisdiction, civil registry and the execution of notarial instruments with which the secretary of the Breede Raad (Broad Council) was charged. In 1809 H.W. Daendels reduced the personnel to a chief factor and some assistants. Finally in 1822 the Factory was closed down.

The archives, to a great extent consisting of papers of the period before 1796, contain:

a. Archives of the Commercial Council, consisting of papers of a general nature including resolutions 1742–1809, papers received 1747–1820, 'Bataviasche instructies' (Batavian instructions) 1739, 1743–1756, 'Patriasche instructies' (instructions from the Netherlands) 1756–1805, reports received from Batavia and the Netherlands up to 1814, papers despatched to Batavia and the Netherlands up to 1822, daily records 1743–1816; papers of a special nature 1779–1808; papers concerning Dutch and English embassies to Peking; financial papers 1797–1823.

b. Archives of the secretary of the Broad Council (charged with the execution of notarial instruments), consisting of protocols 1800–1815.

c. Private correspondence 1800–1810 of the supercarga's.

For the names and subjects mentioned in the inventory see the index.

Entry: J.J.G. Bolduan. *Inventaris van het archief van de Nederlandsche Factorij te Canton*. ['s-Gravenhage, 1939].

Size: *C*. 8 running metres (393 inventory items).

Deposited at the First Section.

**"Algemene Secretarie" te Batavia (General Secretariat at Batavia), 1942–1950** [35]

These papers consist of a great number of random files mainly relating to the Japanese period in Indonesia and the period 1945–1949. The collection is subdivided into:

a. *Eerste zending (First despatch)* for the greater part consisting of papers originating from the Algemene Secretarie and of papers of the Secretarie van de Luitenant-Gouverneur-Generaal (Secretariat of the Lieutenant-Governor-General), the Departement van Justitie (Department of Justice), the Regeringsvoorlichtingsdienst (Government Information Service), Islamitische Zaken (Islamic Affairs), the Departement van Binnenlands Bestuur (Department of the Indies Civil Service), the Kantoor voor Japanse Zaken (Office for Japanese Affairs), the Hooggerechtshof (High Court of Justice), the Procureur-Generaal (Public Prosecutor) and the Departement van Verkeer en Energie (Department of Transport and Power). For the names and subjects mentioned in the file list see the index.

b. *Tweede zending (Second despatch)* also for the greater part consisting of papers originating from the General Secretariat and of papers originating from other services, including the delegation negotiating on behalf of the Dutch government with the Republic Indonesia about the transfer of sovereignty. For the names and subjects mentioned in the lists see the index.

Entries: File list first despatch and inventory on index cards second despatch.

Size: *C.* 64 running metres.

Permission is needed for consultation of the papers.

**Procureur-Generaal bij het Hooggerechtshof (Public Prosecutor with the High Court of Justice), 1942, 1945–1949** [36]

These papers consist of a number of files originating from the Public Prosecutor. Only a very small part of the papers from the collection concerns the Japanese period in Indonesia. The remaining part contains, besides political papers of the period 1945–1949, many files concerning Dutchmen and Indonesians during that period. For the names and subjects mentioned in the file list see the index.

Entry: File list.

Size: *C.* 23 running metres (894 files).

Permission is needed for consultation of the papers.

**Archivalia betreffende Nieuw-Guinea (Archives concerning New Guinea), 1919–1949** [37]

These papers concerning New Guinea originate from various public services, including the General Secretariat. The collection, for the greater part consisting of papers after 1945, has the following subdivision: papers concerning the political situation in South East Asia; papers concerning the government in New Guinea;

papers concerning military affairs. For the names and subjects mentioned in the inventory see the index.

Entry: Inventory.

Size: *C.* 1.50 running metres (202 inventory items).

## Foreign Affairs [38–68]

For the other archives see no. 278–335

### Ministerie van Buitenlandse Zaken (Ministry of Foreign Affairs), BuZa 1795–1810 [38]

By decree of April 3rd 1795 of the Provisioneele Representanten van het Volk van Holland (Provisional Representatives of the Dutch People) the Comité tot de Buitenlandsche Zaken (Committee for Foreign Affairs) was appointed. By decree of March 25th 1796 this was replaced by the Commissie tot de Buitenlandsche Zaken (Commission for Foreign Affairs). The members were appointed by the National Assembly (see no. 4). When the Executive Government was established (see no. 7) the Department of Foreign Affairs was set up on January 25th 1798, subordinate to the Executive Government, and in April 1798 an Agent of Foreign Relations was appointed. December 1st 1801 he got the title of Secretary of State, being subordinate to the State Government (see no. 8) and from 1805 to the Grand Pensionary (see no. 9). In 1806 the Department became a ministry under King Louis Napoleon (see no. 10). By decree of July 15th 1810 of the Prince-Viceregent (see no. 11) the Ministry of Foreign Affairs was abolished.

The archives consist of minutes and "verbalen" (for an explanation see no. 25: classification) April 3rd 1795–July 3rd 1810; papers despatched 1796–1810; papers received 1796–1810 from authorities in the Netherlands, including those of War 1796–1810, Naval Affairs 1796–1808, Committee for the Affairs of the East Indies Trade and Possessions 1796–1800, Council for the Asian Possessions and Establishments 1800–1806, Ministry of Trade and Colonial Affairs 1806–1807, Ministry of Naval and Colonial Affairs 1808–1810; papers received 1796–1810 from diplomatic representatives from Algeria, Baden, Bavaria, Denmark, France, Great Britain, Hanseatic towns, Hessen-Kassel, Italy, Mainz, Austria, Portugal, Prussia, Russia, Spain (i.a. Chr.A. Ver Huell 1807–1808), Tanger, Tripolis, Tunis, Turkey, United States (i.a. from R.G. van Polanen 1797–1802), Westphalia (i.a. from D. van Hogendorp 1807–1808), Württemberg, Sweden and Switzerland; a series of treaties, including one with Spain about the defence of Java 1800.

Entries: Systematic and alphabetic indexes. The most important headings concerning Asia are: "Oost-Indische handel en bezittingen" (East Indies trade and possessions), 1796–1800; "Oost-Indische expeditie" (East Indies expedition), 1796, 1799, 1801–1803; "Aziatische bezittingen en etablissementen" (Asian possessions and establishments), 1802–1806; "Koophandel en koloniën" (Trade and colonies), 1807 and "Marine en koloniën" (Naval and colonial affairs), 1808–1810. In addition to the systematic and alphabetic indexes there is also an inventory in typescript.

Size: *C.* 44.50 running metres (442 inventory items).

**Ministerie van Buitenlandse Zaken (Ministry of Foreign Affairs), BuZa 1813–1870, 1870–1940** [39]

By sovereign decree of December 7th 1813 a Ministry of Foreign Affairs was again established after the French period (see no. 38, 40), headed by a secretary of state, from 1815 forward by a minister. From 1814–1824 the Ministry was composed of a bureau for political, juridical, protocol and secret financial affairs and of a bureau for consular and trading affairs. In 1824 both bureaux were united. In 1861 the Cabinet of the minister was separated from it. From 1876 the bureaux for political affairs and for consular and trading affairs were re-established, and in addition a section for domestic and financial matters was founded.

The archives up to 1870 include series of non secret, secret and very secret correspondence and minutes with postbooks, alphabetic and systematic indexes etc. and files arranged according to section. After 1870 these archives are subdivided into the 'groot archief' (large archives) and the 'kabinetsarchief' (archives of the cabinet). The 'groot archief' includes a series of A-files (concerning political and diplomatic affairs), B-files (concerning consular and trading affairs) and after 1918 DEZ-files (originating from the Direction Economic Affairs and separated from the B-files). The archives up to 1870 are deposited as a whole at the ARA. In addition the ARA keeps a series of ratifications by foreign powers of treaties, conventions etc. concluded with the Netherlands 1814–1900, and a supplement of papers 1814–1940, including the ratifications 1900–1940. The A-files up to 1918 have been transferred to the ARA. For this survey however, these papers are still classed as part of the archives deposited with the Ministry of Foreign Affairs (see no. 280). For the names and subjects mentioned in the inventories see the index.

Entries: Postbooks, systematic and alphabetic indexes etc. Papers concerning Asia are mainly to be found under the headings "consulaire zaken" (consular affairs), "koloniën" (colonies), "koophandel en zeevaart" (trade and shipping), names of the countries and towns where the several legations and consulates were established, the names of the ministries to which Colonial Affairs did belong up to 1848 (see no. 25). In addition to the aforesaid entries also inventories of the sections of the archives 1813–1830, in *VROA*, 1918, p. 291–330; 1831–1848, in *VROA*, 1921, p. 111–195, including also ratifications up to 1900; 1849–1870, in *VROA*, 1923, p. 174–205; inventory in typescript of the supplement 1814–1940.

Size: *C.* 1200 running metres (3579 inventory items, ratifications 593 items).

**Dutch legations and consulates, 1741, 1778, 1786, 1795–1811** [40]

The Dutch legations and consulates were abolished whenever Foreign Affairs (see no. 38) were closed down. Archives are present of the following stations:

a. Antwerp (consulate) 1804–1810. 13 inventory items.
See also no. 41, 285.

b. Baden 1807–1809. 16 inventory items.

c. Denmark 1793–1809. 29 inventory items. Included are papers 1797–1799 of the Committee for the Affairs of the East Indies Trade and Possessions.
See also no. 42, 55, 304, 320.

d. German Empire (Vienna) 1786–1808. 50 inventory items.

e. France 1795–1810. 200 inventory items. Included are papers 1807–1810 of C.H. Ver Huell.
See also no. 44, 293.

f. Hanseatic towns 1795–1809. 21 inventory items.
See also no. 46.

g. Hessen-Kassel 1796–1804. 10 inventory items.
See also no. 43.

h. Upper und Lower Rhine Districts 1778–1800. 9 inventory items.

i. Philadelphia (consulate) 1795–1801. 2 inventory items.
See also no. 54, 317.

j. Portugal 1802–1808. 7 inventory items. Included are correspondence 1802–1806 of C.H. Grasveld, papers 1807 concerning H.W. Daendels, correspondence 1806–1807 of J. Fraser.
See also no. 49, 60, 307.

k. Prussia 1795–1810. 43 inventory items. Included are papers 1803 concerning Chr. Jadich, soldier in Ceylon.
See also no. 50.

l. Russia 1791–1809. 20 inventory items. Included is correspondence 1803–1805 of D. van Hogendorp and 1803–1808 of C.H. Ver Huell.
See also no. 51, 308.

m. Spain 1795–1810. 12 inventory items. Included are papers of Chr. A. Ver Huell and papers concerning the Philippines.
See also no. 52, 311.

n. Turkey 1741–1811. 445 inventory items.
See also no. 53, 313.

o. Württemberg 1797–1809. 19 inventory items.
See also no. 43.

p. Sweden 1795–1806. 22 inventory items.
See also no. 42, 55, 66, 304, 320.

Entry: Inventory.

Size: *C.* 22.50 running metres (939 inventory items).

**Dutch legations after 1813** [41–56]

Present are the archives of the posts in Argentina 1880–1920; Bavaria, Württemberg and Baden 1814–1867; Belgium 1839–1890, 1897; Denmark 1814–1863; German League and Frankfurt, Nassau, Hessen and Electoral Hessen 1816–1867; France 1814–1884; Great Britain and Ireland 1814–1914; Hannover 1838–1866; Hanseatic towns 1815–1841; Italy 1863–1910; Japan 1870–1891; Papal States 1826–1830; Austria 1814–1842; Persia 1890–1906; Portugal 1823–1898; Portugal and Brazil 1815–1821; Prussia 1814–1900; Roumania 1889–1910; Russia 1814–1890; Saxony 1828–1856; Sardinia 1818–1855; Serbia 1903–1911; Spain 1814–1899; Tuscany 1826–1828; Turkey 1814–1874; United States 1814–1909; Sweden and Norway 1814–1862; Sweden, Norway and Denmark 1863–1910; Switzerland 1817–1915. According to the inventories the archives given below contain papers concerning Asia.

**Belgium, 1839–1890, 1897** [41]

Included are correspondence 1839–1890 with postbooks and systematic and alphabetic indexes and papers kept separately up to 1897 i.a. concerning Belgians in the Dutch East Indies army 1853–1889, purchase of arms for Atjeh 1873, C.H.A.F. Catenius 1880–1886, J.R.H. Gisser 1889–1890, J.A. Alartin 1897, C. Roeder and E. Werleman 1893, C.F.W. Wichers van Kerchem 1882–1885, eruption of the Krakatau 1883–1884, archives of the Brussels committee of the fund for support of the relatives of soldiers killed in Lombok 1894 (the Dutch envoy F.H. van Stralen was a member).

Entry: Inventory.

Size: *C.* 19.50 running metres (241 inventory items).

See also no. 40 a, 285.

**Denmark, 1814–1863** [42]

Including correspondence 1814–1862 with postbooks and systematic and alphabetic indexes and papers up to 1863 kept separately, including the Royal Danish Asian Company 1787, commercial treaties 1816–1863 of Denmark and the Hanseatic towns with Japan.

Entries: Inventory. In *VROA*, 1918, I, p. 389–390; supplement inventory on index cards.

Size: *C.* 5 running metres (79 inventory items).

See also no. 40 c, 40 p, 55, 66, 304, 320.

**German League and Frankfurt, Nassau, Hessen and Electoral Hessen, 1816–1867** [43]

Including papers concerning the establishment of consulates of Württemberg (1857–1858) and Hessen (1863–1864) in the Dutch East Indies.

Entries: Inventory. In *VROA*, 1918, I, p. 374–377; supplement inventory on index cards.

Size: *C.* 0.75 running metre (36 inventory items).

See also no. 40 g, 40 o.

**France, 1814–1884** [44]

Including correspondence 1814–1884 with postbooks and papers kept separately i.a. correspondence 1814–1815 with the Ministry of Colonial Affairs concerning Java and the admission of French ships there.

Entries: Inventory. In *VROA*, 1918, I, p. 352–374; supplement inventory on index cards.

Size: *C.* 33.50 running metres (363 inventory items).

See also no. 40 e, 293.

**Great Britain and Ireland, 1814–1914** [45]

Including correspondence 1814–1914 with postbooks and systematic and alphabetic indexes and papers kept separately up to 1914, i.a. correspondence with the Ministry of Colonial Affairs concerning the Dutch and English colonies. For the names and subjects mentioned in the inventories see the index.

Entries: Inventory. In *VROA*, 1918, I, p. 343–351; supplement inventories on index cards.

Size: *C.* 38 running metres (581 inventory items).

See also no. 61, 314.

**Hanseatic towns, 1815–1841** [46]

Including correspondence 1826–1831 about recruiting of soldiers for the Dutch East Indies army.

Entry: Inventory. In *VROA*, 1918, I, p. 387–388.

Size: *C.* 2 running metres (22 inventory items).

See also no. 40 b.

**Japan, 1856, 1870–1890** [47]

The archives consist of general correspondence 1870–1890 with postbooks and of correspondence kept separately from 1856 mainly concerning matters of trade and

Dutchmen in Japan. For the names and subjects mentioned in the inventory see the index.

Entry: Inventory on index cards.

Size: *C.* 2.50 running metres (57 inventory items).

See also no. 62, 67, 301.

**Persia, 1890–1906** [48]

The archives have the following subdivision: papers received and despatched, postbooks, miscellanies.

Entry: Inventory on index cards.

Size: *C.* 2 running metres (44 inventory items).

See also no. 298.

**Portugal, 1823–1828, 1840–1898** [49]

Including correspondence 1823–1828, correspondence 1840–1895 with postbooks, papers under the heading "zaken van strafrecht" (matters of criminal law) concerning the arrest of the Malay crew of a Dutch ship on Madeira 1856–1857, papers under the heading "zaken betreffende koloniën" (matters concerning the colonies) i.a. about Timor, Solor, Sulu Babi 1841–1895, papers under the heading "miscellanies" about J.J. Lopez de Lima 1856.

Entries: Inventory. In *VROA*, 1918, I, p. 380–382; supplement inventory on index cards.

Size: *C.* 5 running metres (101 inventory items).

See also no. 40 j, 60, 307.

**Prussia, 1814–1900** [50]

Including correspondence 1814–1900 with postbooks and alphabetic indexes, after 1876 A- and B-files, papers 1845–1900 concerning Prussian soldiers in the Dutch East Indies army, papers 1881 concerning the establishment of a Chinese legation in the Netherlands, papers concerning the Dutch colonies 1883–1887 and concerning the German colonial policy 1879–1887.

Entries: Inventory. In *VROA*, 1918, I, p. 380–382; supplement inventory on index cards.

Size: *C.* 20 running metres (295 inventory items).

See also no. 40 k.

**Russia, 1814–1890** [51]

Including papers 1876 concerning the promotion of the Dutch interests in Persia by the Russian legation.

Entries: Inventory. In *VROA*, 1918, I, p. 392–396; supplement inventory on index cards.

Size: *C.* 9.50 running metres (154 inventory items).

See also no. 40 L, 308.

**Spain, 1814–1913** [52]

Including correspondence 1814–1900 with postbooks, papers under the heading "handels- en verkeerszaken" (matters of trade and transport) about an exhibition of colonial trade at Amsterdam 1882–1883, papers under the heading "zaken betreffende koloniën" (matters concerning the colonies) about piracy in the Sulu archipelago 1855–1868, government in the Philippines 1863, consuls in the colonies 1863–1866, 1859–1870 (under the heading "algemeen" (general).

Entries: Inventory. In *VROA*, 1918, I, p. 424–428; supplement inventory on index cards.

Size: *C.* 17 running metres (272 inventory items).

See also no. 40 m, 311.

**Turkey (also consulate), 1814–1874** [53]

Including correspondence 1814–1849, letters from the Ministry of Colonial Affairs 1815–1829, papers 1870 concerning the protection of Turks against Persians, commercial treaty Belgium-Persia 1842.

Entries: Inventory. In *VROA*, 1918, I, p. 396–419; supplement inventory on index cards.

Size: *C.* 13.50 running metres (343 inventory items).

See also no. 40 n, 313.

**United States, 1814–1909** [54]

Including correspondence 1814–1909 with postbooks, letters 1816–1830 from i.a. the governors general of the Dutch East Indies [G.A.G.Ph. van der Capellen and L.P.J. du Bus de Gisignies], letters 1825–1830 to the Ministry of Colonial Affairs, papers concerning trade agreements with China 1853–1879 and Japan 1871–1874, papers concerning the import of Sumatra-tobacco 1882–1890, correspondence 1881–1883 about the colonial exhibition in Amsterdam, correspondence 1894–1898 about the appointment of a consular agent in Makasar.

Entries: Inventory. In *VROA*, 1918, I, p. 431–435; supplement inventory on index cards.

Size: *C.* 9.50 running metres (208 inventory items).

See also no. 40 i, 317.

**Sweden, Norway and Denmark, 1863–1910** [55]

Including commercial treaty 1863 of Denmark with China.

Entry: Inventory on index cards.

Size: *C.* 7 running metres (88 inventory items).

See also no. 40 c, 40 p, 42, 66, 304, 320.

**Switzerland, 1817–1915** [56]

Including correspondence 1817–1915 with postbooks, correspondence 1828–1830 of A.L.W. Seyffardt with the Ministry of Colonial Affairs, papers concerning the opium conference 1912–1915, Russo-Japanese War 1904–1905, frontier dispute between the Netherlands and Portugal on Timor 1913, NOT 1915, Swiss soldiers in the Dutch East Indies army 1894–1915.

Entries: Inventory. In *VROA*, 1918, I, p. 421–424; supplement inventory on index cards.

Size: *C.* 10.50 running metres (278 inventory items).

See also no. 58, 59, 68, 321.

**Dutch consulates after 1813** [57–68]

Present are the archives of the posts in Alexandria 1815–1825; Athens and Nauplia 1834–1846; Berlin 1867–1877; Bern consulate general 1818–1901, vice consulate 1865–1889, consulate general for Luxembourg 1880–1890; Bordeaux 1873; Caracas 1859–1877; Coruna 1831–1886; Cyprus 1829–1831; Elmina 1872–1880; Geneva 1850–1915; Guatemala 1853–1859, 1901–1906; Kanea 1828–1833; Lausanne 1892–1894; Lisbon 1814–1861; London 1814–1909; Mahon 1817–1870; Mogador 1845–1897; Mozambique 1874–1876; Nagasaki 1860–1915; Ningpoo 1863–1886; New York 1818–1917; Penang 1872–1908; Philadelphia 1869–1903; Rio de Janeiro 1826–1831; Singapore 1863–1908; Stockholm 1821–1920; Tangier 1830–1907; Truxillo 1826–1834; Tunis 1814–1872; Yokohama 1860–1870; Zürich 1850–1874. According to the inventories the archives given below contain papers concerning Asia.

**Berlin (Consulate general), 1867–1877** [57]

Including correspondence 1867–1868 with the Ministry of Colonial Affairs.

Entry: Inventory on index cards.

Size: *C.* 0.50 running metre (13 inventory items).

**Bern (Consulate general), 1818–1901** [58]

Including papers concerning Swiss soldiers in the Dutch East Indies 1827–1899, Koloniaal Werfdepôt (Colonial Recruiting Depot) 1869–1884, Ministry of Colonial Affairs 1867–1888, consular officials in the Dutch East Indies 1875–1889, import and export duties 1865–1874, decorations in the Atjeh war 1873–1876, deceased's estates 1830–1893, eruption of the Krakatau 1883, rules for residence in Java 1838, Militair Invalidenhuis (Home for Disabled Soldiers) in Bronbeek, diplomatic service in Japan 1861–1870, trade treaties with Japan 1881–1887, medicine in the Dutch East Indies 1879–1890, J. Loudon 1875, E. de Waal 1871, A. Hottinger 1847–1865.

Entry: Inventory on index cards.

Size: *C.* 6.50 running metres (351 inventory items).

See also no. 56, 59, 68, 321.

**Geneva, 1850–1915** [59]

Including papers concerning Swiss soldiers in the Dutch East Indies army 1870–1915.

Entry: Inventory on index cards.

Size: *C.* 3 running metres (159 inventory items).

See also no. 56, 58, 68, 321.

**Lisbon, 1814–1861** [60]

Including letters from the Ministry of Colonial Affairs 1816–1823, [A.A.] Buyskes 1815, papers 1799–1808 of the Committee for the Affairs of the East Indies Trade and Possessions, the Council for Asian Possessions and Establishments and the Ministry of Colonial Affairs.

Entries: Inventory. In *VROA,* 1918, I, p. 438–446; supplement inventory on index cards.

Size: *C.* 3 running metres (74 inventory items).

See also no. 40 j, 49, 307.

**London, 1814–1909** [61]

Including papers 1859–1860 of Reynst & Vinju and J.W. May concerning the sale of copper ore struck at the Western district of Borneo.

Entry: Inventory on index cards.

Size: *C.* 8 running metres (181 inventory items).

See also no. 45, 314.

**Nagasaki (vice consulate), 1860–1915** [62]

These archives consist of correspondence 1860–1908 with postbooks, series of civil registry 1861–1910, register of Dutchmen 1882–1915, notarial instruments 1860–1872, circular letters, files. For the names and subjects mentioned in the inventory see the index.

Entry: Inventory on index cards.

Size: *C.* 2 running metres (94 inventory items).

See also no. 47, 67, 301.

**Ningpo, 1863–1886** [63]

These few archives consist of registers of births, marriages and deaths 1877–1882, 1886, lists of ships arrived 1873–1875, papers concerning the colonial exhibition at Amsterdam 1883, treaty between the Netherlands and China 1863, printed matter 1863.

Entry: Inventory on index cards.

Size: *C.* 0.30 running metre (8 inventory items).

See also no. 290.

**Penang, 1872–1908** [64]

These archives consist of correspondence 1872–1908 with systematic and alphabetic indexes, mainly concerning Atjeh. For the names and subjects mentioned in the inventory see the index.

Entry: Inventory.

Size: *C.* 7 running metres (119 inventory items).

**Singapore (Consulate general), 1863–1908** [65]

These archives consist of correspondence and telegrams 1863–1902 with postbooks (including Atjeh reports 1899–1900), registers of consular rights 1885–1905, registers of signing on and signing off 1900–1908.

Entry: Inventory.

Size: *C.* 12 running metres (156 inventory items).

See also no. 310.

**Stockholm (Consulate general), 1821–1920** [66]

Including letters from the Department of Agriculture, Industry and Trade at Buitenzorg 1916–1919.

Entry: Inventory on index cards.

Size: *C.* 7.50 running metres (168 inventory items).

See also no. 40 c, 40 p, 42, 55, 304, 320.

**Yokohama (Consulate general), 1860–1870** [67]

These archives consist of correspondence 1860–1870 with postbooks, papers 1860–1862 received and despatched during mission journeys to Edo, import and export duties 1862–1869, Nederlands Marinehospitaal (Dutch Naval Hospital)

1864–1869, Marine Kolendepôt (Naval Coaling Depot) 1865–1869, cashbook 1861–1865, storehouse and kambang accounts 1860–1862, Japanese receipts.

Entry: Inventory on index cards.

Size: *C.* 2 running metres (30 inventory items).

See also no. 47, 62, 301.

**Zürich, 1850–1874** [68]

Including papers 1850–1874 concerning Swiss soldiers in the Dutch East Indies army.

Entry: Inventory on index cards.

Size: *C.* 0.20 running metre (9 inventory items).

See also no. 56, 58, 59, 321.

## Naval Affairs [69–70]

**Ministerie van Marine (Ministry of Naval Affairs), 1795–1813** [69]

February 25th 1795 the Comité tot de Zaken der Marine (Committee for the Affairs of the Navy) was established, which became subordinate to the National Assembly (see no. 4) in March 1796. In February 1798 an Agent of Naval Affairs was appointed, who was subordinate to the Executive Government (see no. 7). January 1st 1802 the Raad van Marine (Council of Naval Affairs) was established by the State Government (see no. 8), being subordinate to that body. May 1805 the Grand Pensionary (see no. 9) appointed a Secretary of State. During the reign of Louis Napoleon (see no. 10) in June 1806 the Ministry of Naval Affairs was established, amalgamating with Colonial Affairs in January 1808 (see no. 20). The Ministry of Naval and Colonial Affairs was, after a few rearrangments, finally composed of a General Secretariat i.a. for the accountancy of Naval and Colonial Affairs, a First Division for personnel and military affairs of the Navy, a Second Division for the material of the Navy and a Third and Fourth Division for colonial affairs. After the incorporation of the Netherlands into France the policy concerning Naval and Colonial Affairs was laid down in Paris (see also no. 11). In the Netherlands current affairs were finished under the Prince-Viceregent. From 1801–1808 Naval Affairs have dealt with the affairs concerning the navy in the colonies (see also no. 19–20).

The archives, not easy of access and in 1844 half destroyed and half damaged by fire, consist of 3 parts:

a. Resolutions, "verbalen" (for an explanation see no. 25: classification), minutes 1795–1810 of the Ministry; papers 1795–1813 originating from authorities in the Netherlands and France, including fragments of letters from the Council for the Asian Possesssions and Establishments concerning the government of the colonies 1804; despatches 1795–1813 to authorities in the Netherlands and

France, including letters 1801–1805 to the Asian Council; filed secret and non secret papers, including papers 1800 concerning the military and political organization of the East Indies and Cape of Good Hope, memorandum 1807 of J.J. Melvill van Carnbee concerning the defence of Java, letters 1807 from [J.W.] de Winter, [A.] Kikkert and [A.A.] Buyskes concerning their expedition to the East Indies, papers 1801 concerning G. Masson, reports 1803 of [S.] Dekker, A. Melissen and A.A. Buyskes concerning a Dutch squadron in the East Indies, court martial papers 1797 concerning S.W.B. Lucas, register of papers despatched 1805–1806 of A.A. Buyskes, reports and letters 1802–1804 of S. Dekker about his voyage to the East Indies, journal-minutes 1807 of P. Hartsinck as commander of the naval forces east of Cape of Good Hope, "verbaal" of the proceedings of A.G. Besier 1799–1807 with alphabetic index.

b. List of ship registers, chronologically arranged, of naval craft 1795–1810, including lists of exchanged prisoners of war s.a. after the surrender of Cape of Good Hope and Ceylon by [S.W.B.] Lucas.

c. Logbooks etc. of naval craft, originating from the collection of admiralty boards, including a logbook by A. van der Waal Cornelisz of his voyage through the East Indies archipelago 1800–1801 under the command of I. van Kervel, logbook A.A. Buyskes of his voyage through the East Indies archipelago 1800–1804, logbook J. Hasteke of the voyage from Cape of Good Hope to the East Indies 1802–1806 under the command of J. Claris, logbook W. Courrech of the voyage from Cape of Good Hope to the East Indies 1802–1804 under the command of J. Claris, logbook J.H. Dibbetz of his voyage through the East Indies archipelago 1802–1805 with notes, logbook J.J. Melvill van Carnbee of the voyage from the East Indies to Cape of Good Hope 1802–1805, logbook of the voyage from Cape of Good Hope to the East Indies 1803–1804 under the command of A.A. Buyskes, logbook P. Hartsinck of his voyage through the East Indies archipelago 1804–1806, logbook J.J. Melvill van Carnbee of his voyage from Cape of Good Hope to the East Indies 1804–1806, papers of [J.W.] de Winter including a list of officers and petty officers in the East and West Indies 1798.

Entry: Manuscript inventory with appendix I relating to the ship registers and appendix II relating to the logbooks etc.

Size: *C.* 134 running metres (respectively 968, 1225 (ship registers) and 190 (logbooks) inventory items).

For the maps originating from the Ministry of Naval Affairs see no. 274.

**Ministerie van Marine (Ministry of Naval Affairs), 1813–1900, 1900–1940** [70]

The first minister of Naval Affairs after the French period (see no. 69) was appointed December 7th 1813. At first the Dutch East Indies had their own colonial navy (set up in 1815), but its administration was conducted by the Ministry in the Netherlands. From 1838 the naval forces in the Indies consisted of a Dutch squadron, a colonial and a civil navy. From that time the Dutch squadron would supply the crew of the colonial navy. In 1841 the colonial navy was abolished.

The naval forces in the Dutch East Indies became the responsibility of the ministers of Naval and Colonial Affairs. (Supplements Ph. Kleintjes. *Staatsinstellingen van Nederlandsch-Indië;* 5th ed. Amsterdam, 1927, 2 vol.)

The archives consist of a series of non secret "verbalen" (for an explanation see no. 25: classification) 1813–1900 and secret "verbalen" 1813–1844 with postbooks, systematic and alphabetic indexes. The secret archives 1845–1928 were destroyed by fire, the non secret archives 1901–1928 largely. Of this latter part a number of papers, and the systematic and alphabetic indexes are surviving. Finally the archives contain the "stamboeken" (rolls) 1795–1908 and the logbooks and payrolls of naval craft 1750–1940.

Entries: Postbooks, systematic and alphabetic indexes. Papers concerning the Dutch East Indies may be found i.a. under the headings "Koloniën (Ministerie van)" (Colonial Affairs, Ministry of), "Koloniale troepen" (Colonial troops) and "Koloniale marine" (Colonial navy). Besides these entries there are also alphabetic indexes to the "stamboeken" (rolls), a list (alphabetical to name of the ship) of logbooks, provisional survey of the archives.

Size: *C.* 600 running metres.

For the archives after 1900 see no. 337, for the logbooks after 1940 see no. 359.

## War [71–72]

### Ministerie van Oorlog (Ministry of War), 1795–1813 [71]

By decree of March 24th 1795 of the Provisioneele Representanten van het Volk van Holland (Provisional Representatives of the Dutch People) the Comité tot de Algemeene Zaken van het Bondgenootschap te Lande (Committee for the General Affairs of the Alliance on Land) was established, subordinate to the National Assembly (see no. 4) from March 1796. February 1st 1798 the Executive Government (see no. 7) appointed an Agent of War. The State Government (see no. 8) established the Council of War December 3rd 1801, dissolved May 31st 1803. Then a Secretary of State for the Affairs of War was appointed, who became subordinate to the Grand Pensionary (see no. 9) in 1805. By royal decree of June 29th 1806 Louis Napoleon (see no. 10) set up the Ministry of War. After the incorporation of the Netherlands into France (see no. 11) in 1811 a Secretary General was appointed, charged with the liquidation of the Ministry. From 1801–1810 the Ministry dealt with the military affairs in the colonies (see also no. 19–20).

The archives consist of secret and non secret minutes 1795–1798 with separate annexes; "verbalen" (for an explanation see no. 25: classification) 1798–1810 with separate annexes; "verbalen" and annexes 1810–1813; papers received 1795–1811, including reports 1804 concerning the East Indies troops; regimental "stamboeken" (rolls) of soldiers as a separate series.

Entries: Postbooks, systematic and alphabetic indexes. The most important headings relating to Asia in these are: "Coloniale troupes" (Colonial troops), 1804–1810, in the secret index: "Coloniën" (Colonial Affairs), 1800–1802. Besides postbooks, systematic and alphabetic indexes a manuscript inventory.

Size: *C.* 205 running metres (1547 inventory items).

**Ministerie van Oorlog (Ministry of War), 1813–1913** [72]

November 17th 1813 the first minister of War after the French period (see no. 71) was appointed. Since the Dutch East Indies had its own land forces, for which the minister of Colonial Affairs was responsible, the Ministry of War had little concern with the East Indies army. It arranged only the detailing of soldiers or the sending of expeditionary troops, e.g. during the Java war. (Supplements Ph. Kleintjes. *Staatsinstellingen van Nederlandsch-Indië;* 5th ed. Amsterdam, 1927. 2 vol.)

The core of the archives is a series of "verbalen" (for an explanation see no. 25: classification) 1813–1913 with postbooks and indexes, with in addition "stamboeken" (rolls) of soldiers of the Dutch land forces, including of the Koloniaal Werfdepôt (Colonial Recruiting Depot) at Harderwijk, and moreover the archives contain the "Van Thielen Collection" including papers 1826–1830 about the expeditionary armies to Java.

Entries: Postbooks and indexes. Papers concerning the Dutch East Indies may be found under the headings "Koloniale militaire zaken" (Colonial military affairs), "Koloniale dienst" (Colonial service), Koloniaal Werfdepôt, "Koloniale reserve" (Colonial reserves). Besides these entries there are also alphabetic indexes to the "stamboeken" (rolls) and inventories 1813–1844 and 1813–1913. The inventory 1813–1844 is in cyclostyle under the title: M.D. Lammerts. *Inventaris van het archief van het Ministerie van Oorlog 1813–1844 en van de zogenaamde collectie Van Thielen, 1795–1844.* 's-Gravenhage, 1948.

Size: *C.* 2500 running metres (15,367 inventory items).

For the archives after 1913 see no. 363.

## Algemene Zaken/Kabinet van de Minister-President (General Affairs/ Cabinet of the Prime Minister) [73–75]

**Departement voor Algemeene Oorlogvoering van het Koninkrijk, AOK/Kabinet van de Minister-President (Kingdom's General War Department/Cabinet of the Prime Minister), 1940–1945** [73]

These archives filed by the Dutch government in exile at London, include files concerning Australia (Dutch representation, information, politics, military, school of administration Melbourne, Dutch refugees); India; Burma-road; China (Dutch representation, politics); Far Eastern Commission; Institute of Pacific Relations;

Japan (politics, peace treaty, military, warfare, prisoners of war); Ministry of Colonial Affairs; Dutch East Indies (*Staatsblad van Nederlandsch-Indië*, governor general, pensions, finance, shipping, banking- and monetary system, Javasche Bank, petroleum, military, KNIL, public health, NRK, NICA, government); New Zealand; Philippines; Portuguese Timor; Far East; personal files i.a. concerning H. Colijn, A.H.J. Lovink, J.H. van Mook, R.A.A. Soejono, Ch. [J.I.M.] Welter. The archives are classified according to the UDC.

Entry: File list.

Size: *C.* 12 running metres (91 boxes).

Permission is necessary for consultation of the papers.

**Stukken "Indonesië in overgangstijd"/Nieuw Guinea (Papers "Indonesia during the period of transition"/New Guinea), 1940–1959/1943–1958** [74]

These papers are mainly relating to the recognition of the sovereignty of Indonesia, position of Indonesia after 1949 and the relations with foreign countries. In addition there are a number of papers (15 boxes) concerning New Guinea 1943–1958. The papers are arranged according to the UDC. For the names and subjects mentioned in the file list see the index.

Entry: File list.

Size: *C.* 16 running metres (131 boxes).

Permission is necessary for consultation of the papers.

**Regeringsvoorlichtingsdienst (Government Information Service), RVD 1945–1950** [75]

This collection includes press reports ANEP-ANETA, papers "luisterafdeling" Bandung, Ministerie van Overzeese Gebiedsdelen, Siam, photograph archives with the heading "Indië" (East Indies).

Entry: Provisional survey.

Size: *C.* 1 running metre.

Permission is necessary for consultation of the papers.

## Justice [76–79]

**Krijgsraden te Velde (Army Courts), 1923–1962** [76]

For the history of Asia are especially important the papers from 1942, drawn up by the Army Courts in Australia and Indonesia (soldiers of KL and KNIL), courts martial at sea in Indonesia, temporary courts martial (Japanese war criminals) and the Superior Court Martial in Indonesia. For the names and subjects mentioned in the shelflist see the index.

Entry: Shelflist.

Size: *C.* 30 running metres.

Permission is necessary for consultation of the papers.

**United Nations War Crimes Commission, 1943–1953** [77]

Including some files concerning Asia, viz. Japanese ships; conduct of the war with Japan; Japanese prisoners of war; demobilization; assistance to Japan; Far Eastern Commission; repatriation from China.

The archives are arranged according to the UDC.

Entry: File list.

Size: *C.* 12 running metres.

Permission is necessary for consultation of the papers.

**Gereconstrueerde registers van Burgerlijke Stand van consulaire posten der Nederlanden (Reconstructed registers of births, deaths and marriages of consular posts of the Netherlands), 1860–1969** [78]

Of importance for the history of Asia are the registers of the posts Bangkok, Deshima, Kobe, Nagasaki, Ningpo, Shanghai, Teheran, Tientsin and Yokohama.

Entry: Shelflist.

Size: *C.* 10 running metres.

**Rijksinstituut voor Oorlogsdocumentatie (State Institute for War Records), RIOD 1935–1966** [79]

The Institute belongs to the Ministry of Education and Sciences.

The archives consist of a series of duplicates 1935–1966, of which are relevant for the history of Asia: 1. the "Nederlands-Indische collectie" (Dutch East Indies collection) 1941–1956, including papers NIGIS, ANEP, ANETA, NIBEG, de *Stem van Ambon,* Penjuluh, Australian press reports, Japanese intelligence service; 2. International Military Tribunal for the Far East, Tokyo, against Araki e.a. For the remaining papers of the Institute see no. 561.

Entry: Shelflist.

Size: *C.* 51 running metres.

## Economic Affairs [80–81]

**Netherlands Shipping and Trading Committee, NSTC 1940–1952** [80]

The NSTC was set up during the war with the object of promoting Dutch shipping and trading interests in time of war, with offices in the United States and Great Britain. The committee was dissolved about 1952.

The archives consist of reports, correspondence and other papers received and despatched 1940–1952, including papers of sub-committees like the Netherlands East Indies Section, papers concerning the ships on commission, papers from and/or about i.a. the Ministry of Colonial Affairs, Th. Cremer, Internatio 'Rotterdam', NHM, Nederlandsch-Indische Handelsbank (Dutch East Indies Trading Bank), Holland-China Trading Co., Java, China, Japan, Calcutta, tobacco, coffee, jute.

Entry: Provisional survey.

Size: *C.* 35 running metres.

**Directie Indische Zaken; souvereiniteitsoverdracht (Direction East Indian Affairs; transfer of the souvereignty), 1946–1950** [81]

This collection relating to Indonesia consists of minutes, reports, memorandums, telegrams and other papers 1946–1950 from and/or concerning i.a. the Coördinatie College Nederland-Nederlands-Indië (Co-ordinating Body the Netherlands-Dutch East Indies), the Commissie-Generaal (Committee General), Directoraat Overzeese Zaken (Directorate Overseas Affairs), Economische Commissie voor Azië en het Verre Oosten (Economic Committee for Asia and the Far East), [H.J.] van Mook, [J.A.] Jonkman, Commissie Goede Diensten (Committee of Good Offices), Java Bank, Ronde Tafel-Conferentie (Round Table Conference), Linggardjati, Eastern Indonesia, New Guinea, [J.M.] de Booy, Rubber Study Group, military affairs, Surinam and the Antilles, with register and table of contents.

Entry: Provisional survey.

Size: *C.* 2 running metres.

Permission is necessary for consultation of the papers.

## Private Organizations in the Economic Field [82–143]

For the history of the companies mentioned below was used *Handboek voor cultuur- en handelsondernemingen in Nederlandsch-Indië.* Amsterdam, 1888–1935.

**Nederlandsche Handel-Maatschappij (Dutch Trading Company), NHM 1824–1957** [82]

The NHM was founded by Royal Decree of March 29th 1824, partly by the initiative and monetary assistance of King Willem I. The object of the Company was 'to promote national trade, shipping, fishing and agriculture, factories and business; . . .'. For that purpose several agencies were established in the Netherlands and in foreign countries, including in the United States, Japan, China, Malaya and Burma. In 1826 the Factory of Batavia was set up as head office in the Dutch East Indies. Starting there several agencies were founded, including in 1826 Semarang, Surabaja and Padang (Padang abolished in 1887), and then successively Menado, Makasar and Palembang (later all three of them abolished again). In

1858 the agency of Singapore was established with offices at Medan and Penang. Up to *c.* 1880 the NHM in the Dutch East Indies was principally concerned with the cultivation of and trade in cash-crops such as coffee, tea, tobacco, sugar. Moreover it had a share in the trade between the various islands. After *c.* 1880 it became more and more engaged in banking affairs. (*Gedenkboek van de Nederlandsche Handel-Maatschappij, 1824–1924*. Amsterdam, 1924. W.M.F. Mansvelt. *Geschiedenis van de Nederlandsche Handel-Maatschappij;* uitgegeven ter gelegenheid van het honderdjarig bestaan. Haarlem, 1925, 1926. 2 vol. *Verzameling geschriften;* samengesteld ten kantore van de Nederlandsche Handel-Maatschappij, ten einde te dienen als leidraad bij de voorlichting van Prins Bernhard van Lippe Biesterfeld omtrent het economische leven van Nederland en Nederlandsch-Indië. Amsterdam, 1936).

The archives consist of general correspondence, papers drawn up and received by the Board of Supervisory Directors, the President, the Managing Board, the Secretariat, the First Section to the Fourth inclusive, accounting papers. The papers of the Second and Third Section are especially relevant for the history of Asia, because these include the papers concerning the various agencies and estate companies. For the names and subjects mentioned in the shelflist see the index.

Entry: Shelflist.

Size: *C.* 1000 running metres.

Permission is necessary for consultation of the papers.

**Nederlandsch-Indische Handelsbank (Dutch East Indies Trading Bank), 1863–1967** [83]

The Trading Bank was founded in 1863 in Amsterdam to meet the urgent need of credit institutions in the Dutch East Indies. Agencies were set up in Batavia (1864), Semarang and Surabaja (1865), Indramaju (1900), Singapore (1901), Bandung (1904), Hongkong (1906), Probolinggo and Ampenan (1911), Tjilatjap (1912), Cheribon, Tegal and Pekalongan (1913), Medan (1914), and after 1914 in Amoy, Bangkok, Banjuwangi, Bombay, Calcutta, Djambi, Gorontalo, Kobe, Makasar, Malang, Manilla, Menado, Osaka, Palembang, Pasuruan, Pontianak, Sjanghai, Swatow, Teluk Betung, Tokio and Yokohama. The Bank maintained close relations with the Nederlandsch-Indische Landbouw Maatschappij (Dutch East Indies Agricultural Company). In 1950 the name was changed into Nationale Handelsbank (National Trading Bank).

The archives consist of papers 1863–1967, including articles of association, rules and regulations, minutes, accounting papers, correspondence with the several agencies, files. For the names and subjects mentioned in the shelflist see the index.

Entry: Shelflist.

Size: *C.* 89 running metres (1751 items).

Permission is necessary for consultation of the papers.

## Koloniale Bank (Colonial Bank) [84–91]

Included are the archives of the actual Bank (see no. 84) and of a number of undertakings which were operated by the Bank or were connected with the Bank (see no. 85–91).

### Koloniale Bank (Colonial Bank), 1881–1969 [84]

The Colonial Bank was established at Amsterdam in 1881. Its object included 'to provide operating capital, in the first place in the Dutch East Indies, to agricultural and industrial undertakings', to make loans, 'to administer property of third parties', 'to acquire, operate and alienate agricultural and industrial undertakings on their own account or on joint account with third parties'. The Bank operated mainly sugar undertakings, and it participated in tea-, coffee-, cocoa-, quinine-, rubber-, indigo-, tobacco- and rice-cultures. In 1949 its name was changed into Cultuur-, Handel- en Industriebank, Cultuurbank (Agriculture-, Trade- and Industry Bank, Agriculture Bank) and in 1969 it was liquidated.

The archives consist of articles of association, annual reports, minutes with annexes, correspondence, circulars, accounting papers; annual reports of the companies Gondang Lipoero 1898–1931; Estate Company Kalie Djompo 1919–1956; Kawi-Coffee-Estate Company 1894–1934; Agricultural Company Manishardjo 1910–1939; Estate Company Maron (in 1934 acquired by Gending) 1893–1935; Estate Company Medarie 1907–1953; Estate Company Ophir 1927–1934; Estate Company Randoe Goenting/Tjandi Sewoe 1897–1953; Estate Company Sedatie 1902–1932; Estate Company Seloredjo 1904–1949; Coca Producenten Vereeniging (Coca Producers' Association) 1925–1950; circulars, articles of association, annual reports and other papers of and concerning the companies MONI 1927–1960, SUMCAMA 1930–1960, LEIDWEST 1927–1960, Estate Company Batoe Lempit 1914–1960, NEDINRUB 1938–1960, Cultuur Maatschappij tot Exploitatie der d'Abo-Landen (Estate Company for the Exploitation of the d'Abo-Lands), with parts of Lerep and Rowo Seneng 1924–1954, ONRANI 1922–1969, BENISO 1918–1967, Federatie van Nederlandsch-Indische Bergcultuurondernemingen (Federation of Dutch East Indies Upland Estate Companies) 1954–1969, Vereeniging voor de Theecultuur (Association for Tea Culture) 1955–1968, Vereeniging voor de Kapok- en Cacaocultuur (Association for Kapok- and Cocoa Culture) 1955–1969, Internationale Vereeniging voor de Rubbercultuur (International Association for Rubber Culture) 1954–1969, Koffieproducenten Vereeniging (Coffee Growers' Association) 1955–1968, Vereeniging van Thee-Importeurs (Association of Tea Importers) 1950–1964, Vereeeniging van Kinabast Producenten (Association of Cinchona Bark Producers) 1928–1966, Estate Company Soloredjo 1904–1949; photographs.

Entry: Shelflist.

Size: *C.* 70 running metres.

Permission is necessary for consultation of the papers.

**Cultuur Maatschappij 'Gending' (Estate Company 'Gending'), 1888–1969** [85]

When Gending was established at Amsterdam in 1888, the Colonial Bank possessed 2/5 of the shares, next year half of the shares. The Company operated the agricultural undertaking Gending near Probolinggo. The principal product was sugar. In 1934 it acquired the Estate Company Maron (established in 1893).

The archives consist of annual reports 1892–1957; correspondence i.a. with the deputy at Surabaja 1888–1937; papers concerning the financial records 1888–1962; reports, production records etc.; papers of i.a. J. Kreyger, C. Visser, D. Feikema and the Kartasoera Estate Company; preliminary papers from 1877; maps.

Entry: Shelflist Colonial Bank (see no. 84).

Size: *C.* 10 running metres.

**Cultuur-Maatschappij 'Kanigoro' (Estate Company 'Kanigoro'), 1891–1969** [86]

In 1891 the sugar undertaking Kanigoro came into the possession of the Colonial Bank. The head office was established at Amsterdam. The undertaking was situated in Madiun.

The archives consist of articles of association; annual reports 1911–1951; minutes 1891–1952; correspondence 1896–1905; papers concerning the financial records 1892–1958; reports, production records etc.; papers of i.a. P.C. Brutel de la Rivière; concession Ponorogo; delimitation of the boundaries Purwodadi i.a. of . . . Lagaay; papers Pagotan.

Entry: Shelflist Colonial Bank (see no. 84).

Size: *C.* 6 running metres.

**Cultuurmaatschappij Belirie (Estate Company Belirie), 1892–1969** [87]

When the Estate Company Belirie was established in 1892 at Amsterdam, the Colonial Bank possessed half of the shares. The company operated the coffee- and rubber undertaking Silosanen in Besuki.

The archives consist of annual reports 1892–1958; minutes 1892–1919; private correspondence 1890–1892; correspondence 1892–1933 including of the deputy at Surabaja with annexes 1890–1919, Ruhaak & Co., J.W. Mac Donald, . . . Huber; papers concerning the financial records 1891–1958; reports coffee- and rubber crop 1920–1958.

Entry: Shelflist Colonial Bank (see no. 84).

Size: *C.* 4 running metres.

**Cultuur Maatschappij Klein Getas (Estate Company Klein Getas), 1900–1958** [88]

Established in 1900 at Bodja with the object of operating the coffee- and rubber undertaking Klein Getas.

The archives consist of annual reports, minutes, correspondence i.a. with G.L. d'Abo and the deputy at Surabaja, circulars, crop reports, accounting papers.

Entry: Shelflist Colonial Bank (see no. 84).

Size: *C.* 2.25 running metres.

**Landbouw Maatschappij 'Pager Alam' (Agricultural Company 'Pager Alam'), 1902–1969** [89]

When Pager Alam was established in 1902 at Amsterdam, the Colonial Bank possessed half of the shares. The Company operated three undertakings in Palembang, viz. the coffee undertaking Pager Alam, the tea undertaking Goenoeng Dempo and the quinine undertaking Goenoeng Agoeng, of which the last two undertakings as such were separated from Pager Alam in 1926.

The archives are from 1927–1940 divided according to the three undertakings and consist of annual reports 1929–1940; monthly reports 1902–1958; correspondence 1903–1933, 1958–1959 with A.T. Hughes, papers concerning the financial records 1902–1958; reports, production records etc., coffee-, tea- and quinine crop reports; papers of A. van Karacson 1896, the tea undertaking Koleberes 1929, the acquisition (1955) of the tea undertaking Tandjoeng Keling.

Entry: Shelflist Colonial Bank (see no. 84).

Size: *C.* 6 running metres.

**Landbouw Maatschappij 'Soekamadjoe' (Agricultural Company 'Soekamadjoe'), 1903–1969** [90]

The tea- and rubber undertaking Soekamadjoe was established in 1903 in Surabaja. In 1909 the Colonial Bank purchased 420 of the 500 shares.

The archives consist of annual reports 1927–1938; minutes 1910–1932; papers concerning the financial records 1918–1959; tea- and rubber crop reports 1932–1959, also of Tjipeter Mangala 1955/1956; sundries 1909–1959; lists of products not paid for 1909–1911.

Entry: Shelflist Colonial Bank (see no. 84).

Size: *C.* 2 running metres.

**Bandjaratma, 1910–1969** [91]

In 1910 concession was granted by the East Indian Government for the building of the sugar factory Bandjaratma near Brebes. In 1913 production was started. The factory was in the possession of the Colonial Bank.

The archives consist of annual reports 1927–1937; papers concerning the financial records 1918–1958; planting reports 1910–1958; records concerning machinery and production; technical and chemical reports; papers concerning Petaroekan, Soemberhardjo, Wadoek Pendjalin and Wadoek Pemalie; maps also of Wadoek Pemalie.

Entry: Shelflist Colonial Bank (see no. 84).

Size: *C.* 6 running metres.

**De Ondernemersraad voor Nederlandsch-Indië (The Employer's Council for the Dutch East Indies), ONRANI 1921–1968** [92]

The Council was established August 24th 1921 at The Hague. Its object was to promote the social interests of the undertakings and industries working in the Dutch East Indies.

The archives consist of annual reports, circulars, documentation and correspondence with the associated members. For the names and subjects mentioned in the shelflist see the index.

Entry: Shelflist.

Size: *C.* 28 running metres.

**Federatie van Vereenigingen van Nederlandsch-Indische Bergcultuurondernemingen (Federation of Associations of Dutch East Indies Upland Estate Companies)** [93–99]

Included are the archives of the actual Federation, the five federated associations (see no. 94–98) and the Rubber-Stichting (Rubber Foundation) (see no. 99).

See also no. 620.

**Federatie van Vereenigingen van N.I. Bergcultuurondernemingen (Federation of Associations of Dutch East Indies Upland Estate Companies), 1935–1968** [93]

The Federation was established at Amsterdam in 1935 and started its activities January 1st 1936. In this connection the Vertegenwoordiging van Bergcultures in Nederland (Agency of Upland Cultures in the Netherlands), established in 1926 at The Hague, was abolished. Members of the Federation became the five associations of estate companies mentioned below. The object of the Federation was 'to conduct the administrations, with which it had been entrusted, of the federated associations' and 'to promote the common interests of two or more federated associations'. In 1949 its name was changed into 'Federatie van Verenigingen van Bergcultures in Indonesië' (Federation of Associations of Upland Cultures in Indonesia). From 1961 up to its liquidation in 1968 it was called 'Federatie van Verenigingen van Cultuurondernemingen in Indonesië' (Federation of Associations of Estate Companies in Indonesia).

The archives consist of articles of association; annual reports 1936–1943; budgets 1936–1966; correspondence 1952–1957; minutes 1936–1968; circulars 1934–1968; papers Agency of Upland Cultures in the Netherlands 1928–1933; papers 'agraria' (agriculture) 1918–1961, including legislation, agricultural policy, hereditary lease, papers . . . Sonius, Sumatra Committee; papers 'arbeidsaangelegenheden' (labour matters) 1922–1958, including planters regulations, coolies, recruiting and crimping of labourers, child labour; papers 'crisis cultuurfondsen' (crisis culture stocks);

papers concerning the development after the second world war; papers 'nationalisatie Nederlandse bedrijven' (nationalization Dutch industries); papers 'general subjects', including correspondence with W.J. de Jonge in Batavia 1947.

Entry: Shelflist.

Size: *C.* 30.50 running metres.

Permission is necessary for consultation of the papers.

**Internationale Vereeniging voor de Rubber- (en Andere) Cultures in Nederlandsch-Indië (International Association for the Rubber (and Other) Cultures in the Dutch East Indies), 1913–1970** [94]

The Association was established in 1913 at The Hague. The object of the Association was to promote the rubber cultures in the Dutch East Indies and the interests of its members. Through the intermediary of the Rubber Growers' Association British companies became members too. From 1923–1935 it was called "Internationale Vereeniging voor Rubber- en Andere Cultures" (International Association for Rubber and Other Cultures). In 1935 the Association became a member of the Federation of Associations of Dutch East Indies Upland Estate Companies and its registration was transferred to Amsterdam. In 1949 its name was changed into "Internationale Vereniging voor de Rubber-Cultures in Indonesië" (International Association for Rubber Cultures in Indonesia). From 1961 up to its liquidation in 1970 it was called "Internationale Vereniging voor de Rubber-Cultuur" (International Association for the Rubber Culture).

The archives consist of articles of association; annual reports 1914–1950; budgets 1936–1952; minutes 1914–1955; circulars 1936–1965; papers "rubber restriction" 1920–1957, including trade, prices, legislation in Ceylon and Malaya, native rubber; papers "ordinances, by-laws, books" 1931–1944, including dealings with the rubber restriction in the Volksraad (People's Council), legislation in British territories in Asia; miscellaneous papers.

Entry: Shelflist.

Size: *C.* 8.70 running metres.

**Vereeniging voor de Theecultuur in Nederlandsch-Indië (Association for the Tea Culture in the Dutch East Indies), 1918–1968** [95]

The Association was established in 1918 at Amsterdam. The object was to promote the tea culture in the Dutch East Indies and the interests of its members. In 1935 the Association became a member of the Federation of Associations of Dutch East Indies Upland Estate Companies. In 1949 its name was changed into "Vereniging voor de Theecultuur in Indonesië" (Association for the Tea Culture in Indonesia). From 1961 up to its liquidation in 1968 it was called "Vereniging voor de Theecultuur" (Association for the Tea Culture).

The archives consist of articles of association; annual reports 1918–1955; budgets 1937–1952; correspondence 1936–1956; minutes 1919–1960; circulars 1936–1960;

papers "theerestrictie" (tea restriction) 1932–1958, including legislation, trade contracts, dealing with the tea restriction in the Volksraad (People's Council) by W.J. de Jonge, W.J.Ph. van Waning, papers tea restriction by Th.A. Meister and F. Kramer.

Entry: Shelflist.

Size: *C.* 5 running metres.

**Koffie- (en Cacao) Producenten Vereeniging (Coffee- (and Cocoa) Producers' Association), 1926–1968** [96]

The Association was established in 1926 at Amsterdam as a Coffee and Cocoa Producers' Association. The object was to promote the coffee and cocoa culture in the Dutch East Indies and the interests of its members. In 1935 the Association became a member of the Federation of Associations of Dutch East Indies Upland Estate Companies. From 1937 up to its liquidation in 1968 it was called "Koffieproducenten Vereeniging" (Coffee Producers' Association).

The archives consist of articles of association; annual reports 1926–1943; budgets 1936–1952; correspondence 1935–1957; minutes 1926–1948; circulars 1936–1958; papers concerning i.a. the coffee consumption on Java and Madura.

Entry: Shelflist.

Size: *C.* 3.50 running metres.

**Vereeniging voor de Kinacultuur (Association for the Cinchona Culture), 1935–1968** [97]

The Association was established in 1935 at Amsterdam with the object to promote the cinchona culture in the Dutch East Indies and the interests of its members. From its establishment the Association was a member of the Federation of Associations of Dutch East Indies Upland Estate Companies and was liquidated in 1968.

The archives consist of articles of association; annual reports 1936–1943; budgets 1937–1957; correspondence 1936–1956; minutes 1936–1947; circulars 1936–1954; papers concerning i.a. quinine and anti-malaria campaign.

Entry: Shelflist.

Size: *C.* 1 running metre.

**Vereeniging voor de Kapok- en Cacaocultuur in N.I. (Association for the Kapok and Cocoa Culture in the Dutch East Indies), 1936–1968** [98]

The Association was established in 1936 at Amsterdam with the object to promote the kapok and cocoa culture in the Dutch East Indies and the interests of its members. From its establishment the Association was a member of the Federation of Associations of Dutch East Indies Upland Estate Companies and was liquidated in 1968.

The archives consist of articles of association; annual reports 1936–1943; budgets 1936–1955; correspondence 1936–1953; minutes 1936–1947; circulars 1936–1957; papers 1936–1954 concerning kapok, including trade, prices, Saigon kapok; papers 1946–1954 concerning cocoa.

Entry: Shelflist.

Size: *C.* 1.50 running metres.

**Rubber-Stichting (Rubber Foundation), 1936–1970** [99]

The Foundation was established in 1936 by the East Indian government and the Internationale Vereeniging voor de Rubber-Cultuur (International Association for the Rubber Culture) in the Dutch East Indies. The Foundation took the place of the technical section of the Association for the Rubber Culture. The Foundation was housed in the office of the Federation of Associations of Dutch East Indies Upland Estate Companies at Amsterdam. The object of the Foundation was to stimulate the use of rubber by means of research and propaganda. The Foundation was liquidated in 1970.

The archives consist of the papers originating from the techno-commercial department, viz. of annual reports 1936–1955; budgets 1936–1968; minutes 1936–1970; circulars 1936–1967; minutes 1937–1940 of the editorial board of the periodical *Rubber;* miscellaneous papers; photographs.

Entry: Shelflist.

Size: *C.* 7 running metres (108 items).

**Java Suikervereeniging (Java Sugar Association), 1917–1918** [100]

The Association was established in 1917 at Semarang through the efforts of the East Indies government. The object was to stimulate the trade of sugar. The Association was abolished in 1918 and replaced by the Vereenigde Java Suiker Producenten (Associated Java Sugar Producers) at Amsterdam (see no. 101). The small amount of archives consists of correspondence, minutes, reports.

Entry: Inventory is forthcoming.

Size: *C.* 0.40 running metre.

**Vereenigde Java Suiker Producenten (Associated Java Sugar Producers), VJSP 1918–1932** [101]

The VJSP is the continuation of the Java Sugar Association (see no. 100), in 1918 established at Amsterdam and abolished in 1932. The Nederlandsch-Indische Vereeniging voor de Afzet van Suiker (Dutch East Indies Association for the Sale of Sugar) (see no. 102) took over its duties.

The archives consist of minutes, correspondence, articles of association, annual reports, production figures, accounting papers.

Entry: Inventory is forthcoming.

Size: *C.* 17 running metres.

**Nederlandsch-Indische Vereeniging voor de Afzet van Suiker (Dutch East Indies Association for the Sale of Sugar), NIVAS 1932–1957** [102]

The NIVAS is the continuation of the Associated Java Sugar Producers (see no. 101), in 1932 established at Surabaja through the efforts of the East Indies government. The object was to promote the interests of its members regarding the sale of sugar and to give information about the growing of sugar. Membership was obligatory for all sugar producers.

The archives consist of minutes, correspondence, articles of association, annual reports, production figures, accounting papers.

Entry: Inventory is forthcoming.

Size: *C.* 22 running metres.

**Cultuurmaatschappij Kadhipaten (Estate Company Kadhipaten), 1858–1959** [103]

The Company was established in Madjalengka (Cheribon) and was concerned with the growing of indigo, after 1873 mainly with the growing of sugar. Originally the Company was the property of the Nederlandsch-Indische Maatschappij van Levensverzekering en Lijfrente (Dutch East India Company of Life Insurance and Life Annuity); 1873 bought up by J.G.Ch.A. de Vogel, F. van der Ven and J.S. Bowles; 1905 limited company; 1920 represented by the firm of Tiedeman and Van Kerchem at Surabaja; 1957 nationalized, up to 1965 investment company.

The archives consist of papers 1858–1959 of the head office at The Hague: articles of association, minutes, annual reports 1901–1957; correspondence with representatives in the Principalities, Bandung, Surabaja, Batavia, Algemeen Syndicaat voor Suikerondernemingen (General Syndicate for Sugar Undertakings), Algemeen Landbouw Syndicaat (General Agricultural Syndicate), South and West Sumatra Syndicate, Ondernemersbond voor Indonesië (Employers' Union for Indonesia), Nederlandsch-Indische Vereeniging voor Afzet van Suiker (Dutch East Indies Association for the Sale of Sugar), Bond van Eigenaren van in Nederlands-Indië gevestigde Suikerondernemingen (Union of Owners of Sugar Undertakings established in the Dutch East Indies), departments in Indonesia, Java Werkgeversbond (Java Employers' Union), Tiedeman and Van Kerchem, Proefstation voor de Java Suiker Industrie (Experimental Station for the Java Sugar Industry) at Pasuruan, Stork and Co.; reports about crops, machines etc.; personnel papers; ledgers and other accounting papers 1858–1959; maps, photographs.

Entry: Inventory.

Size: *C.* 30 running metres (317 inventory items).

**Dusseldorp & Co. Cultuurmaatschappijen (Dusseldorp & Co. Estate Companies), 1884–1966** [104]

No data are known.

The archives consist of annual reports of the estate companies Lodaja, Goenoeng Malang, Pasir Malang, Soekawana, Balapoelang, Pasir Nangha, the Agricultural Company Preanger, with some notarial papers, circulars of the Algemeene Vereeniging van Rubberplanters van Sumatra (General Association of Rubber Growers of Sumatra), papers about the estate company Soekanagra and photographs of various companies.

Entry: Shelflist.

Size: *C.* 2.50 running metres.

**Maatschappij tot Exploitatie der Vereenigde Majang-Landen (Company for the Development of the United Majang-Lands), VML 1885–1914, 1914–1970** [105]

In 1914 the Company was established by A. Janssen at Amsterdam in order to solve the financial difficulties of the estate companies Goenoeng Majang, Kali Badjing/Glantangan and Wono Djati, of which he had the management. The object of the company was to run estate companies of coffee, rubber, cocoa and other crops. The company was liquidated in 1970.

The archives consist of private and business papers 1885–1918 originating from A. Janssen, papers 1894–1951 drawn up and received by P.W. and C.W. Janssen, papers about and originating from undertakings of which A. Janssen had the management since 1892, and the archives of the VML. For the names and subjects mentioned in the shelflist see the index.

Entry: Shelflist.

Size: *C.* 40 running metres.

**Maatschappij tot Exploitatie van het Land Widodaren (Company for the Development of the Estate Widodaren), 1885–1912** [106]

The Company was established in 1885 at The Hague. The object was to grow coffee in Besuki.

The small amount of archives consists of minute books, a waste book, journal and ledger.

Entry: Shelflist.

Size: *C.* 0.50 running metre.

**Klattensche Cultuurmaatschappij (Klaten Estate Company), KCM 1887–1962** [107]

The Company was established in 1887 at The Hague. The object was to run five agricultural undertakings in Surakarta, viz. Gondang Winangun and Djethis (including Djogonallan) Gajamprit, Papringan-Kebon Arum, Gantiwarno, Demangan and two in Djokjakarta, viz. Sorogedoog and Wanudjojo. The Company was under the management of the Vereenigde Klattensche Cultuur-Maatschappij (United Klaten Estate Company).

The archives consist of annual reports, correspondence, reports, accounting papers. For the names and subjects mentioned in the shelflist see the index.

Entry: Shelflist.

Size: *C.* 29 running metres.

**Maatschappij tot Exploitatie van Rijstlanden op Java Michiels-Arnold (Company for the Development of Rice Lands on Java Michiels-Arnold), Micharnold 1887–1957** [108]

The Company was established in 1887 at The Hague. The object was to run undertakings, on terms of letting or on its own account, in Buitenzorg, viz. the fields Tjibaroesa, Tjipamingkis, Tjileungsi, Klapanoenggal, the tea undertakings Tinggardjaja and Selawangi and the rubber undertakings Klapanoenggal and Menteng.

The archives consist of correspondence, circulars, annual reports, minutes and accounting papers. For the names and subjects mentioned in the shelflist see the index.

Entry: Shelflist.

Size: *C.* 26 running metres.

**Cultuurmaatschappij Ngredjo (Estate Company Ngredjo), 1887–1958** [109]

The Company was established in 1887 at Amsterdam. The object was to develop the coffee lands Soekoredjo and Tretes Panggoeng at Malang.

The archives consist of minutes, annual reports, articles of association, correspondence and accounting papers. Besides there are papers originating from and about the Nederlandsch-Indische Cultuurmaatschappij (Dutch East Indies Estate Company) (undertaking Wonokerto), Java-Caoutchouc Company (undertaking Leuweung Belembang and Leuweung Kolot I and II), Amsterdam Langsa Rubber Company, Estate Company Goenoeng Boeleud, Estate Company Kali Bantar and the Koffiecultuurmaatschappij (Coffee Estate Company) Taloeg Goenoeng.

Entry: Shelflist.

Size: *C.* 2.50 running metres.

**Maatschappij tot Exploitatie van de Suikeronderneming Tjomal (Company for the Running of the Sugar Undertaking Tjomal), 1889–1951** [110]

The Company was established in 1889 at The Hague. The object was to run the undertaking Tjomal in Pekalongan with the arrack distillery belonging to it.

The archives consist of minutes, articles of association, annual reports, correspondence, accounting papers, photographs, papers Internatio, Nederlandsche Handel-

Maatschappij (Dutch Trading Company), Estate Company Tjiranggon, Bond van Eigenaren van Nederlandsch-Indische Suikerondernemingen (Union of Owners of Dutch East Indies Sugar Undertakings), M.G. van Heel, . . . Huese, Ong Liang Swie.

Entry: Shelflist.

Size: *C.* 5 running metres.

**Cultuurmaatschappij Wonolongan (Estate Company Wonolongan), 1895–1958** [111]

The Company was established in 1895 at The Hague. The object was the acquistion and running of the sugar undertakings Wonolongan, Oemboel (both in Probolinggo) and Wringin-Anom (Panarukan).

The archives consist of reports, minutes, correspondence and accounting papers. For the names and subjects mentioned in the shelflist see the index.

Entry: Shelflist.

Size: *C.* 30 running metres.

**Cultuur Maatschappijen "Kremboong en Toelangan" en "Sindang-Wangi" (Estate Companies "Kremboong and Toelangan" and "Sindang-Wangi"), 1903–1965** [112]

Both companies were established in 1903 at The Hague under the same management. The object of "Kremboong and Toelangan" was to grow sugar at the undertakings of that name, the object of the other to grow tea at the Sindangwangi undertaking.

The archives consist of minutes, annual reports, drafts of outgoing letters and accounting papers.

Size: *C.* 6 running metres.

**Oost-Java Rubbermaatschappij (East Java Rubber Company), 1910–c. 1970** [113]

The Company was established in 1910 at Rotterdam after the association of the estate companies Soember Bokor, Soember Djeroe and Pantjoer Sarie. The object was the production and sale of rubber, coffee and other plantation products. The company was liquidated *c.* 1970.

The small amount of archives consists of minutes 1940, minutes 1939–1940 of the Rotterdamsche Kina-Maatschappij "Tjikembang" (Rotterdam Cinchona Company "Tjikembang"), some financial papers and the printed annual reports of the Oost-Java Rubbermij (East Java Rubber Company) 1910–1960, Coffee Estate Company Soember Bokor 1898–1901, Coffee Estate Company Soember Djeroe 1898–1901, Estate Company Pantjoer Sarie 1899–1901, Coffee Estate Company Soember Kerto 1898–1939, Coffee Estate Company Kali Klepo 1899–1939.

Size: *C.* 0.20 running metre.

**Tabaksmaatschappij Sumatra (Tobacco Company Sumatra), 1888–1894** [114]

The Company was established in 1888 at Rotterdam. The object was to develop, cultivate and run grounds in Siak for the growing of tobacco. In 1894 the Company was liquidated.

The archives consist of the memorandum of association, correspondence, accounting papers, liquidation papers, papers H.J.G. Tuning, A. and E. de Haas, minutes, annual reports, papers about coolies: Chinese, Javanese, Banjarese.

Entry: Shelflist.

Size: *C.* 2.50 running metres.

**Cultuuronderneming Way Lima (Estate Company Way Lima), 1891–1911** [115]

The Company was established in 1892 at Amsterdam. The object was to develop the plot of land Way Lima in the Lampongs, where coffee was grown.

The archives consist of correspondence, accounting papers, articles of association, working lists, papers about coolies, rice, quinine, Kedongdong, Way Ratai, Tandjong Karang, letters G.M. and J.R. Andreas, A.W.J. Graafland, E.G.A. Halewijn, H. Muller Szn, S. Muller Hzn, D. Burger Dzn, A. Waardenburg.

Entry: Shelflist.

Size: *C.* 3.50 running metres.

**Lampongsche Districten (Lampong Districts), 1892–1900** [116]

Established in 1892 at Amsterdam.

The small amount of archives consists of private letters about the Lampongs and drafts of outgoing letters.

Entry: Shelflist.

Size: *C.* 0.50 running metre.

**Cultuurmaatschappij Kedongdong (Estate Company Kedongdong), 1894–1905** [117]

The Company was established in 1894 at Amsterdam. The object was to develop the plot of land Kedongdong in the Lampongs, where coffee, gambier, cocoa and nutmeg was grown.

The archives consist of correspondence, articles of association, annual reports, minutes, accounting papers, papers West-Telok, nutmeg, mace, letters . . . Broekhuyzen, . . . Piepers, . . . Suermondt, S. Muller Hzn, A. van Rijckevorsel Hzn, M.C. van Rijckevorsel-Sint Martin.

Entry: Shelfist.

Size: *C.* 1 running metre.

**Langkat Tabaksbouw Maatschappij (Langkat Tobacco Growing Company), 1898–1902** [118]

The Company was established in 1898 at Amsterdam.

The small amount of archives consists of minutes, drafts of outgoing letters and accounting papers.

Entry: Shelflist.

Size: *C.* 0.50 running metre.

**Nederlandsche Rubbermaatschappij (Dutch Rubber Company), 1909–1923** [119]

The Company was established in 1909 at Amsterdam. The object was to develop agricultural undertakings on Sumatra and also in other parts of Indonesia, with the exception of Djokjakarta and Surakarta.

The archives consist of annual reports, articles of association, drafts of outgoing letters, accounting papers, drawings, papers Liberta, Kotangan, Soengei Mangkei, Soengei Poetih, Serangdjaja, letters H.J.W. Westenberg and J.F. de Beauforill.

Entry: Shelflist.

Size: *C.* 6 running metres.

**Borneo Tabak Maatschappij (Borneo Tobacco Company), 1901–1904** [120]

The Company was established in 1900 at Amsterdam to run the undertakings Karan and Mahé.

The small amount of archives consists of minutes, copy book and ledger.

Entry: Shelflist.

Size: *C.* 0.20 running metre.

**Oliefabrieken Insulinde (Oil Factories Insulinde), 1916–1932** [121]

The factories were engaged in crushing oil from copra.

The archives consist of letters, reports, accounting papers, papers from and/or about B. Zwets, J. Hijering, . . . Dumas, W. Swerts, B. Streefland, Singapore, Padang, Menado, Kebumen, Sumatra, Rangkasbetung, Kediri, Rembang and Semarang.

Entry: Shelflist.

Size: *C.* 2 running metres.

See also no. 412.

**Nederlandsche Maatschappij voor Nieuw-Guinea (Dutch Company for New Guinea), 1928–c. 1940** [122]

The Company was established in 1928 at Amsterdam. The object was to collect and study data about New Guinea in order to obtain and develop undertakings of agriculture and industry there and to foster cultures, also of forest and sea produce, but not of gold and oil.

The archives consist of files concerning several undertakings, annual reports, minutes, correspondence and accounting papers.

Size: *C.* 8 running metres.

**Mijnbouw Maatschappij Salida (Mining Company Salida), 1910–1930** [123]

The Company was established in 1910 at The Hague. The object was to develop 4 mining concessions containing gold, silver, lead, zinc and coal in Painan.

The archives consist of correspondence, articles of association, annual reports, minutes and accounting papers. For the names and subjects mentioned in the shelflist see the index.

Entry: Shelflist.

Size: *C.* 3.50 running metres.

**Petroleummaatschappijen en mijnbouwmaatschappijen (Petroleum and mining companies), 1864–1950** [124]

The collection consists of annual reports, articles of association, minutes, profit and loss accounts of several companies. For the names mentioned in the shelflist see the index.

Entry: Shelflist.

Size: *C.* 2 running metres.

**Railway and tramway companies** [125–140]

For the history of the companies mentioned below have been consulted: *De tramwegen op Java;* gedenkboek samengesteld ter gelegenheid van het vijfentwintig-jarig bestaan der Samarang-Joana Stoomtram-Maatschappij. 's-Gravenhage, 1907. A.W.A. Weyerman. *Geschiedkundig overzicht van de totstandkoming der aan spoor- en tramwegmaatschappijen in Nederlandsch-Indië verleende concessie-voorwaarden en der op grond van die concessievoorwaarden gesloten overeenkomsten.* 's-Gravenhage, 1916, 1917. 2 vol. (Algemeene mededeelingen [van de] dienst der Staatsspoorwegen op Java; no 11). S.A. Reitsma. *Korte geschiedenis der Nederlandsch-Indische spoor- en tramwegen.* Weltevreden, 1928.

**Vereeniging van Nederlandsch-Indische Spoor- en Tramwegmaatschappijen (Association of Dutch East Indies Rail- and Tramway Companies), 1917–c. 1950** [125]

The Association was established in 1917 at The Hague. The object was to promote the general interests of the rail- and tramway system in the Dutch East Indies. All East Indies companies were members of the association.

The archives consist of minutes, reports, correspondence, documentation about the associated companies, accounting papers.

Size: *C.* 11 running metres.

**Nederlandsch-Indische Spoorweg Maatschappij (Dutch East Indies Railway Company), NISM 1863–1940** [126]

The Company was established in 1863 at The Hague. The object was to construct and work a railway, running from Semarang via Tempuran through the Sukawati lands to Surakarta and from there to Djokjakarta with a branch line via Tuntang to Willem I. In 1867 the first section (Semarang-Tanggung) was opened.

The archives consist of (annual) reports, technical reports, minutes, correspondence, accounting papers. For the names and subjects mentioned in the shelflist see the index.

Entry: Shelflist.

Size: *C.* 100 running metres.

**Dutch East Indies steam tramway companies**

Under this heading are mentioned the joint archives of the four associated companies SJS, OJS, SDS and SCS (see no. 127), the separate archives of the associated companies (see no. 128–131) and a number of allied companies (see no. 132–136).

**Gemeenschappelijk archief SJS, OJS, SDS en SCS (Joint archives SJS, OJS, SDS and SCS), 1881–1950** [127]

In the Dutch East Indies the SJS, OJS, SDS and SCS were independent companies. However, in the Netherlands there was one joint management. Moreover there were a number of identical regulations concerning the four companies, viz. regulations for the management, instruction for their joint general agent in the Dutch East Indies, rules for the personnel (the personnel could i.a. be transferred from one company to another), technical regulations concerning the department of civil engineering, regulations for the maintenance of rolling stock and to secure the rights to the ground.

The archives consist of the correspondence of the management in the Netherlands, in so far as that correspondence is not exclusively concerning one of the four associated companies. The archives are arranged according to the file system. Besides there are data about members of the personnel, articles of association, regulations, instructions, papers concerning the construction, working and rolling stock, photographs, drawings. For the names and subjects mentioned in the subjectlist see the index.

Entry: Subjectlist.

Size: *C.* 35 running metres.

**Semarang-Joana Stoomtram-Maatschappij (Semarang-Joana Steam Tram Company), SJS 1881–1952** [128]

Established in 1881, after J.F. Dijkman, W. Walker and G.H. Clifford had obtained concessions in 1879. 1882 opening of the first line Semarang-Djombang.

The archives consist of correspondence, reports, regulations, papers about personnel, bookkeeping and rolling stock arranged according to files 1881–1952. For the names and subjects mentioned in the subjectlist see the index.

Entry: Subjectlist.

Size: *C.* 27 running metres.

**Oost-Java Stoomtram-Maatschappij (East Java Steam Tram Company), OJS 1888–1944** [129]

Established in 1888, after concessions had been granted to W.A. Zilver Rupe and A.J. Snouck Hurgronje in 1886. 1889 opening of the first line Udjong-Fort Prins Hendrik.

The archives consist of correspondence, reports, regulations, papers about the personnel, bookkeeping and rolling stock arranged in files 1888–1944. For the names and subjects mentioned in the subjectlist see the index.

Entry: Subjectlist.

Size: *C.* 16 running metres.

**Serajoedal Stoomtram-Maatschappij (Serajoedal Steam Tram Company), SDS 1894–1940** [130]

Established in 1894, after R.H. Eysonius de Waal and O.J.A. Repelaer van Driel had obtained concessions in 1883, which were granted in 1893 and in 1894 were taken over by the Financiële Maatschappij van Nijverheidsondernemingen in Nederlandsch-Indië (Financial Company of Industrial Undertakings in the Dutch East Indies). 1897 opening of the first lines Maos-Purwokerto, Purwokerto-Sukaradja.

The archives consist of correspondence, reports, regulations, papers about the personnel, bookkeeping and rolling stock arranged according to files 1894–1940. For the names and subjects mentioned in the subjectlist see the index.

Entry: Subjectlist.

Size: *C.* 20 running metres.

**Semarang-Cheribon Stoomtram-Maatschappij (Semarang-Cheribon Steam Tram Company), SCS 1895–1946** [131]

Established in 1895, after concessions had been granted in 1884 for the first time, which were taken over in 1893 by the Financiële Maatschappij van Nijverheidsondernemingen in Nederlandsch-Indië (Financial Company of Industrial Undertakings in the Dutch East Indies). 1897 opening of the first lines Semarang-West-Kaliwungu, Kalibodri-Weliri.

The archives consist of correspondence, reports, regulations, papers about the personnel, bookkeeping and rolling stock arranged in files 1895–1946. For the names and subjects mentioned in the subjectlist see the index.

Entry: Subjectlist.

Size: *C.* 28 running metres.

**Java-Spoorweg-Maatschappij (Java Railway Company), 1884–1914** [132]

The Company operated the line Tegal-Balapulang, which was taken over in 1895 by the Semarang-Cheribon Steam Tram Company.

The archives consist of files, copy books and correspondence.

Entry: Shelflist.

Size: *C.* 2 running metres.

**Probolinggo Stoomtram-Maatschappij (Probolinggo Steam Tram Company), PbSM 1894–1950** [133]

Established in 1894, after C.S. Hien had obtained concessions in the same year. 1897 opening of the first lines Djati-Gending, Gending-Djabung, Probolinggo-Djati.

The archives consist of minutes, annual reports, concessions, correspondence, accounting papers.

Entry: Shelflist.

Size: *C.* 2 running metres.

**Pasoeroean Stoomtram-Maatschappij (Pasoeroean Steam Tram Company), 1895–1951** [134]

Established in 1895, after in 1891 concessions had been granted to J.A. Boulet, which were taken over in 1892 by D. Mournier. 1896 opening of the first lines Pasuruan-Warundowo, Pasuruan-Boom.

The archives consist of annual reports, contracts, correspondence, accounting papers.

Entry: Shelflist.

Size: *C.* 4 running metres.

**Modjokerto Stoomtram-Maatschappij (Modjokerto Steam Tram Company), 1891–1928** [135]

Established in 1896. The object was to construct and operate rail- and tramway lines for general traffic in the Dutch East Indies.

The small amount of archives consists of petitions.

Entry: Shelflist.

Size: *C.* 0.50 running metre.

**Madoera Stoomtram-Maatschappij (Madoera Steam Tram Company), 1897–1966** [136]

Established in 1897. The object was to construct and operate the steam tramway Kamal-Bangkalan-Sempang, Pamakasan-Sumenep-Kalianget.

The archives consist of annual reports.

Entry: Shelflist.

Size: *C.* 1 running metre.

**Bataviasche Ooster-Spoorweg-Maatschappij (Batavian Eastern Railway Company), 1884–1898** [137]

The Company was established in 1884 at Amsterdam. The object was the transport of passengers and goods on the lines Batavia-Weltevreden, Meester Cornelis-Bekasi. In 1895 a line to Krawang was opened. In 1898 the Company was bought up by the Staats Spoorwegen (State Railways).

The small amount of archives consists of articles of association, regulations, annual reports, accounting papers and papers about the liquidation.

Entry: Shelflist.

Size: *C.* 0.70 running metre.

**Kediri Stoomtrammaatschappij (Kediri Steam Tram Company), KSM 1895–1970** [138]

The Kediri Steam Tram Company was established in 1895 at Amsterdam. The Company worked in close association with the Malang Stoomtrammaatschappij (Malang Steam Tram Company) (see no. 139), which is evidenced by the board of managers. The first manager was C.E. van Kesteren, supervisory directors J.K. Kempees, P.C. Huyser and A. Janssen. Head of the construction and operation was J. Bakker. The first tramlines Djombang-Pare-Kediri, Kediri-Wates and Pare-Papar were opened in 1897. The Company was liquidated in 1970.

The archives consist of printed annual reports 1895–1969, correspondence 1895–1970; minutes 1896–1948; series and papers i.a. concerning the bookkeeping and the personnel. Besides the Kediri archives (111 inventory items) and the Malang archives (95 inventory items) there is a joint part (47 inventory items). For the names and subjects mentioned in the inventory see the index.

Entry: Adam, H.B.N.B. *Inventaris van de archieven van de Kediri-Stoomtrammaatschappij en de Malang-Stoomtrammaatschappij, 1895–1970.* 's-Gravenhage, 1976.

Size: *C.* 7 running metres (253 inventory items).

**Malang Stoomtrammaatschappij (Malang Steam Tram Company), MS 1896–1970** [139]

The Malang Steam Tram Company was established in 1896 at Amsterdam. From the beginning the Company worked in close association with the Kediri Steam Tram Company (see no. 138), which is apparent from the board of managers. The first manager was C.E. van Kesteren, supervisory directors J.K. Kempees and A. Janssen. Administrator was J. Bakker. The first tramline Malang-Bululawang was opened in 1897. The Company was liquidated in 1970.

The archives consist of printed annual reports 1896–1956; correspondence 1896–1970; minutes 1896–1950; series and papers i.a. concerning the bookkeeping and the personnel. Besides the Malang archives (95 inventory items) and the Kediri archives (111 inventory items) there is also a joint part (47 inventory items). For the names and subjects mentioned in the inventory see the index.

Entry: Adam, H.B.N.B. *Inventaris van de archieven van de Kediri-Stoomtrammaatschappij en de Malang-Stoomtrammaatschappij, 1895–1970.* 's-Gravenhage, 1976.

Size: *C.* 7 running metres (253 inventory items).

**Deli Spoorweg Maatschappij (Deli Railway Company), DSM 1883–1970** [140]

The DSM was established in 1883 at Amsterdam. The first manager was J.T. Cremer, supervisory directors P.W. Janssen, B. Heldring and I.J. van Santen. Agent in the East Indies was W.H.J. Dates. The first railway line Labuan-Medan was opened in 1886. The Company was liquidated in 1970.

The archives consist of printed annual reports 1883–1969; correspondence 1883–1970; minutes books; travel records; series and papers concerning the bookkeeping 1883–1952; various files including contracts, papers concerning Belawan, Perlanaan, Teluknibung, Asahan, the Vereeniging Nederlandsch-Indische Spoor- en Tramwegmaatschappijen (Association of Dutch East Indies Rail- and Tramway Companies), papers concerning the rolling stock; photographs and glass negatives. For the names and subjects mentioned in the shelflist see the index.

Entry: Shelflist.

Size: *C.* 25 running metres.

**Handelsvennootschap v/h Maintz & Co (Trading Company late Maintz & Co), 1874–1970** [141]

The Trading Company, established in 1917, was the continuation of the trading firm Maintz & Co, since 1874 the property of S. and E. Maintz. The establishment was effected with the assistance of the Nederlandsch-Indische Escompto Mij

(Dutch East Indies Discount Company) and the Allgemeine Elektrizitäts Gesellschaft by intermediary of H.R. du Mosch. In 1917 its head office was transferred from Paris to Amsterdam. The object of the Company was to trade in the fullest sense of the word. Its management included the Algemeene Nederlandsch-Indische Electriciteit Mij (General Dutch East Indies Electric Supply Company), the Nederlandsch-Indische Waterkracht Exploitatie Mij (Dutch East Indies Water Power Operating Company), later Waterkracht Exploitatie Mij Indonesië (Water Power Operating Company Indonesia), the Solosche Electriciteits Mij (Solo Electric Supply Company), the Gemeenschappelijk Electriciteits Bedrijf Bandoeng en Omstreken (Joint Electric Supply Company Bandoeng and District), continuation of the Bandoengsche Electriciteit Mij (Bandoeng Electric Supply Company), the Company "Electra", the Electric Supply Company Banjoemas, the East Java Electric Supply Company, the Electric Supply Company Rembang, the Electric Supply Company Sumatra, the Electric Supply Company Bali and Lombok, the Leerlooierij "Wonokromo" (Tanning Industry "Wonokromo"), the Tapioca flour factory Paree at Surabaja, the Nederlandsch-Indische Kunstmestfabriek (Dutch East Indies Fertilizer Factory) at Batavia, the Tea Undertaking "Patoeahwattee", the Nederlandsch-Indische Mij voor Chemische Producten (Dutch East Indies Company for Chemical Products) at Surabaja, the Amsterdamsch Kantoor voor Indische Zaken (Amsterdam Office for East Indies Affairs) and the Nederlandsche Turfgasgenerator Mij (Dutch Peat Gas Producer Company) at Amsterdam. The Company was agent for the Estate Company Moetoran, the Allgemeine Elektrizitäts Gesellschaft, the Maatschappij voor Ondernemingen in Nederlands-Indië (Company for Undertakings in the Dutch East Indies), the Deutsch-Australische Dampfschiffs Gesellschaft and the Oostenrijkse Lloyd (Austrian Lloyd).

The archives consist of the archives of the Maintz & Co and Trading Company, 1874–1941; the General Dutch East Indies Electric Supply Company, 1909–1970; the Dutch East Indies Water Power Operating Company, 1926–1939; the Mij tot Exploitatie van Waterleidingen in Nederlandsch-Indië (Company for Operating Water Works in the Dutch East Indies), 1929–1939; the Joint Electric Supply Undertaking Bandoeng and District, 1918–1927; the Amsterdam Office for East Indies Affairs, 1911–1939; the Tea Undertaking "Patoeahwattee", 1910–1926; the Teakhout Import Mij (Teakwood Import Company), 1929–1930; the Dutch Peat Gas Producer Company, 1939–1945; concessions, contracts and concessions of various electric supply companies, 1900–1953; papers originating from the Ondernemersbond voor Indonesië (Employers' Union for Indonesia), 1948–1956, the Ondernemersraad voor Nederlandsch-Indië/Indonesia (Employers' Council for the Dutch East Indies/Indonesia), 1947–1966, the Lembaga Alat-Alat Pembajaran Luar Negeri (Currency Exchange Office), 1956–1958; photographic material, negatives, slides, tape-recordings, 1910–1968. For the names and subjects mentioned in the shelflist see the index.

Entry: Shelflist.

Size: *C.* 51.50 running metres (990 inventory items).

**Nederlandsche Overzee Trustmaatschappij (Dutch Overseas Trust Company), NOT 1914–1919, 1920–1937** [142]

The NOT was established in November 1914 at the initiative of C.J.K. van Aalst, president of the Commissie voor den Nederlandschen Handel (Committee for the Dutch Trade). The object aimed at by the Company was to secure the Dutch import and export during the First World War. In July 1919 its activities came to an end. (Ch.A. van Manen. *De Nederlandsche Overzee Trustmaatschappij;* middelpunt van het verkeer van onzijdig Nederland met het buitenland tijdens den wereldoorlog 1914–1919. 's-Gravenhage, 1953). This book for the greater part founded on the archives of the NOT, shows that the Company was also concerned with the trade to the Dutch East Indies.

The archives consist of:

a. The archives of the Algemene Secretarie (General Secretariat), including minutes; papers concerning C. van Vollenhoven, G.W. van der Goes, the Merauke Company; files concerning shipping, coffee, tobacco, cocoa, the NOT and foreign countries.

b. The archives of the Centraal Bureau (Central Office), including correspondence (alphabetically arranged per annum according to names of persons and firms); correspondence with foreign legations; reports, statistics etc.

c. Archival documents of i.a. the press service, the accounts and clerical service, the legal office, the committee for the navigation, Tabaksbelangen (Tobacco Interests).

d. Papers 1920–1937 in connection with the discontinuance of the NOT; papers of the Stichting Archieven der NOT (Foundation Archives of the NOT), with a separate inventory.

e. Secret archives, including series of the bookkeeping 1915–1926 and papers concerning penalty cases, with a separate inventory.

f. Library of the NOT, with a catalogue (*c.* 12 running metres).

Entry: Inventories A–C, D, E.
Size: Together with the library mentioned sub f *c.* 80 running metres.

**Holland-Bombay Handelsmaatschappij (Holland-Bombay Trading Company), 1890–1939** [143]

The Company was established in 1890 at Amsterdam. The object was to trade with British India, both for its own account or on commission.

The archives consist of correspondence with Calcutta, Bombay, Karachi, Indonesia, Persia, letters from M.W.C. Jansen, A. Tetzner, D. van Wijngaarden, G.A. Becker, A. Cagi, L. Marques, . . . van Obost-De Corrido, H. Poppe, papers about J.G. Bendien, opium, accounting papers.

Entry: Shelflist.
Size: *C.* 6.50 running metres.

## Other Private Organizations [144–151]

**Commissie tot de Zaken der Protestantsche Kerken in Nederlandsch-Oost- en West-Indië (Committee for the Affairs of the Protestant Churches in the Dutch East and West Indies), 1815–1946** [144]

In 1820 the Committee for the Affairs of the Protestant Churches in the Dutch East and West Indies was established at The Hague, after a provisional committee, which had been established already in 1815. The Committee, also called Indische Commissie (Indies Committee), was composed of seven members, including the secretaries of the Algemene Hervormde en Luthers-Evangelische Synode (General Reformed and Evangelic-Lutherian Synod) and of the Provinciaal Kerkbestuur, PKB (Provincial Clerical Board) of South Holland. The Crown laid down the conduct of affairs and appointed the ministers on nomination by the Committee. The responsibility of the Committee was to promote the interests of the church in the East and West Indies. For that purpose it carried on a regular correspondence with the churches in the colonies and it also kept in regular touch with the bible and missionary societies there. It represented at the Algemene Synode der Hervormde Kerk (General Synod of the Reformed Dutch Church) the Protestantsch Kerkbestuur (Protestant Clerical Board) established in 1844 at Batavia. The ministers sent to the Dutch East Indies were appointed by the governor general on the recommandation of the PKB. In 1935 the PKB established the Commissie tot Behartiging in Nederland van de Belangen der Protestantsche Kerken in Nederlandsch-Indië (Committee for Promotion in the Netherlands of the Interests of the Protestant Churches in the Dutch East Indies). From that time this Committee, also called Nieuwe Commissie (New Committee), or Oost-Indische Commissie (East Indies Committee), was in charge of the direct contact between the Netherlands and the churches in the Dutch East Indies. Henceforth the PKB appointed, besides the members of the East Indies Committee, also the ministers. After 1946 the aforesaid committees were no longer working for Indonesia

The archives consist of papers 1815 and 1820 of the establishment of the Committee; minutes 1815–1949 of the Indies Committee, after 1935 only West Indies affairs, with annexes including letters received and drafts of papers despatched; reports 1816–1940 of the Indies Committee about the situation of the Protestant churches in the East and West Indies; reports 1893–1905 of the PKB; applications 1891–1908 received by the Indies Committee; cash books 1868–1934 of the Indies Committee; papers 1935–1946 of the West Indies Committee; papers 1935–1946 of the East Indies Committee: postbooks of papers received and despatched, papers internal affairs, Protestant Clerical Board at Batavia, ministers, curates, financial matters, Zendingsschool (Protestant Missionary Trainings College) at Oegstgeest, Nederlandse Zendingsraad (Dutch Protestant Missionary Council), Missionary conferences at Oxford and Edinburgh, Indische Kring (East Indies Circle); deposited papers before 1815; printed matter such as government gazettes, reports, notes etc. concerning the Dutch East Indies Church, books, pamphlets, newspapers, magazines, church journals, missionary journals, a few maps.

Entry: Provisional survey; tables to the annexes of the minutes from 1847.

Size: *C.* 11.50 running metres.

Deposited at the ARA by the Reformed Dutch Church, see no. 574 a.

**Oost- en West-Indische Opvoedingsmaatschappij (East and West Indies Educational Society), 1850–1885** [145]

The Society was established in 1850 at Delft. The object was the education in the home country of Dutch children whose parents or guardians lived in the colonies. The Society was abolished in 1886.

The small amount of archives consists of articles of association, regulations, annual reports, minutes, drafts of outgoing letters, papers about the pupils, Jongens Wezeninrichting (Boys' Orphanage) at Surabaja and accountancy.

Entry: Shelflist.

Size: *C.* 0.50 running metre.

**Nederlandsche Roode Kruis (Dutch Red Cross), NRK 1867–1945** [146]

By Royal Decree of July 19th 1867 the NRK was established. October 8th 1870 the Bataviaasch Comité (Batavian Committee) of the Red Cross was established, being the first section in the Dutch East Indies. In 1872 the name was changed into Centraal Comité Nederlandsch-Indië van de Nederlandsche Vereeniging tot het Verleenen van Hulp aan Zieke en Gewonde Krijgslieden in Tijd van Oorlog (Central Committee Dutch East Indies of the Dutch Association for Rendering Assistance to Sick and Wounded Soldiers in Times of War). In 1895 its name was changed again into Centraal Comité in Nederlandsch-Indië van de Vereeniging van het Nederlandsche Roode Kruis (Central Committee in the Dutch East Indies of the Association of the Dutch Red Cross). After the alteration of the articles in 1920 3 policlinics were established in Batavia and Buitenzorg and in 1929 also a hospital. The Committee in the Dutch East Indies was mainly concerned with nursing, assistance to victims of calamities, welfare work and training of personnel.

The archives 1867–1945 consist of:

A. General series and papers, including minutes, correspondence, annual reports.

B. Series and papers of a special nature, viz.:
   a. The organism, including, relating to the organization in the Dutch East Indies, papers concerning the articles of association 1920–1939; correspondence 1915–1940; personal notes; internal service; finance.
   b. Contact with third parties, including the international Red Cross; information.
   c. Execution of the NRK duties, including duty in time of war; social field; assistance in case of calamities; assistance in foreign countries.

C. Deposited archives, including the papers of the Indies section of the chief committee of the NRK. This small amount of archives includes articles of association; correspondence 1870–1875, with postbook; lists of goods sent to the Indies 1873–1879; lists of letters received from the Central Committee at Batavia 1874–1878; lists of hospital orderlies 1875–1891.

Entry: J.H. Rombach. *Inventaris van het archief van het hoofdbestuur van het Nederlandsche Roode Kruis, 1867–1945*. 's-Gravenhage, 1966.

Size: *C.* 18 running metres (187 inventory items).

See also no. 728 l.

**Vereeniging Het Buitenland (Association Foreign Countries), 1888–1938** [147]

The Association was established in 1888 at Amsterdam. The object was to promote the settlement in foreign countries of young Dutchmen in business positions.

Relevant for the history of Asia are the minutes, annual reports, correspondence and papers from and/or about Peking 1900, consulate Shanghai 1887–1891, Bangkok 1892–1897, Japan 1912, Bombay 1909, China 1890, Persia 1890, photographs Formosa.

Entry: Shelflist.

Size: *C.* 3.25 running metres.

**Orde van de Ster (Order of the Star), 1912–1931** [148]

The Orde van de Ster in het Oosten (Order of the Star in the East) was established in 1911 in Benares by A. Besant. Head of the Order was the world teacher J. Krishnamurti. 1924 headquarters in Ommen, from 1926 called Orde van de Ster, 1929 abolished by Krishnamurti, but the meetings continued.

The archives consist of correspondence 1912–1931 of the headquarters with i.a. Australia, India, Dutch East Indies and New Zealand; drafts of articles 1928–1929 for *The Star Bulletin;* lists of addresses and a declaration of abolition with correspondence 1929–1931.

Entry: Inventory.

Size: *C.* 11 running metres (32 inventory items).

**Koloniale School voor Meisjes en Vrouwen (Colonial School for Girls and Women), 1920–1949** [149]

The School was established in 1920 and opened in 1921 at The Hague. The object was to prepare girls and women for the life in the East and West Indies by means of quarterly courses. They were instructed in maternity nursing, infant welfare work, care of the sick, hygiene, Indies and Dutch cooking, Malay, and the geography and anthropology of the East and West Indies. The school was abolished in 1949. (*Koloniale School voor Meisjes en Vrouwen*. 's-Gravenhage, 1923).

The archives include papers concerning the establishment; rules and regulations; minutes of the board and members' meetings 1920–1949; papers received and despatched 1921–1953 after 1949 relating to the abolition; papers concerning personnel and pupils, including papers received and despatched by the headmistress 1920–1928, lists of pupils, letters from former pupils, photographs of pupils; papers concerning financial matters 1920–1946; papers concerning propaganda 1920–1946; documentation.

Entry: Inventory.

Size: *C.* 2 running metres (80 inventory items).

**Stichting het Nationaal Nieuw-Guinea Comité (Foundation the National New Guinea Committee), 1949–1962** [150]

The Committee was established December 22nd 1949 by G.L. Tichelman, W. Mendes da Costa, G.C. Schrader, P.M. Verbruggen, and J.F.J. Huybers. The object was to collect and distribute documentation about New Guinea, to support the inhabitants of New Guinea both in social and cultural matters and to promote the common interests of the Netherlands and New Guinea. One tried to achieve this object i.a. by means of lecture meetings, promotion of education on New Guinea and the publication of the bimonthly periodical *Nederlands Nieuw-Guinea*, published from 1952–1962. In 1950 the Foundation established the Emigratie- en Transmigratieraad (Emigration and Transmigration Council). The Committee was abolished in 1962.

The archives include minutes 1949–1962, annual reports 1950–1962, papers received and despatched 1950–1962, accounting papers 1950–1962, papers concerning memberships, papers concerning the Emigration and Transmigration Council, papers originating from the editorial staff of *Nederlands Nieuw-Guinea* 1952–1962, papers concerning the social and medical work of the Committee, documentation, propaganda films.

Entry: Inventory.

Size: *C.* 10 running metres (332 inventory items).

**Comité Nationale Actie Steunt Spijtoptanten Indonesië (Committee National Campaign for Support of Spijtoptanten Indonesia), NASSI 1960–1969** [151]

The Committee was established in 1960 at The Hague by K. Stolk, Hein Buitenweg, E.A.C. den Hamer, S.L.F. Catalani and Tjalie Robinson. W.L.G. Lemaire became president of the board, vicepresident S.H. Spoor-Dykema. The object of the NASSI was to intermediate with obtaining visa, residence permits etc. for people who had decided in favour of the Indonesian nationality, but on second thoughts wanted to go to the Netherlands. The Committee stopped its activities in 1969. (*Tong-Tong.* 4th vol., 1960).

The archives consist for the greater part of the 3782 files of persons who had sent in an application for residence in the Netherlands; besides non disposed files;

granted applications for visa; papers originating from [F.H.J.M.] Daams; papers originating from [J.J.Th.] ten Broecke Hoekstra; files 1968–1969.

Entry: Inventory.

Size: *C.* 25 running metres (317 items).

Permission is necessary for consultation of the papers.

## Families and Private Persons [152–270]

### Van Alphen/Engelhard [152]

The collection Van Alphen/Engelhard consists of the papers originating from various members of the Van Alphen family and of papers of N. Engelhard and W.A. Alting, who legated their archives to D.F. van Alphen. Papers concerning Asia after 1795 are originating from:

*D.F. van Alphen, August 30th 1774 Utrecht – October 16th 1840 Voorschoten*
After having been student at Leyden University for a short time Van Alphen went on his first voyage to the Dutch East Indies in 1791; 1793–1796 second lieutenant with the navy in the Indies; 1797 in the service of the East India Company; 1798 at Surabaja; 1801–1808 administrator at Surabaja; 1808 leaves Java, taken prisoner by the English and arrives in the Netherlands in 1809 via the United States; 1815 till his death member of the Second Chamber.

The papers originating from D.F. van Alphen are pertaining to the period 1660–1840. A large portion of these concerns the period before 1795. The papers after 1795 are mainly concerned with private matters (landed property on Java) and his membership of the Second Chamber.

*F.E.M. van Alphen, December 5th 1822 Leyden – January 14th 1899 Brummen*
1846 junior clerk with the Ministry of Colonial Affairs; 1851 clerk; 1856 principal clerk; 1861 referendary, head section Oost-Indische Algemene Zaken (East Indies General Affairs); 1876 secretary general Ministry of Colonial Affairs; 1880 member of the Raad van State (Council of State); 1894 retired.

*H. van Alphen, died 1899*
1826 clerk with the Ministry of Colonial Affairs; 1844 referendary. Originating from F.E.M. van Alphen and H. van Alphen are reports of the Raad van State 1828 concerning the cultures in Java; departmental papers concerning the East Indies 1834–1880; documentation from periodicals, pamphlets etc. 1849–1893; economic journals 1881–1893; budget of the Dutch East Indies 1894; English blue books concerning Sulu, Borneo, New Guinea and the Nisero affair 1882–1890; reports of the meetings of the Indisch Genootschap (East Indies Society) 1866–1891.

*N. Engelhard, December 1st 1761 Eelde – May 31st 1831 Buitenzorg*
1778 in the service of the East India Company; 1780 under merchant; 1781 second administrator of the warehouses on the Waterpoort; 1784 second administrator of

the warehouses on Onrust; 1785 first administrator there; 1789 sabandhaar and licentmeester (master of the duties); 1791–1800 delegate for and about the affairs of the Native; 1800 president of the heemraden (polder boards); 1801 extraordinary councillor of the East Indies; 1801 till his dismissal in 1808 governor of the north east coast of Java; up to 1816 retired; 1816–1817 member of the Commissie tot Overneming van de Etablissementen (Committee for Acquisition of the Establishments) on Bandjarmasin; 1817–1818 president of the Commissie tot Overneming (Committee for Acquisition) of the Moluccas; from 1818 retired; 1823–1824 residence in the Netherlands; 1826 member of the Commissie die de Hamangkoe Boeana of Sepoeh in zijn waardigheid van sultan moet herstellen (Committee for Retrieving the Dignity of Sultan for the Hamangkoe Boeana or Sepoeh); after this Engelhard occupied no more offices.

Originating from Engelhard are papers concerning his activities from 1787 with early papers from 1615; a large part was drawn up and/or received by him after 1795.

For the names and subjects mentioned in the inventory (concerning D.F. van Alphen and N. Engelhard) about the period after 1795, see the index.

Entries: The collection was given to the ARA by instalments. Of these separate inventories have been made, printed in *VROA*, 1896, p. 112–115; 1900, p. 29–57; 1916, p. 358–378. Of the acquisition mentioned in *VROA*, 1941, p. 20 an inventory has been made.

Size: *C.* 9 running metres (63, 374, 193, 1 inventory items).

**W.A. Alting, November 11th 1724 Groningen – June 7th 1800 Kampung Melaju** [153]

1741–1750 law student at Groningen; 1750 under merchant for the VOC at Batavia; 1754 merchant and geheimschijver (secret writer); 1756 second secretary of the Hooge Regeering (Supreme Government); 1759 first secretary; 1762 extraordinary member of the Raad van Indië (Council of the Indies); 1772 member of the Council of the Indies; 1776 collector general; 1777 director general; 1780 temporarily governor general; 1797 dismissal.

The collection, consisting of papers from 1596, contains only a few papers (5 inventory items) about the period after 1795, viz. private reports and other papers 1778–1796; papers 1781–1796 concerning the Chinese; papers 1786–1796 of [C.F.] Reimer; letters etc. 1788–1796 of P. Sluyskens, commander of Galle; description of the events in Ceylon by J.G. van Angelbeek, 1796.

Entry: Inventory. In *VROA*, 1900, p. 29–36; 1916, p. 377–378.

Size: *C.* 2.50 running metres (93 inventory items).

Deposited at the First Section.

**F.M. van Asbeck, February 23rd 1889 Nieuwediep – ?** [154]

Law student at Leyden and Lausanne, 1916 thesis; 1919–1924, 1927–1933 official at the disposal of the Algemene Secretarie (General Secretariat), lastly with the grade of secretary of the government, from 1925 also extraordinary professor of international law and of foreign colonial law at the Rechtshogeschool (University of Law) at Batavia; 1935–1940 member of the Permanente Mandatencommissie van de Volkenbond (Standing Mandates Commission of the League of Nations); 1939 professor of overseas constitutional and administrative law at Leyden; 1947 of international law and political history; 1954 member of the Permanent Hof van Arbitrage (Standing Court of Arbitration); 1959 judge at the Europees Hof voor de Rechten van de Mens (European Court for the Rights of Men).

The collection mainly consists of papers 1920–1933 and papers 1945–1960, viz. notes, advices etc. General Secretariat about the legal status of Dutch subjects and the Chinese and Japanese, papers Indonesian question 1945–1949 (Malino, Den Pasar, Allied Forces, Pangkal Pinang, freedom of religion, Linggardjati, "politiële" (military) actions, H.J. van Mook, New Guinea), besides private correspondence, papers League of Nations, New Guinea up to 1960, Siam.

Entry: List drawn up at the transfer.

Size: *C.* 0.50 running metre.

Permission is necessary for consultation of the papers.

**C.J. Asselbergs, March 28th 1869 Bergen op Zoom – December 11th 1949 Breda** [155]

1886 training at the KMA (Royal Military Academy); 1891 as second lieutenant to the Dutch East Indies; 1899 as first lieutenant back to the Netherlands; 1899–1900 officer in South Africa; 1904 retired.

Only a small part of the collection (7 inventory items) concerns the Dutch East Indies, viz. papers concerning the expedition to Samalanga, operations in Atjeh, descriptions of the population of Atjeh, lists of salaries and notes concerning the Malay language 1877–1899.

Entry: Inventory.

Size: *C.* 2.25 running metres (65 inventory items).

**J.C. Baud, October 24th 1789 The Hague – June 27th 1859 The Hague** [156]

1803 training artillery school at Breda; 1804–1810 naval cadet; 1811 second clerk with the English government at Batavia; 1812 second translator at Batavia; 1813 retired; 1814 principal clerk with the government secretariat at Batavia; 1816 secretary of the government of the Dutch East Indies; 1819 general secretary of the Supreme Government; 1821–1824 retired in the Netherlands; 1824 i.a. concerned in the establishment of the NHM (Dutch Trading Company) and the Stoombootmaatschappij (Steamship Company) at Rotterdam; 1824 director for the affairs of the East Indies possessions at the Ministry of Nationale Nijverheid en

Koloniën (National Industry and Colonial Affairs); 1825–1832 director for the colonies at the Ministry of Naval and Colonial Affairs; 1825–1830 extraordinary member of the Commissie tot Zaken der Protestantsche Kerken in Oost- en West-Indië (Committee for the Affairs of the Protestant Churches in the East and West Indies); 1827 concerned in the establishment of the Javasche Bank (Java Bank); 1827–*c*. 1830 member of the Commissie tot het Dempen van de Java-oorlog (Committee for Quelling the Java War); 1832 vice president of the Supreme Government; 1833 intermediary governor general of the Dutch East Indies; after his return to the Netherlands in 1836 up to 1840 councillor of state in extraordinary service and adviser for the Colonies; 1838–1839 member of the Raad van State (Council of State); 1840–1848 minister of Colonial Affairs; up to 1850 retired; 1850–1858 member of the Tweede Kamer (Second Chamber); 1851 co-founder of the KITLV; 1853 president of the Staatscommissie van Onderzoek (State Committee of Inquiry) in connection with the Emancipation of Slaves in the Dutch Colonies; 1853–1854 concerned in the draft bill monetary system in the Dutch East Indies and the draft government regulations. (P. Mijer. *Jean Chrétien Baud geschetst.* Utrecht, 1878).

The collection consists of papers drawn up and/or received during Baud's years of office from 1811. The greater part of the papers are from the periods when he was director for colonial affairs (174 inventory items), intermediary governor general (182 inventory items), Minister of Colonial Affairs (204 inventory items) and member of the Second Chamber (107 inventory items). Besides there are papers drawn up as director of the widows' and orphans' fund and director of institutes for deaf-mutes and blinds in the Dutch East Indies; moreover the collection contains an autobiography and a service record and miscellanies from 1596 concerning i.a. judicial system, history, linguistics, geography and ethnology of the Dutch East Indies. In principle the functions are divided as follows: general; finance; trade and cultures; banking and monetary system; military affairs. The greater part of these papers concerns the Dutch East Indies, a much smaller part concerns the Netherlands and the West Indies. For the names and subjects mentioned in the inventory see the index.

Entry: Inventory. In *VROA,* 1917, p. 497–621.

Size: *C.* 8.50 running metres (1072 inventory items).

See also J. van den Bosch and J.C. Baud. *Briefwisseling, 1829–1832 en 1834–1836;* vol. 1. Brieven van Van den Bosch; uitgegeven door J.J. Westendorp Boerma. Utrecht, 1956. (WHG; serie 3; no. 80).

**Van Beresteyn** [157]

The collection consists of papers originating from the Van Beresteyn family and related families. For the history of Asia the papers originating from the following persons are relevant:

a. *P.A. van Beresteyn, June 21st 1824–1904*
   Official at the offices of the Provincial States at Utrecht; 1847 official Algemene

Secretarie (General Secretariat) Batavia, later secretary and director of the Oost-Indische Maatschappij van Administratie en Lijfrente (East Indies Company of Administration and Life Annuity), also co-administrator of the Oost-Indische Zee- en Brandassurantie Maatschappij (East Indies Sea and Fire Insurance Company), the Javasche Zee- en Brandassurantie Maatschappij (Java Sea and Fire Insurance Company) and the Nederlandsche Lloyd (Dutch Lloyd).

The papers originating from him concern his activities with the various insurance companies from 1847 forward. For the names and subjects mentioned in the provisional inventory see the index.

b. *O.J.B. van Beresteyn, 1872–1919*
In 1892 as journalist to Indonesia for the *Soerabajasche Courant;* 1895 war correspondent for that newspaper in the Chinese-Japanese War, advocate for good Chinese-Dutch trade relations; 1896 war correspondent for the *Locomotief* in the Atjeh-war; 1910 concerned in the foundation of the *Gazette de Hollande.*

Originating from him are diaries and papers connected with his journalistic activities. For the names and subjects mentioned in the provisional inventory see the index.

c. *E.A. Forsten, 1811–1843*
Biologist.

Originating from him are papers about a biological expedition in Celebes, 1839–1842.

d. *F.M.S. van Zuylen, 1820–1861*
KNIL officer.

Originating from him are notes about the Fort Erfprins and notes about a journey through Hitu.

e. *G.E.V.L. van Zuylen, 1837–1895*
1857 second lieutenant KNIL; 1859 first lieutenant; 1865 captain; 1872 major; 1876 up to *c.* 1878 lieutenant colonel; founder of the Bukit Gompong estate company in Padang.

Originating from him are papers concerning his military career and papers concerning Bukit Gompong. For the names and subjects mentioned in the provisional inventory see the index.

Entry: Provisional inventory.

Size: *C.* 23 running metres (2040 inventory items, of which *c.* 70 originating from the aforesaid persons).

**A.G. Besier, May 19th 1758 Deventer – March 13th 1829 Deventer** [158]

1775 law student at Leyden where he graduated in 1780; 1780 lawyer at Zwolle, 1783 alderman, 1787 burgomaster; 1796 commissioner director of the Navy; 1799 member of the Executive Government; 1801–1805 member of the State Government, concerned with the reorganization of the Aziatische Raad (Asia Council) as member of the section for Naval and Colonial Affairs; 1805 retired at Amsterdam; 1810 Deventer; 1814 member of the municipal council of Deventer; 1821 member of the Provincial States of Overijssel.

The collection consists of papers 1794–1805 concerning naval affairs, Ceylon, China, Cochin-China, India, Cape of Good Hope, Malaya, East Indies, Penang, West Indies. For the names and subjects mentioned in the very extensive inventory, see the index.

Entry: F.J.J. Besier. *Voorlopige inventaris van het archief van de familie Besier;* nader bewerkt door W. Klaasen. [Deventer], 1963.

Size: *C.* 0.30 running metre (104 inventory items).

**Bezemer** [159]

The collection consists of the papers originating from P. Bezemer and A. Bezemer. All papers concerning the Dutch East Indies are originating from the latter.

*A. Bezemer, 1783–1852*
1795 naval cadet; 1801 naval lieutenant; 1808 first naval lieutenant; 1814 on leave sailing for the merchant navy; 1816 on half-pay sailing for the merchant navy; 1821 naval captain lieutenant; 1831 naval captain; 1832–1850 inspector of the equipment, victuals and uniforms of the navy; 1850 retired as rear admiral.

The collection includes journals 1802–1810 of his voyages from the Mediterranean to the Dutch East Indies and through the Indian waterways; journals 1819–1826 of his voyages from Amsterdam and Antwerp to the Dutch East Indies and Japan, with registers of letters despatched, missives and papers concerning his voyage in 1825 from Batavia to Japan and back; register of letters despatched to several authorities 1814–1853. For the names and subjects mentioned in the inventory see the index.

Entry: Inventory. In *VROA,* 1907, p. 110–116.

Size: *C.* 0.80 running metre (48 inventory items).

**P.A. Bik, July 18th 1798 Amsterdam – April 10th 1855 Breda** [160]

1816 official Bandjarmasin; 1817 scriba (secretary); 1818 official West Coast of Borneo; 1819 employed at the secretariat of the general direction of Finance; 1822 postmaster Batavia; 1839 ontvanger van het collateraal (collector of the collateral succession duties) at Batavia, member of the orphans' court; 1841–1845 chief factor of the trade in Deshima; 1846 compulsory retired; 1851 pension.

Besides the papers concerning his official career the collection consists of papers drawn up or received by him as chief factor in Deshima. For the names and subjects mentioned in the provisional inventory see the index.

Entry: Provisional inventory.

Size: *C.* 0.80 running metre (*c.* 50 inventory items).

**M. van Blankenstein, June 13th 1880 Oudekerk aan de IJssel – September 18th 1964 Wassenaar** [161]

1902–1906 student of the Dutch language at Leyden, 1911 graduated; 1906–1937 journalist with the *NRC*, 1906 correspondent at Copenhagen, 1908 Berlin, 1920 travelling correspondent, 1931 foreign external editor; 1937 journalist with the *Utrechts Nieuwsblad;* up to 1940 also contributor to i.a. the *Soerabaiasch Handelsblad, De Haagsche Post* and *Indépendence Belge;* 1940 chief editor of *Vrij Nederland* at London, then of *De Stem van Nederland;* 1946 contributor of *Het Parool,* from 1959 foreign external editor; from 1952 contributor of several Indonesian newspapers. (Supplements *Keesings Historisch Archief.* 1964).

The collection consists for the greater part of letters received 1940–1961, including many concerning Indonesia, and also papers concerning his activities as journalist and papers of a private nature. For the names and subjects mentioned in the inventory see the index.

Entry: Inventory.

Size: *C.* 0.90 running metre (631 inventory items).

Permission is necessary for consultation of the papers.

**M. Boon** [162]

Official at the Ministry of Foreign Affairs, from 1965 up to 1966 entrusted with the collection of sources concerning the history of Indonesia 1945–1950.

The collection consists of copies and record copies of papers originating from L.J.M. Beel, M. Brouwer, W.F.L. van Bylandt, W. Drees, H.J. Kruls, C. Lion Cachet, L. Neher, A.S. Pinke, S. Posthuma, E.M.J.A. Sassen, J.R.H. van Schaik and E.M. Stok.

Size: *C.* 0.10 running metre.

Permission is necessary for consultation of the papers.

**J. van den Bosch, February 2nd 1780 Herwijnen – January 28th 1844 The Hague** [163]

1798 first lieutenant with the Comité voor de Oost-Indische Handel en Bezittingen (Committee for the East Indies Trade and Possessions); 1801 captain adjutant to the governor general; 1807 lieutenant colonel and adjutant general; 1808 discharge, 1810 compelled to leave Java, arrives in the Netherlands after English captivity in 1812; 1813–1816 officer in the Dutch army; 1818 co-founder (and up to 1843 member of the board) of the Maatschappij van Weldadigheid (Society of Charity); 1825 administrator of the Nationale Militie en Schutterijen (National and Civic Militia); 1827 as commissioner general to Curaçao, 1828 to Paramaribo; 1828 in the Netherlands appointed governor general of the Dutch East Indies, arrives at Batavia in 1830, introduction 'cultuurstelsel' (cultivation system) in co-operation with the NHM; 1832 commissioner general of the Dutch East Indies; 1833, when J.C. Baud becomes governor general, Van den Bosch continued to direct finance, the cultures and defence; 1834 returns to the Netherlands and becomes minister of Colonial Affairs (up to 1839); 1839–1842 retired; 1842 till his death member of the Second Chamber. (J.J. Westendorp Boerma. *Een geestdriftig Nederlander;* Johannes van den Bosch. Amsterdam, 1950).

The collection is the residue of the functions of Van den Bosch from 1807. The greater part of the East Indies papers consists of papers drawn up and/or received when he was governor general (295 inventory items) and when he was minister of Colonial Affairs (234 items). In addition of the official papers the collection also contains papers concerning his private life, documentation relating to Van den Bosch and papers originating from relations, i.a. from his son H. van den Bosch and his grandson W.J.P. van den Bosch. For the names and subjects mentioned in the inventory see the index.

Entry: G.J.W. de Jongh. *Beschrijving van een verzameling stukken afkomstig van Johannes van den Bosch en enige van zijn nakomelingen.* 's-Gravenhage, 1968.

Size: *C.* 10 running metres (901 inventory items).

See also J. van den Bosch en J.C. Baud. *Briefwisseling, 1829–1832 en 1834–1836;* vol. 2. Brieven van Baud; uitgegeven door J.J. Westendorp Boerma. Utrecht, 1956. (WHG; serie 3; no. 81). J. van den Bosch en D.J. de Eerens. *Particuliere briefwisseling, 1834–1840, en enige daarop betrekking hebbende andere stukken;* uitgegeven door F.C. Gerretson en W.Ph. Coolhaas. Groningen, 1960. (WHG; serie 3; no. 83).

**J.F.R.S. van den Bossche, September 4th 1819 Bergen – January 12th 1889** [164]

1848 comptroller Palembang; 1852 assistant resident Tebing Tinggi; 1855 on leave; 1857 governor Coast of Guinea; 1859 resident Banka; 1861 resident Besuki; 1862 governor West Coast of Sumatra; 1868–1871 member of the Raad van Indië (Indies Council); when retired private estate proprietor in Borneo where he had acquired mining concessions; died during the voyage to the Netherlands.

The collection consists of papers 1850–1889, including letters, regulations, Malay law books, papers G.M.W. Zuur, sugar manufactures Besuki, coffee cultures in the Preanger, coal mining in Kutai, singkep tin, rice growing, Tobacco Undertaking Senang Hati, Mining Company Tambang Salida.

Entry: Shelflist.

Size: *C.* 1 running metre.

**J.P. van Braam, October 27th 1737 Werkhoven – July 16th 1803 Zwolle** [165]

1748 naval cadet Admiralty of Amsterdam, voyages to Curaçao; 1753 lieutenant; 1758 commander; 1764 in the service of the VOC to the East Indies; 1767 equipagemeester (master of the equipment) in Bengal; 1773 discharge and in 1776 to the Netherlands; 1782 as commander in chief of a squadron to the Dutch East Indies, 1784 expedition against Riouw; 1786 back in the Netherlands; 1788 rear admiral; 1792 vice admiral.

The collection consists almost entirely of papers of before 1796, only two inventory items are pertaining to Asia after 1796, viz. observation of the councillor of the Indies W.H. van IJsseldijk about the government and the trade in Asia (1803) and an instruction from the governor of the North East Coast of Java N. Engelhard (1803), and also a supplement of papers 18th and 19th century.

Entries: Inventories. In *VROA*, 1915, p. 393–421; 1916, p. 478–481; 1950, p. 18.

Size: *C.* 1.50 running metres (273 inventory items).

Deposited at the First Section.

**Brugmans** [166]

The collection consists of the papers originating from P.A. Brugmans, A. Brugmans, D.A. Meerman van der Goes and Ph.H. de Haart. Only the papers of P.A. Brugmans (120 inventory items) contain papers concerning Asia after 1795.

*P.A. Brugmans, July 26th 1769 Groningen – May 2nd 1851 Amsterdam*
1786 law student at Groningen, 1790 at Leyden, 1792 graduated at Groningen; 1793–1796 second lawyer VOC of Amsterdam; subsequently up to *c.* 1838 lawyer at Amsterdam; 1839 staatsraad (councillor of state) in extraordinary service; 1840 member of the Council of State; 1841–1848 member of the First Chamber.

Originating from him are papers drawn up or received as lawyer of the VOC, only a few concerning the period after 1795, viz. minutes 1792–1796, letters 1796 to P. de Grandiour, repertory of charters, regulations etc. of the VOC 1602–1796, papers 1795–1796 concerning the liquidation of the VOC, papers 1795–1796 concerning J.C. Hartsinck at Lisbon, discourse 1796 about a possible attack of the English on Java; besides papers 1796–*c.* 1838 drawn up and received as lawyer at Amsterdam, including papers concerning the liquidation of the VOC, papers concerning J.C. Schultz, J. Pieters, A. Kuvel, S.C. Nederburgh and J.F. Spoenlin, papers concerning trade and cultures i.a. in connection with opinions of D. van Hogendorp, discourse by L. Manneron about Cape of Good Hope and Ceylon, lists of estimated receipts and expenses in Amboina, Banda and Ternate.

Entry: Inventory. In *VROA*, 1919, p. 347–366.

Size: *C.* 1.50 running metres (176 inventory items).

Deposited at the First Section.

**L.P.J. du Bus de Gisignies, February 28th 1780 Dottignies – May 31st 1849 Oostmalle** [167]

Training in Belgium for jurist; 1802 in the service of the French Empire; 1815 member of the Second Chamber; 1818 president of the Second Chamber; 1820 governor of Antwerp; 1823 governor of South Brabant; 1825 commissioner general of the Dutch East Indies, where he arrives in 1826; returns to Europe in 1830; after 1830 retired. (H. van Wijck. *De Nederlandsche Oost-Indische bezittingen onder het bestuur van den kommissaris-generaal Du Bus de Gisignies (1826–1830).* 's-Gravenhage, 1866).

Besides a small number of papers concerning i.a. his landed estates in Belgium and his offices in the Netherlands and Belgium, the collection consists completely of papers pertaining to his function as commissioner general of the Dutch East Indies, including documentation 17th century up to 1825, journal and other papers concerning his outward voyage, papers drawn up and/or received as commissioner general (minutes, correspondence, papers concerning officials, financial matters, trade and cultures, military matters, regencies etc.), papers concerning his homeward voyage, papers 1830–1847 drawn up and/or received as ex-commissioner general. For the names and subjects mentioned in the inventory see the index.

Entry: Inventory.

Size: *C.* 12 running metres (550 inventory items).

**D.C. Buurman van Vreeden, June 16th 1902 Surakarta – June 12th 1964 Paris** [168]

1921 cadet KMA; 1925 second lieutenant KNIL; 1928 first lieutenant; 1931 on leave in the Netherlands, student Hogere Krijgsschool (Army-Staff College); 1934 with the marechaussee (mounted constabulary) in Atjeh; 1935 Magelang; 1936 Tarakan; 1938 headquarters staff at Bandung, 1939 captain; 1942 liaison officer British headquarters at New Delhi, major, after that to Australia; 1943 military attaché Tsjungkin, then to New Delhi and liaison officer with Mountbatten; 1944 lieutenant colonel Dutch staff at Kandy; 1946 colonel, chief of the general staff at Batavia; 1949 (temporary) lieutenant general, (interim) army commander as successor of S.H. Spoor; 1950 returns to the Netherlands, where he was territorial commander till his discharge in 1957; 1957 deputy director of armaments with the West European Union at Paris.

Besides a number of appointment decrees the collection consists completely of papers 1945–1950 concerning Indonesia, drawn up and received by him as officer of the general staff and army commander. For the names and subjects mentioned in the inventory see the index.

Entry: Inventory of the S.H. Spoor (see no. 257) and D.C. Buurman van Vreeden collections.

Size: *C.* 0.60 running metre (154 inventory items).

Permission is necessary for consultation of the papers.

**A.A. Buyskes, January 21st 1771 Enkhuizen – January 23rd 1838 Loosduinen** [169]

1783 naval cadet; 1788 naval lieutenant; 1789 to the Dutch East Indies for survey of harbours and bays; 1792 first naval lieutenant; 1793 to the Netherlands; 1798 naval captain lieutenant; 1801 naval captain; 1802 to the Dutch East Indies, charged with taking over the Moluccas; 1804 to the Netherlands; 1807 rear admiral, commander of the naval forces in the Dutch East Indies; 1808 vice president of the Indies Council; 1809 to the Netherlands; 1814 commissioner general of the Dutch East Indies, arrives in Batavia in 1816; 1817 expedition to Amboina and Saparua; 1818 to the Netherlands; 1827 vice admiral.

The collection consists almost entirely of journals, i.a. of his voyages 1788–1819 from the Netherlands to the East Indies and through the Indian archipelago and of his voyage from Surabaja to Amboina and back; furthermore a report as commissioner general about the restoration of peace in the Moluccas 1817–1818 and a service record 1829.

Entry: Inventory. In J. de Hullu. *De archieven der admiraliteitscolleges.* 's-Gravenhage, 1924 (p. 301–303).

Size: *C.* 0.20 running metre (4 inventory items).

Deposited at the First Section.

**Canneman** [170]

This collection consists for the greater part of papers originating from the Dutch official E. Canneman. A much smaller number of papers, originating from W. Canneman, the son of E. Canneman, are pertaining to the Dutch East Indies.

*W. Canneman, born at The Hague in 1817*
In 1836, when he was 19 years old, Canneman went to the East Indies where he became a student of the Instituut voor de Javaansche Taal (Institute for the Javanese Language) at Surakarta; 1839 comptroller of the national revenue; 1845 on two years' leave to the Netherlands; 1848 secretary and vendumeester (auctioneer) in Besuki; 1849 discharge; 1851 assistant resident Banjumas (Probolinggo); 1854 assistant resident Bagelen (Ledok); 1855 inspector of the cultures in Java; 1864 to the Netherlands; 1866 retired.

Originating from W. Canneman: papers concerning the Institute for the Javanese Language 1834–1838; papers about Makasar 1833–1846; letters received as comptroller at Wonosobo; papers inspection of the cultures 1839; contract with Buleleng (Bali) 1841; letters from Indonesians from Bagelen and Pekalongan 1842–1863; papers concerning the affair of the demang of Situbondo 1844–1845; report of the cultures 1852; articles from the *Javasche Courant* about i.a. rice and sugar cultures 1854–1863.

Entry: Inventory. In *VROA,* 1904, p. 71–79. Supplement (only papers originating from E. Canneman) in *IRAA;* III. 's-Gravenhage, 1930, p. 264–266.

Size: *C.* 1.75 running metres (91, 20 inventory items).

**J. Cock Blomhoff, August 5th 1779 Amsterdam – October 13th 1853 Amersfoort** [171]

Up to 1802 soldier in the Netherlands, then into trade at Bremen; 1805 to Java; 1808 first lieutenant of H.W. Daendels; 1809 warehousekeeper of the Dutch factory in Deshima; 1813 as temporary chief factor to Java to conclude a commercial treaty with the British, is taken prisoner and arrives then, via England, in the Netherlands; 1816 chief factor of Deshima; 1824 retired in the Netherlands.

The collection consists mainly of papers 1817–1824 concerning Japan, viz. financial papers, papers concerning the trade, letters etc. from authorities at Batavia, Nagasaki, Osaka and Edo, papers concerning J.F. van Overmeer, papers concerning products, institutions, customs etc. in Japan; in addition papers concerning Gujerat and Korea.

Entry: Inventory. In *VROA*, 1907, p. 108–109.

Size: *C.* 0.10 running metre (14 inventory items).

**W. Cool, April 2nd 1877 Utrecht – January 30th 1947 The Hague** [172]

1894 student civil engineer Delft; 1898 engineer of public works of Rotterdam; 1914 member of the Commissie Kraus-De Jongh ter Advies Havenwerken in Nederlandsch-Indië (Advisory Committee Kraus-De Jongh for Harbour Works in the Dutch East Indies); 1915–1920 general consultant for the harbours in the Dutch East Indies; 1921 president of the Comité van Bestuur der Nederlandsch-Indische Spoorweg Mij. (Committee of Management of the Dutch East Indies Railway Company) at Semarang; 1931 discharge and to the Netherlands. He was also a member of i.a. the Koninklijk Instituut van Ingenieurs (Royal Institute of Engineers), of the Board of Governors of the Polytechnische Hogeschool (Polytechnic University) of Bandung, member and (vice) president of the Board of Supervisory Directors of the Trading Company late Maintz & Co., idem of the Tokohandel (Retail Trade) G.A. van de Pol.

The collection consists of papers drawn up and received in all his functions and additional functions mentioned above, and also of private papers and documentation about harbours, shipping, dredging works, construction of bridges and roads, water management i.a. in Japan, China, Singapore and Indonesia. For the names and subjects mentioned in the inventory see the index.

Entry: Inventory.

Size: *C.* 4.70 running metres (252 inventory items).

**Couperus** [173]

The collection consists of papers originating from the Couperus family. Many papers are pertaining to genealogical research, only a small part to Asia. For the names and subjects mentioned in the inventory see the index.

The papers concerning Asia are originating from:

a. *P.Th. Couperus, c. August 12th 1787 Malacca – 1823*
Son of A. Couperus, governor of Malaya (the collection contains also papers of him). In the *RAvNI* of 1816 P.Th. Couperus is mentioned as secretary of the Commercial Committee, in the one of 1817–1822 as assistant collector of the import and export duties.
b. *J.R. Couperus, February 24th 1816 Batavia – 1902*
Son of P.Th. Couperus. Training for jurist in the Netherlands; 1838 to the Dutch East Indies; 1839 at the disposal of the commissioner for Sumatra; 1840

second comptroller of the national revenue and cultures; 1843 secretary and fiscal of the Rechtbank van Ommegang (court) at Batavia; 1844 fiscal of the Council of Justice at Padang; 1846 member of the Council of Justice at Batavia; 1847 temporary travelling judge of the Western District of Borneo; [1848] member of the Council of Justice at Batavia; 1849 griffier (clerk) of the High Court of Justice; 1850 raadsheer (justice) of the High Court of Justice; 1854 also member of the High Court Martial; 1859 discharge, to the Netherlands. (*Het Regt in Nederlandsch-Indië;* regtskundig tijdschrift; XX (1862), p. 360).

c. *F.E. Couperus, 1857–1910*
1884 extraordinary deputy clerk of the landraad (district council) of Pasuruan (principal town and Bangil); 1887 clerk of the district council of Batavia (Meester Cornelis); 1889 member of the Council of Justice Surabaja; 1892 also second lieutenant secretary of the court martial; 1895 president of the district council of Madura (Pamekasan); 1896 on leave to the Netherlands; 1899 president of the district council of Bantam (Serang); 1901 member of the Council of Justice of Semarang; 1909 president of the Council of Justice of Semarang. *(RAvNI)*

d. *Other members of the family.* The papers concerning Asia are originating from A. Couperus (1752–1813) about the capitulation of Malacca in 1795, by himself, P. Walbeeck, J.R. Couperus and C.R.G. Vlielander Hein-Couperus; G.G. Couperus (1819–1876) concerning the sugar contracts in Java, 1860; G.W. Couperus (1842–1891) concerning coffee in Java, *c.* 1880; J.R. Couperus jr. (1853–1940) records of service, papers about rinderpest in West Java and about the Principalities by himself as resident of Djokjakarta and by M. van Geuns at Surabaja 1884–1908.

Entry: Inventory.

Size: *C.* 2 running metres (124 inventory items).

Deposited at the First Section.

**J. Couvreur, 1900–?** [174]

1926 assistant commander Sawahlunto (Fort van der Capellen); 1927 idem Batipuh (Pariaman); 1928 comptroller/commander Painan (Balaiselasa); 1929 idem Fort de Kock (Pajakumbo); 1930 idem Indragiri (Kuantan districts); 1932 idem Buton and Laiwu (Muna); 1935 idem Pare-Pare; 1938–1942 comptroller for Makasar; after 1945 i.a. assistant resident Sumba.

The collection consists almost completely of papers 1948 concerning self government in Sumba. These include as preliminary papers a note of A. Couvreur of 1913 and *Het eiland Soemba in 1759;* by H. Erasmusz. Besides these papers the collection also contains photographs of East Java and the Ronde Tafel Conferentie (Round Table Conference) 1947–1949.

Entry: Inventory. In *VROA*, 1964, p. 43.

Size: *C.* 0.15 running metre (7 inventory items).

**Cremer** [175]

For the history of Asia are especially relevant the papers originating from J.Th. Cremer; in addition the collection contains letters from J.Th. Cremer to his son H. Cremer.

*J.Th. Cremer, June 30st 1847 Zwolle – August 14th 1923 Amsterdam*
1863 junior clerk Firm of Balck; 1865 Firm of Ellerman; 1867 Nederlandsche Handel-Maatschappij (Dutch Trading Company), 1868 sent to Indonesia; 1871 employed by the Deli-Maatschappij (Deli Company), 1879 founder of the Deli-Plantersvereeniging (Deli Planters' Association); 1880 to the Netherlands; 1881–1894 member of the Second Chamber; 1897–1901 minister of Colonial Affairs; 1907–1918 president of the Dutch Trading Company; 1921 member of the First Chamber; after 1880 he travelled to Indonesia in 1883, 1905–1907 and 1913; he was founder and director of i.a. the Deli Spoorweg Maatschappij (Deli Railway Company), the Nederlandsch-Indische Landbouw Maatschappij (Dutch East Indies Agricultural Company), the Koninklijke Paketvaart Maatschappij (Royal Packet Boat Company) and the Koloniaal Instituut (Colonial Institute) at Amsterdam (the present Koninklijk Instituut voor de Tropen (Royal Institute for the Tropics).

The papers originating from J.Th. Cremer run to 165 inventory items and consist of private papers, financial papers, papers about real estate, papers about his functions from 1868, correspondence from 1866 and publications and speeches. For the names and subjects mentioned in the inventory see the index.

Entry: Inventory.

Size: *C.* 6 running metres (496 inventory items).

**H.W. Daendels, October 21st 1762 Hattem – May 2nd 1818 St. George d'Elmina** [176]

Law student at Harderwijk where he graduated in 1783; in 1788 he fled to France as patriot (anti-Orangeman); 1792 commander of the Batavian legion; 1794 brigade general, campaign through the Netherlands; 1795 lieutenant general; 1799 fights against the English; 1802 discharge; 1802 extraordinary councillor of state; 1807–1810 governor general of the East Indies, took over the government there January 1808; 1812–1813 campaign to Russia; 1815 governor on the Coast of Guinea where he died.

The collection consists of papers 1783–1818 with some papers of after 1818 originating from other persons. 48 inventory items concern papers drawn up or received by Daendels as governor general of the Indies. For the names and subjects mentioned in the inventory see the index.

Entry: Inventory. Herein are also included the acquisitions mentioned in *VROA*, 1894, p. 72–74 and *VROA*, 1927, p. 112–114.

Size: *C.* 1.75 running metres (189 inventory items).

**M.L. van Deventer, March 10th 1832 The Hague – January 22nd 1892 Prinsenhage** [177]

1848 official Ministry of Foreign Affairs, 1872 principal clerk; 1875–1879 consul general Rio de Janeiro; after his return to the Netherlands publicist in the field of the history of the Dutch East Indies.

The collection consists of papers *c.* 1870–1892, including correspondence and notes to the publications *Geschiedenis der Nederlanders op Java.* 1887; *De opkomst van het Nederlandsch gezag in Oost-Indië.* 1888; and *Het Nederlandsch gezag over Java.* 1891; the collection contains also photographs.

Size: *C.* 1 running metre.

**Diemont-Van Alphen** [178]

The collection consists almost completely of journals, viz. by H.P.N. 't Hooft of his voyages to and residences in the Dutch East Indies 1814–1826 and a journal by S. 't Hooft of his voyage to the Dutch East Indies 1815–1816.

Entry: Inventory. In *VROA,* 1897, p. 44–45.

Size: *C.* 0.20 running metre (6 inventory items).

**H. Doeff, December 2nd 1777 Amsterdam – October 19th 1835 Amsterdam** [179]

1796 to Batavia; 1799 dispensier and scriba (secretary) in Japan; 1801 warehouse-keeper; 1803–1817 chief factor, was also engaged in scientific research; 1819 from Batavia to the Netherlands; 1824 concerned in the foundation of the NHM.

The collection consists of official papers concerning Japan, colonial trade and the foundation of the NHM, private papers and papers about his scientific work, and also some papers originating from his wife H. Doeff-Jacobs and his great-grandson H. Doeff jr. For the names and subjects mentioned in the inventory see the index.

Entry: Inventory.

Size: *C.* 1.75 running metres (97 inventory items).

**A.J. Duymaer van Twist, February 20th 1809 Deventer – December 3rd 1887 Diepenveen** [180]

Training for jurist at Leyden; 1832 lawyer at Deventer; 1843 member of the Second Chamber; 1850 president of the Second Chamber; 1851 governor general of the Dutch East Indies; 1856 to the Netherlands, retired, appointed minister of State; 1858–1862 member of the Second Chamber; 1865–1881 member of the First Chamber. (J. Zwart. *A.J. Duymaer van Twist;* een historisch-liberaal staatsman, 1809–1887. Utrecht, 1939. (Utrechtsche bijdragen tot de geschiedenis, het staatsrecht en de economie van Nederlandsch-Indië; no. XVI)).

The collection consists for a large part of papers concerning the Dutch East Indies. They were drawn up and received mainly by Duymaer van Twist during and after his governor generalship, up to *c.* 1880. For the names and subjects mentioned in the inventory see the index.

Entry: Inventory.

Size: *C.* 0.50 running metre (*c.* 120 inventory items).

See also Brieven aan den gouverneur-generaal [A.J.] Duymaer van Twist van vrienden in het vaderland; medegedeeld door J. Zwart. In *BMHG;* vol. 60 (1939), p. 1–86.

**De Eerens** [181]

The collection consists of papers originating from several members of the De Eerens family. Indies papers are originating from:

*D.J. de Eerens, March 17th 1781 Alkmaar – May 30st 1840 Buitenzorg*
1798 soldier; 1800 cadet; 1801 second lieutenant; 1803 adjutant major; 1806 first lieutenant; 1807 captain; 1809 lieutenant colonel; 1812 adjutant commander of the general staff; 1813 chief of the general staff; 1815 general major; 1826 lieutenant general and inspector general of the infantry; 1829 director general of the Ministry of War; 1834 lieutenant governor general of the Dutch East Indies, arrives in 1835 in Batavia and succeeds in 1836 J.C. Baud as governor general.

Originating from him are letters from J. van den Bosch about i.a. the cultivation system and financial matters, and also letters from J.C. Baud, J.B. Cleerens, J.P. Cornets de Groot, . . . Hoogeveen, L. Launy, P. Merkus, H.G. Nahuys van Burgst, J. Penning Nieuwland, J.C. Reynst, L. Vitalis, W. de Vogel and papers concerning the Indies Council, the susuhunan of Surakarta, the sultan of Sumenep, regents, military expeditions on the West Coast of Sumatra and Bondjol.

The inventory is forthcoming.

Size: *C.* 0.65 running metre.

See also J. van den Bosch en D.J. de Eerens. *Particuliere briefwisseling, 1834–1840, en enige daarop betrekking hebbende andere stukken;* uitgegeven door F.C. Gerretson en W.Ph. Coolhaas. Groningen, 1960. (WHG; serie 3; no. 3).

**C.Th. Elout, March 22nd 1767 Haarlem – May 3rd 1841 The Hague** [182]

Training for jurist at Leyden; 1788 lawyer at Amsterdam; 1795 schout-crimineel (criminal sheriff), dijkgraaf (dikereeve), opperstrandvoogd (chief receiver of wreck) Texel; 1796 councillor in the Court of Holland and Zeeland; 1799 member of the departmental court of justice Texel; 1802 procureur-generaal (public prosecutor) of the national court of justice; 1805 commissioner general of the Dutch East Indies for the improvement of the government there; 1806 voyage to the Indies is interrupted when Daendels is appointed governor general; 1808 member of the Staatsraad (State Council); 1813 member of the committees for the drawing up of the constitutions of 1814 and 1815; 1814 member of the Raad van State (Council of State); 1814 commissioner general of the Dutch East Indies, arrives in Batavia in 1816 and returns to the Netherlands in 1819; 1820 member of the negotiating committee for the settlement of East Indian affairs with the British; 1824 minister of Nationale Nijverheid en Koloniën (National Industry and Colonial Affairs); 1829 discharge, minister of State.

Papers concerning the Indies have been drawn up and/or received by Elout mainly when he was commissioner general, negotiator with the British and Minister of Colonial Affairs. Besides a number of preliminary papers from 1788 for the purpose of the negotiations, the collection also contains a number of papers *c.* 1850–1858 originating from his son P.J. Elout van Soeterwoude (i.a. member of the Council of State 1864–1874; member of the Second Chamber 1853, 1879–1880 and of the First Chamber 1886–1887) concerning British interference in Borneo and the Londens tractaat (London Treaty). For the names and subjects mentioned in the inventories see the index.

Entries: Inventories, viz. an inventory in typescript of the acquisitions 1853, 1869, 1874, a printed inventory in *VROA*, 1908, p. 59–60 and 1917, p. 437–468 and an inventory in typescript of his private correspondence.

Size: *C.* 2 running metres (74, 15, 174, 366 inventory items).

See also *Bijdragen tot de kennis van het koloniaal beheer;* getrokken uit de nagelaten papieren van [C.Th.] Elout; [door P.J. Elout van Soeterwoude]. 's-Gravenhage, 1851.

*Bijdragen tot de geschiedenis van het koloniaal beheer;* getrokken uit de nagelaten papieren van [C.Th.] Elout; [door P.J. Elout van Soeterwoude]. 's-Gravenhage, 1861.

*Bijdragen tot de geschiedenis der onderhandelingen met Engeland;* betreffende de overzeesche bezittingen, 1820–1824; getrokken uit de nagelaten papieren van [C.Th.] Elout; [door P.J. Elout van Soeterwoude]. 's-Gravenhage, 1863.

*Bijdragen betrekkelijk koloniale en andere aangelegenheden in den Raad van State behandeld;* getrokken uit de nagelaten papieren van [C.Th.] Elout; [door P.J. Elout van Soeterwoude]. 's-Gravenhage, 1874.

**Fabius** [183]

The collection consists of the papers of several members of the Fabius family. Papers concerning the Indies are originating from:

a. *J. Fabius, February 11th 1776 Amsterdam – November 11th 1850 Amsterdam*
   Training for jurist at Utrecht; 1797 lawyer at Amsterdam; 1802 notary at Amsterdam; 1838 member of the Criminal Court of Justice of Amsterdam; 1842 raadsheer (justice) of the Provincial Court of Justice of North Holland.

J. Fabius kept up correspondence with various important East Indies officials, viz. with J. van den Bosch, G.A.G.Ph. van der Capellen, H.J. van de Graaff, W. Kocken, I.J. van Sevenhoven (with a memorandum of Van Sevenhoven about the cultures in Java). The letters were sent to Fabius between 1799 and 1842. (5 inventory items).

b. *G. Fabius, December 13th 1806 Lisse – March 24th 1888 The Hague*
1821 ordinary seaman; 1825 naval cadet; 1838–1842 merchant captain; 1844 second naval lieutenant; 1845–1848 adjutant of rear admiral E.B. van den Bosch in the Dutch East Indies; 1854 naval captain lieutenant in Japan; 1860 naval captain; 1861 commander of the navy in the West Indies; 1864 rear admiral, up to 1868 commander of the navy in the Dutch East Indies; 1867 vice admiral; 1869 retired, then in the Netherlands member of the Second Chamber.

Besides (auto)biographic notes and a record of service there are papers concerning the East Indies originating from G. Fabius, drawn up and/or received when he was adjutant of E.B. van den Bosch (letters from him) and when he was commander of the navy in the East Indies; moreover papers J.H. Donker Curtius, A.A. Buyskes, W.J.C.H. Kattendijk, papers about his voyage to the East Indies and stay in Japan 1850–1851, articles about naval yards in Java and the steam ship service in the Indies. (14 inventory items).

c. *Papers concerning the voyages from and to the East Indies and the residence there of several other members of the Fabius family*
G.C. (letters to J. Fabius, papers military career), G.Th.J. (sugar cultures in Java), J.A.P. (papers military career), A. and K. Fabius (about their voyages and about members of the family). (12 inventory items).

Entries: Inventories J. Fabius in *VROA*, 1913, p. 248–252, *VROA*, 1916, p. 234–238; other members of the family in inventory in typescript.

Size: *C.* 3.50 running metres (15, 32, 728 inventory items).

**Falck** [184]

The collection consists of papers originating from several members of the Falck family. Papers concerning Asia after 1796 are originating from O.W. Falck and his son A.R. Falck.

*O.W. Falck, 1738–1814*
Went in 1756 to the East Indies as constabelsmaat (quarter-gunner) and returned in 1775 as merchant.

Originating from him are some papers (2 inventory items) concerning Asia after 1796, viz. concerning the Council for the Asian Possessions and Establishments 1800 and letters 1802 from S.C. Nederburgh and [Chr.A.] Ver Huell concerning the East Indies trade.

*A.R. Falck, March 19th 1777 Utrecht – March 16th 1843 Brussels*
Training for jurist at Leyden; 1799 lawyer; 1801 member of the municipal government of Amsterdam; 1802 in diplomatic service, up to 1806 at Madrid; 1806

department head at the Ministry of Foreign Affairs, then secretary general of the Ministry of Naval and Colonial Affairs; 1810 lawyer at Amsterdam; 1813 general secretary to Willem I, then secretary of state; 1818 minister of Education, Industry and Colonial Affairs; 1823 member of the negotiating committee for the settlement of East Indies Affairs with the British; 1824 ambassador to the British court; 1832 retired; 1839–1843 ambassador to Brussels.

Concerning the Dutch East Indies the collection contains, besides a number of diaries, papers (13 inventory items) drawn up and/or received by Falck when he was minister of Colonial Affairs and negotiator with the British (London Treaty), with letters 1814–1824 from J.C. Baud (originating from his wife), G.A.G.Ph. van der Capellen (also originating from his wife), C.Th. Elout, H.M. de Kock, J.D. Kruseman, H.W. Muntinghe as commissioner of Palembang and Banka concerning piracy in the Straits of Malaya, J.J. Rochussen, J. Schneither concerning Timor, F. Wappers Melis, memorandums 1814, 1817 about Java of D.F. van Alphen, extract from the Nederlandsche Bezittingen (Dutch Possessions) by J. van den Bosch.

Entry: Inventory. In *VROA,* 1913, p. 187–202, supplements in typescript.

Size: *C.* 2 running metres (133, 4 inventory items).

See also A.R. Falck. *Ambtsbrieven, 1802–1842;* [uitgegeven door O.W. Hora Siccama]. 's-Gravenhage, 1878.

A.R. Falck. *Gedenkschriften;* uitgegeven en met een appendix voorzien door H.T. Colenbrander. 's-Gravenhage, 1913. (RGP; kleine serie; no. 13).

**F.W. Fennekol, c. 1761 St. George d'Elmina – January 23rd 1837 Noordwijk** [185]

1784 finished law studies at Harderwijk, 1788 graduated at Leyden; 1795 member of the Provisioneele Representanten van het Volk van Nederland (Provisional Representatives of the People of the Netherlands); 1796–1800 member of the Comité tot de Zaken van de Oost-Indische Handel en Bezittingen (Committee for the Affairs of East Indies Trade and Possessions); 1800 collector; 1806 retired; 1813–1817 archivist at the Ministry of Colonial Affairs; 1819–1822 official at the Ministry of Justice.

The collection consists of papers 1782–1831, the most important part being a number of private letters 1795–1831 from J.B. Bicker, . . . Bijleveld, A.R. Falck, W. Fijnje, H. Gevers, I.J.A. Gogel, J. Goldberg, J.G.H. Hahn, W. Irhoven van Dam, F. Kumsius, B.F. von Liebeherr, A.J. la Pierre, C.J. Prediger, P.P.J. Quint Ondaatje, C. Scheffer, J.C. Smissaert, . . . van Vlierden, S.I. Wiselius and P. van Zonsbeek.

Entry: Inventory. In *VROA,* 1913, p. 326–327.

Size: *C.* 0.50 running metre (27 inventory items).

**W.K.H. Feuilletau de Bruyn, July 11th 1866 Palembang – ?** [186]

1904–1907 training KMA at Breda; 1907 as second lieutenant to the Dutch East Indies; 1908–1910 expedition in West Ceram; 1910–1916 in New Guinea, 1911 first lieutenant; 1919 captain; 1920–1926 Southern and Eastern district of Borneo; 1919–1922 geology student at Lausanne, thesis on the geology of New Guinea, 1926 second time to Lausanne; 1929 major; 1931 lieutenant colonel; 1933 colonel; 1931–1935 member of the Volksraad (People's Council), 1935 member of the College van Gedelegeerden (Board of Deputies), representative of the Vaderlandsche Club (Fatherland Club); 1935 retired as colonel, interned in the Netherlands during the second world war; after 1945 opponent of the government policy concerning Indonesia; publicist i.a. *De Vrije Pers* (Supplements from *Persoonlijkheden in het Koninkrijk der Nederlanden in woord en beeld.* Amsterdam, 1938).

The collection contains papers concerning the economic defence of the East Indies of the period when he was a member of the Volksraad (People's Council) and when he was an opponent of the government policy concerning Indonesia 1945–1950; Parlementaire Enquêtecommissie (Parliamentary Committee of Inquiry) 1945–1961 with letters from H. Baudet, J. de Courcy, H. Fraser, J. van Hulzen, . . . Hyma, L. de Jong, H.A. Korthals, J.W. Meyer Ranneft, H.J. van Mook, O. Reuchlin, W. Schermerhorn and H.J. de Vries.

Entry: Inventory.

Size: *C.* 0.15 running metre (10 inventory items).
Permission is necessary for consultation of the papers.

**R. de Filliettaz Bousquet, September 2nd 1801 Delft – July 26th 1878 The Hague** [187]

1815–1819 naval cadet; 1819 clerk at the department of Finance at Batavia; 1821 clerk to the inspector of finance; 1824 clerk at the Landelijke Inkomsten en Directe Belastingen (Inland Revenue and Direct Taxes); 1825 secretary Cheribon; 1829 assistant resident Buitenzorg; 1830 resident Ledok; 1831 resident Pekalongan; 1834 governor Makasar; 1841 discharge and to the Netherlands; 1842 returns to the East Indies and in 1843 resident Kedu; 1845 resident Djokjakarta; 1848 discharge; 1849 resident Banjumas; 1850 resident Bagelen; 1854 discharge and to the Netherlands.

The collection consists of papers concerning his career 1841–1856; a memorandum of justification concerning his discharge as governor of Makasar 1839, 1840; papers concerning his discharge (as a result of the accusation of . . . Prehn) as resident of Djokjakarta and as temporary resident of Banjumas 1848–1851; papers concerning land tax and salt duty at Bagelen 1852; papers concerning the rumours about malpractices of R. van Noord Borski, assistant resident of Ledok 1853.

Entry: Inventory. In *VROA*, 1939, p. 22–23.

Size: *C.* 0.20 running metre (5 inventory items).

**I.D. Fransen van de Putte, March 22nd 1822 Goes – March 3rd 1902 The Hague** [188]

After a broken off training for naval cadet from 1838 in the merchant navy, many voyages to the Dutch East Indies; 1849 administrator of the sugar factory Pandji (Besuki); 1857 owner of the factory; 1862 member of the Second Chamber, expert in East Indies Affairs; 1863 minister of Colonial Affairs; 1866 member of the Second Chamber; 1872 Minister of Colonial Affairs, 1873–1874 also minister of Naval Affairs, 1874 member of the Second Chamber; 1880 till his death in 1902 member of the First Chamber.

The collection consists of papers from 1838, viz. letters 1838–1848 about the Indies; papers concerning Pandji and other undertakings; articles by Fransen van de Putte; papers drawn up and received when he was minister and member of the Second and First Chamber; moreover a number of portraits of himself, J.C. Baud, W.R. van Hoëvell, L.W.C. Keuchenius, T.H. der Kinderen, J.J. Rochussen, C.P.K. Winckel. For the names and subjects mentioned in the inventory see the index.

Entry: Inventory, with supplement.

Size: *C.* 0.55 running metre (177 inventory items).

Permission is necessary for consultation of some of the papers.

See also I.D. Fransen van de Putte. *Parlementaire redevoeringen (van Mei 1862–September 1865).* Schiedam, 1872, 1873. 2 vol.

I.D. Fransen van de Putte. *Atjeh;* parlementaire redevoeringen (van Februari 1873–December 1885); met een voorwoord. Schiedam, 1886.

Brieven van I.D. Fransen van de Putte uit diens planterstijd; medegedeeld door R. Reinsma. In *BMHG;* vol. 71(1957), p. 160–185.

**W. Fijnje, January 24th 1750 Zwolle – October 2nd 1809 Amsterdam** [189]

1771 student Faculty of Arts at Leyden, 1774 graduated; 1774 teacher Deventer; 1779 editor *Hollandsche Historische Courant* at Delft; active member of the Patriottische (anti-Orange) movement, 1787 refugee; 1795 president of the assembly Provisioneele Representanten (Provisional Representatives); 1795 member of the Comité tot de Zaken van de Oost-Indische Handel en Bezittingen (Committee for the Affairs of the East Indies Trade and Possessions); 1798 member of the Uitvoerend Bewind (Executive Government); 1798 retired; 1805 chief editor of successively the *Bataafsche Staatscourant,* the *Koninklijke Staatscourant* and the *Koninklijke Courant en Moniteur van Amsterdam.*

The collection consists of papers of himself and his descendants 1772–1852, including papers 1795–1797 concerning the Committee for the Affairs of the East Indies Trade and Possessions.

Entry: Inventory. In *VROA,* 1943, p. 17–18.

Size: *C.* 0.10 running metre (12 inventory items).

**W.O. Gallois, 1848–1917** [190]

From 1871 till 1891 Gallois was employed at the Algemene Secretarie (General Secretariat) at Batavia (1871 second clerk, 1873–1876 first clerk, 1879 referendary, 1883 secretary, 1886 first secretary, 1887 general secretary); 1891 member of the Raad van Indië (Indies Council); 1895–1898 vice president of the Indies Council.

The collection consists entirely of papers concerning the mission of Gallois to the Minahasa in 1892 in connection with a government inquiry into the situation there, viz. letters to the governor general; advice and reports of himself; letters, petitions, memorandums etc. of Z.B. Hoekzema, A.B. Kalenkongan, A. Lasut Wawo Runto, E. Ludu, . . . Reinking (at Amurang), J. Rotinsulu (at Tonsea), M.C.E. Stakman (resident of Menado), reverend J.F. Verhoef and the Zendingsvereeniging (Missionary Society) concerning leasehold lands in Tinombo and Poso, tobacco, import of coolies, statute labour and taxes.

Entry: Inventory. In *VROA*, 1918, p. 796–797.

Size: *C.* 0.10 running metre (20 inventory items).

See also W.O. Gallois. *Rapport nopens den staat van zaken in de Minahassa;* uitgebracht; in voldoening aan de hem bij besluit van 8 februari 1892, no. 1 gedane opdracht. Batavia, 1892.

**P.R.W. van Gesseler Verschuir, May 10th 1883 Groningen – May 27th 1962 The Hague** [191]

Student of indology at Leyden; 1905 official Algemene Secretarie (General Secretariat); 1905 aspirant comptroller Surabaja; 1906 comptroller Banjuwangi; 1908 Panarukan; 1910 Bondowoso; 1912 Sumenep; 1913 on leave; 1915 comptroller Kuningan; 1917 Batavia; 1919 specially charged with food supply; 1921 assistant resident; 1922 on leave; 1924 assistant resident Kudus; 1925 resident at the disposal of the governor of West Java; 1929 governor of Djokjakarta; 1933 retired.

The collection consists mainly of papers drawn up and received from 1905 when he was an official in the Indies. For the names and subjects mentioned in the inventory see the index.

Entry: Inventory.

Size: *C.* 0.30 running metre (61 inventory items).

**Gey van Pittius** [192]

The collection consists of papers originating from several members of the Gey van Pittius family. Papers concerning the Indies are originating from:

a. *A.R.W. Gey van Pittius sr., November 25th 1787 Willemstad – March 6th 1865 Zutphen*
   1803 second lieutenant; 1808 first lieutenant; 1813 captain; 1821 major, expedition Palembang; 1825 lieutenant colonel, fightings Celebes, command of artillery in the Java war; 1829 retired to the Netherlands.

Originating from him are records of service etc.; a biography; journal 1821 of the expedition to Palembang, with annexes; letters 1811–1841 from N.C. Ampt, B.G.M.V. Baltus, J.H. Beer, L.C. du Bus, L.P.J. du Bus de Gisignies, W. Koopman, C.F. Krahmer, E. van Löben Sels, J.D. de Roock and H.R. Trip. (27 inventory items).

b. *A.R.W. Gey van Pittius jr., August 13th 1838 The Hague – January 16th 1896 The Hague*
Son of Gey sr.; 1857 second lieutenant, took part in the expedition to Palembang 1858–1859; 1860 first lieutenant; 1868 captain; 1875 major, took part in the expedition to Atjeh 1878–1881; 1880 lieutenant colonel; 1884 colonel; 1885–1893 chief of the general staff; 1888 general major; 1893 till his discharge in 1895 lieutenant general, commander of the East Indies army, took part in the Lombok expedition 1894.

Originating from him are records of service etc.; appointments as member of the KITLV and the Central Committee of the Red Cross; report expedition Djambi and Lematang Ulu 1859; reports interview with A.W.P. Weitzel and O. van Rees 1884; speeches of Gey at the taking over and surrender of the command of the East Indies army; articles about Atjeh and Lombok expeditions; letters 1884–1895 from W.H. van Capelle, W.K. van Dedem, A. Ernste, J.J.K. Moulin, K.L. Pfeiffer, C. Pijnacker Hordijk, B.T.C.F. Schmidt, J.D. Six, G.M. Verspijck, J.H.P. van der Willige von Schmidt auf Altenstadt, C.H.A. van der Wijck; necrologies; plan of Batavia. (46 inventory items).

c. *A.C.A.P. Gey van Pittius-Blanken*
Wife of Gey jr., viz. diary sheets of her husband about his appointments as lieutenant general and commander of the East Indies army, papers 1904–1935 about the rehabilitation of Gey jr. in consequence of his part in the Lombok expedition. (2 inventory items).

d. *A.R.W. Gey van Pittius*
Son of Gey jr., viz. papers 1904–1937 concerning his career in the East Indies army. (13 inventory items).

Entry: Inventory.

Size: *C.* 0.80 running metre (250 inventory items).

**F.P.A. van Gheel Gildemeester, 1875–?** [193]

Officer infantry KNIL, 1897 second lieutenant; 1900 first lieutenant; 1911 captain; 1919 major; 1921 lieutenant colonel; 1924 colonel.

The collection consists of papers 1900–1913 about the war in Atjeh, and also some papers about Semarang. For the names and subjects mentioned in the shelflist see the index.

Entry: Shelflist.

Size: *C.* 0.10 running metre.

**J. Goldberg, May 8th 1763 Amsterdam – April 25th 1828 Voorschoten** [194]

1793 insurer at Amsterdam; 1795 member of the municipality of Amsterdam, from 1796 also substitute member of the provincial government of Holland and of the National Assembly; 1799–1801 agent for the National Economy; 1805–1810 member of the Staatsraad (State Council) (president of the section finance), 1807–1809 also ambassador to Berlin; 1812 president of the law court of mercantile affairs at The Hague; 1814 member of the Raad van Koophandel en Koloniën (Council of Commercial and Colonial Affairs); 1815 director general of Commercial and Colonial Affairs; 1818–1827 member of the Raad van State (Council of State).

The papers concerning Asia after 1796 (*c.* 140 inventory items) were drawn up and/or received by Goldberg when he was (substitute) member of the provincial government of Holland, of the Staatsraad (Council of State), of the Council of Commercial and Colonial Affairs and when he was director general of Commercial and Colonial Affairs. For the names and subjects mentioned in the inventory see the index.

Entry: Inventory. In *VROA*, 1913, p. 253–303 and *VROA*, 1914, p. 483.

Size: *C.* 31 running metres (391 inventory items).

**Van de Graaff/Van Vinceler** [195]

Papers about the East Indies in this collection are all originating from:

*H.J. van de Graaff, September 7th 1782 Surat – March 1st 1827 near West-Capelle*
Training for jurist at Groningen, Berlin and Utrecht; 1801 settles at Alkmaar where he was subsequently town clerk, judge and town councillor; 1816 to the Dutch East Indies, member of the Council of Finance at Batavia; 1817 asssistant inspector general of the inland revenue; 1820 senior inspector of Finance, member of the East Indies Council; 1826 discharge, drowned when he was shipwrecked off the coast of West-Capelle. Van de Graaff accompanied G.A.G.Ph. van der Capellen on his travels through Java (1816, 1819) and to the Moluccas (1824) and in 1821 he also made a tour of inspection to the Moluccas himself.

The papers originating from Van de Graaff (19 inventory items, *c.* 0.70 running metre) are all concerning the East Indies, letters to him 1816–1827 (alphabetically arranged), papers concerning the monetary system, a despatch from the resident of Japara and Joana D.W. Pinket van Haak 1825 about the Java war, letters kept separately 1796–1828 from and to J. Bousquet, A.A. Buyskes, G. Canning at Calcutta, G.A.G.Ph. van der Capellen, C.Th. Elout, his brother in law G. Fontein Verschuir, Gobson & Co in Ceylon, J.C. Goldman, F.J. van de Graaff, his wife A. van de Graaff-de Dieu, J.W. Janssens, H.M. de Kock, G.F. Meylan, D.W. Pinket van Haak, N. de Vogel and F. Wappers Melis.

Entry: Inventory.

Size of the complete collection: *C.* 0.80 running metre (61 inventory items).

See also P.H. van der Kemp. *Brieven van en aan H.J. van de Graaff, 1816–1826;* een bijdrage tot de kennis der Oost-Indische bestuurstoestanden onder de regeering van G.A.G.P [h.] van der Capellen. Batavia etc., 1901, 1902. (VBG; vol. 52)

**Gunning** [196]

The collection consists of papers originating from J.H. Gunning and his daughter C.H.C. Gunning.

*J.H. Gunning, July 13th 1859 Utrecht – March 5th 1951 Hilversum*
Classicist; 1882 teacher classic languages gymnasium Utrecht; 1888 (con) rector ((vice) principal) gymnasium Zwolle; 1898 study of pedagogy at Jena; 1900 pedagogy tutor university Utrecht; 1904 idem Amsterdam; 1917 idem Utrecht; 1923–1929 extraordinary professor Utrecht; 1915–1918 member of the Provincial States of North Holland; 1925–1935 municipal councillor of Hilversum; 1922 (according to the *RAvNI*) member of the Commissie van Bijstand in Nederland van het Genootschap voor In- en Uitwendige Zending (Committee of Support in the Netherlands of the Society for Internal and External Protestant Mission).

*C.H.C. Gunning, c. 1889 Zwolle – ?*
Born in 1889, see *Zendingsblad der Nederlands Hervormde Kerk*, July, August 1959; in 1915 to the East Indies (see correspondence from 1915); in the *RAvNI* she is mentioned as district nurse of the Genootschap voor In- en Uitwendige Zending 1920–1922; matron of the reformatory institutions at Sukabumi 1925–1931; vice president of the Society for Internal and External Protestant Mission 1939–1940.
The collection contains correspondence 1915–1946 between C.H.C. Gunning and her parents; papers 1919–1942 received by J.H. Gunning; papers 1921–1956 received by C.H.C. Gunning; papers originating from C.H.C. Gunning concerning the Christelijke Jonge Vrouwen Federatie (Federation of Christian Young Women), the Society for Internal and External Protestant Mission, reformatory institutions at Sukabumi.

Entry: Provisional inventory.

Size: *C.* 0.30 running metre (19 inventory items).

**H.H.A. van Gybland Oosterhoff, May 26th 1887 Batavia – January 21st 1937 The Hague** [197]

Study of law and political science at Leyden, Amsterdam and Utrecht; 1910 assistant clerk Ministry of Agriculture, Trade and Industry; 1913 factory inspector The Hague; 1914 assistant secretary Koninklijk Nationaal Steuncomité (Royal National Relief Committee); 1916 BPM, from 1918 private secretary to H.Colijn, 1920 director petroleum company La Corona in Mexico; 1924 secretary of the board of governors of the faculty of indology at Utrecht; was also secretary of the council of employers for Surinam, president of the Dutch-Mexican Chamber of Commerce; travelled much i.a. to the Dutch East Indies 1916–1917; publicist, i.a.

chief editor of *De Rijkseenheid* established in 1929; from 1933 vice president of the Verbond voor Nationaal Herstel (League for National Recovery) which acted together with the Indo-Europeesch Verbond (Indo-European League) and the Vaderlandsche Club (Fatherland Club); member of the Entente Internationale Anti-Communiste; propagandist Dutch culture in South Africa.

Concerning Asia the collection contains papers of and about the Ondernemersraad van Nederlands-Indië (Employers' Council of the Dutch East Indies), the Faculty of Indology, Dutch-Indonesian political relations, correspondence 1927–1937, and also papers originating from some relatives. For the names and subjects mentioned in the inventory see the index.

Entry: W.L.H. Roessingh. *Inventaris van het archief van Mr. Dr. H.H.A. van Gybland Oosterhoff en enige van zijn familieleden.* 's-Gravenhage, 1975.

Size: *C.* 7 running metres (353 inventory items).

Permission is necessary for consultation of the papers.

**R.L. de Haes, April 13th 1818 Zutphen – 1884** [198]

1832 naval cadet; 1837 voyage to the West Indies; 1841 second naval lieutenant; 1841 voyage to the East Indies; 1846 to the West Indies; 1849 to the East Indies; 1852 first naval lieutenant; 1857 to the West Indies, from there to the East Indies; 1858 stationed in the Western district of Borneo, expedition in 1859 against the kampong Saribanao in Sapora; 1861 to the Netherlands, returns to the East Indies in 1865, chief of staff Department of Naval Affairs at Batavia, naval captain; 1869 to the Netherlands, returns to the East Indies in 1871; 1872 King's commissioner on the coast of Guinea in connection with the transference to Great Britain; 1873 to the Netherlands; 1874 director and commander of the Navy at Hellevoetsluis; 1875 rear admiral.

Besides the papers concerning his appointments and promotions the collection contains about the East Indies journals of voyages 1851 and 1861 through the archipelago, a journal 1852 of the voyage from the Indies to the Netherlands, a register of letters despatched and reports 1858–1860, a report of the expedition against Saribanao.

Entry: Inventory.

Size: *C.* 0.10 running metre (15 inventory items).

**W.J. Haisma Rahder** [199]

Jurist, employed with the Algemeene Vereeniging van Rubberplanters ter Oostkust van Sumatra (General Association of Rubber Planters on the East Coast of Sumatra).

The collection consists mainly of printed matter, i.a. of the Deli Plantersvereeniging (Deli Planters' Association) and the AVROS, and also some letters 1939 to K.A. de Weerd.

Size: *C.* 0.20 running metre.

**Hasselman** [200]

The collection consists of the papers originating from J.J. Hasselman and his nephew C.J. Hasselman.

a. *J.J. Hasselman, October 21st 1815 Nederhemert – March 27th 1895 Tiel*
1834 as official of the General Secretariat to Batavia, where he was subsequently second clerk, first clerk 1837 and principal clerk 1839; 1840 assistant resident Kediri (Ngrowo); 1845 on leave; 1847 assistant resident Bantam (Lebak and Pandeglang); 1850 Surakarta; 1851–1855 resident of Djokjakarta; 1855 to the Netherlands, where he became burgomaster of Tiel; 1867–1868 minister of Colonial Affairs; after this i.a. member of the Second Chamber 1869–1871 and member of the Raad van State (Council of State) 1874–1880.

Originating from him are letters 1868 from J. Heemskerk Azn. about the departure of C. Busken Huet to the East Indies.

b. *C.J. Hasselman, October 21st 1861 Herwijnen – February 10th 1944 The Hague*
Training at the Indische Instelling (East Indies Institution) at Delft; 1883 official at the disposal of the Binnenlands Bestuur (Indies Civil Service), rinder pest service; 1885 comptroller Banjumas; 1888 Semarang; 1893 comptroller and 1896 referendary at the General Secretariat at Buitenzorg; 1898 on leave and drafted off to the Ministry of Colonial Affairs; 1902 back to the East Indies and assistant resident in commission for the municipal- and statute labour; 1905 inspector of the inland revenue and forced services; 1907 to the Netherlands, where he was general secretary of the Koloniaal Instituut (Colonial Institute) at Amsterdam 1910–1912; 1912–1940 member of the Raad van State (Council of State).

Originating from him are papers 1902–1914 concerning dessa services and the investigation into prosperity in Java and Madura; papers 1911–1918 concerning the training of East Indies officials; articles 1893–1928 i.a. concerning the Indies Civil Service, colonial policy and British India.

Entry: Inventory.

Size: C. 0.10 running metre (8 inventory items).

**D.H. Havelaar, 1852 Rotterdam – November 22nd 1918 Rotterdam** [201]

1876 aspirant engineer with the Department of Water Management at Batavia; 1877 third class engineer at Padang; 1879 Fort de Kock; 1880 Semarang; 1882 second class engineer; 1883 Bantam (Serang); 1884 Batavia; 1890 first class engineer; 1892 to the Netherlands; 1893 to Surinam; 1899 member and then vice president of the Raad van Bestuur (Council of Governors) of Surinam; 1903 senior engineer second class; 1906 to the Netherlands, where he became alderman of the municipal enterprises of Rotterdam.

The collection contains records of service etc., and concerning the Dutch East Indies a small number of papers (7 inventory items), relating to the dry dock company Tandjung Priok 1887–1891; irrigation works of the Solo valley 1893,

1896; theft of government money in Padang 1879–1880; plan forced labour depot Semarang 1880; economy in management and maintenance of civil and military buildings 1894–1895; shipment of Javanese and British Indian immigrants to Surinam 1900.

Entry: Inventory. In *VROA*, 1919, p. 330–336.

Size: *C*. 1.30 running metres (55 inventory items).

**J.B. van Heutsz, February 3rd 1851 Coevorden – July 10th 1924 Montreux** [202]

1867 military training Kampen; 1872 second lieutenant; 1873 to the Dutch East Indies (Surabaja); 1874 Atjeh, 1875 Kota Radja; 1876 first lieutenant; 1877 local adjutant at Surabaja; 1880 back to Atjeh; 1881 on leave to the Netherlands; 1884 adjutant to the commander of the garrison at the East Coast of Sumatra; 1886 captain; 1887 with the general staff of the East Coast of Sumatra; 1889 chief of the staff in Atjeh; 1891 major; 1893 to the Netherlands; 1894 lieutenant colonel; 1896 returns to the East Indies, military commander of the East Coast of Sumatra; 1896 at the disposal of the civil and military governor of Atjeh; 1897 colonel; 1898 general major, civil and military governor of Atjeh; 1900 lieutenant general, 1904–1909 governor general, during his governor generalship expeditions to Central Sumatra, Central Borneo, Celebes, Bali, Sumbawa, Sumba, Flores, Timor; 1910 to the Netherlands; 1922 to Switzerland (Montreux). (J.C. Lamster. *J.B. van Heutsz als gouverneur-generaal, 1904–1909*. Amsterdam, [1946]).

The collection consists of essays, memorandums, reports 1882–1894 concerning Atjeh, native government in Java and Madura, C. Snouck Hurgronje, Perlak, with a paper by Laging Tobias about the sultanate of Atjeh; speeches 1916 by Van Heutsz; correspondence 1905–1909 about government affairs with H. Colijn, Th.F.A. Delprat, D. Fock, A.W.F. Idenburg, C.A. Kroesen, A.P. Melchior, D.F.W. van Rees, A.F. de Savornin Lohman, C. de Vries and J.C. van der Wijck; papers 1910–1914 concerning the Anglo-Dutch Plantations of Java Ltd.; papers 1921 about the establishment of the Nederlandsch-Indisch Landsyndicaat (Dutch East Indies Land Syndicate) in 1910, with a letter to . . . Verster, the editor of the *Nieuws van den Dag*.

Entry: Inventory. In *VROA*, 1927, p. 115–116.

Size: *C*. 0.10 running metre (6 inventory items).

Permission is necessary for consultation of the papers.

**J.E. Heeres, May 30st 1858 Zuidhorn – February 16th 1932 The Hague** [203]

1877 law study at Groningen, where he graduates in 1885, then study of literature up to 1888; 1888–1897 employed at the Algemeen Rijksarchief (General State Archives, 1888 junior clerk, 1891 clerk, 1894 assistant state archivist), where his task was to make an inventory of the VOC archives; 1897 professor of history of the Indische Instelling (East Indies Institution) at Delft, retired after its abolition in 1900; 1902 professor of history of the Dutch East Indies and the history and method of the protestant mission at the State University of Leyden; 1913–1918 member of the Second Chamber; 1921 founder of the Democratische Partij (Democratic Party) after he had left the Liberale Unie (Liberal Union); many publications about the Dutch colonies, when he was professor in Delft he wrote the column "Indische overzichten" (East Indies surveys) in the *Nieuwe Courant*, was also member of the board of the KITLV. (Supplements from *Handelingen en levensberichten van de Maatschappij der Nederlandsche Letterkunde te Leiden;* 1931–1932, p. 137).

The collection consists mainly of papers collected and copied for his publications about the colonies, including papers purchased from the estates of U.G. Lauts and J.C. van der Muelen. For the names and subjects mentioned in the inventory see the index.

Entry: Inventory.

Size: *C.* 1.25 running metres (43 inventory items).

Deposited at the First Section.

**P. van der Heim, February 8th 1753 Brussels – April 6th 1823 The Hague** [204]

1768 law study at Leyden; up to 1795 secretary of the admiralty of Rotterdam; 1795–1806 retired; 1806 director general for East Indies affairs and trade and in the same year minister of Trade and Colonial Affairs; 1807 also minister of Naval Affairs; 1808 minister of Naval and Colonial Affairs; 1810 sea prefect of Holland and temporary minister of Foreign Affairs; 1811 discharge; after 1813 member of the First Chamber.

The major part (194 inventory items) of this collection consists of papers drawn up and received by Van der Heim when he was minister of Trade and Colonial Affairs/Naval and Colonial Affairs. Of these are relevant for the history of Asia the headings 'algemeen' (general), 'koloniën algemeen' (colonies general) and 'Oost-Indië' (East Indies). For the names and subjects mentioned in the inventory see the index.

Entry: Inventory. In *VROA*, 1923, p. 232–258.

Size: *C.* 2 running metres (277 inventory items).

**H.Th. Hoffmann, July 18th 1898 Amsterdam – December 7th 1960 The Hague** [205]

1915–1935 naval officer, several voyages through the Indonesian archipelago; 1940 official Ministry of Economic Affairs, occupied with the Indonesian-Dutch relations, advocate of the Republic of the South Moluccas.

The collection consists of papers from 1950 mainly concerning the South Moluccas and New Guinea affairs.

Entry: Inventory.

Size: *C.* 0.70 running metre (92 inventory items).

**Van Hogendorp (first collection)** [206]

The main part of this collection consists of the collected papers of:

a. *G.K. van Hogendorp, October 27th 1762 Rotterdam – August 5th 1834 The Hague*
Son of the resident of Rembang Willem van Hogendorp; military training at Berlin; 1786 graduated at Leyden; from 1787 active part in Dutch political life; 1795 retired, publications about the reform of the East India Company; from 1813 again active in Dutch political life, i.a. minister of Foreign Affairs 1813–1814, vice president of the Raad van State (Council of State) 1814, member of the Second Chamber 1815–1825.

The greater part of the papers originating from G.K. van Hogendorp concerns his career in the Netherlands and in addition there are diaries 1777–1834, letters (alphabetically arranged) 1773–1834, letters kept separately 1777–1822 from his brother D. van Hogendorp sr. (see also no. 206 b and 207 a) and letters 1825–1830 from his son W. van Hogendorp jr. (see also no. 206 c and 208 c) and in addition papers concerning Asia written by various persons. For the names and subjects mentioned in the inventory see the index. See also no. 208 a.

b. *D. van Hogendorp sr.*
Brother of G.K. van Hogendorp. See also no. 206 a and 207 a.

Originating from him appointments, papers concerning the Eastern part of Java and East Indies affairs 1773–1822, letters 1777–1820 from his brothers G.K. van Hogendorp and W. van Hogendorp sr. and from his son C.S.W. van Hogendorp, with papers concerning his Brazilian estate Novo Sion.

Furthermore the collection contains letters 1776–1825 originating from W. van Hogendorp sr. from his brothers G.K. van Hogendorp and D. van Hogendorp sr.

c. *W. van Hogendorp jr., March 17th 1795 Amsterdam – July 10th 1838 The Hague*
Son of G.K. van Hogendorp, jurist, 1819 graduated at Leyden; 1819 attaché at Naples; *c.* 1821 referendary Council of State; 1825 first secretary of L.P.J. du Bus de Gisignies; 1829 discharge, travels through Java; 1830 to the Netherlands; 1838 member of the Raad van State (Council of State).

Besides a diary 1826–1827 the papers originating from W. van Hogendorp jr. consist of letters 1818–1830 from G.K. van Hogendorp (with annexes), his mother and other relatives and letters 1818–1825 from various persons (alphabetically arranged; see also no. 208 c).

Entry: Inventory. In *VROA*, 1913, p. 203–247.

Size: *C.* 11 running metres (205 inventory items, 113 items originating from G.K. van Hogendorp).

**Van Hogendorp (second collection)** [207]

This collection consists for the greater part of papers concerning the Dutch East Indies. They are originating from:

a. *D. van Hogendorp sr., c. October 11th 1761 Heenvliet – October 29th 1822 Brazil*
Military training at Berlin; 1782 to the Netherlands; 1784 to the East Indies, expeditions to Malaya and Riouw; 1786 under merchant East India Company at Patna; 1798 idem in Onrust; 1791 resident Japara; 1794 chief merchant and commander Eastern Part of Java; 1798 imprisoned at Batavia for his agitation against the VOC and slavery, escaped via Benkulen to the Netherlands; 1802 member of the Commissie tot de Oost-Indische Zaken (Committee for the East Indies Affairs) and the Staatscommissie ten behoeve van het Regeringsreglement (State Committee for the Government Regulations); 1803 ambassador in Russia; 1805 to the Netherlands; 1806 member of the Raad van State (Council of State), minister of War; 1807 ambassador to Vienna; 1809 to the Netherlands; 1810 member of the Commissie voor de Reünie (Committee for the Reunion); 1811 lieutenant general; 1812 governor of Königsberg, subsequently Lithuania, Breslau, Hamburg; 1813 discharge; 1815 in French state service; 1816 to Brazil. Many publications on the East Indies. See also no. 206 b.

The papers of D. van Hogendorp (63 inventory items) consist of official papers from 1784 and non official papers, including his memoirs from 1789. For the names and subjects mentioned in the inventory see the index.

b. *C.S.W. van Hogendorp, August 15th 1788 Casimbazar – October 29th 1856 Utrecht*
Son of D. van Hogendorp sr.; military training Sorrère; 1806 naval cadet, then naval lieutenant; 1808 auditor at the embassy in Vienna; 1809 captain adjutant; 1811 cavalry captain; 1813 lieutenant colonel in French service; 1817 second class official Dutch East Indies; 1818 member of the Algemene Rekenkamer (General Audit Department); 1818 resident Buitenzorg; 1823 resident Batavia; 1827 to the Netherlands, 1830 discharge, many publications on the Dutch East Indies; 1837 member of the Indies Council, 1839–1841, 1850–1851 temporary vice president, 1840–1841 temporary governor general; 1843–1848 president of the Hoofdcommissie van Onderwijs (Chief Committee of Education) in the Dutch East Indies; 1849 president of the Protestantsch Kerkbestuur (Protestant Clerical Board); 1852 to the Netherlands.

Originating from C.S.W. van Hogendorp are papers drawn up and/or received when he was a member of the Audit Department, resident of Buitenzorg and Batavia, member and temporary vice president of the Indies Council, temporary governor general, member of the Protestant Clerical Board, in addition private papers 1793–1857 including his memoirs and letters (for the names and subjects mentioned in the inventory see the index; 75 inventory items), and also papers originating from his wife C.C. van Hogendorp-Olivier, viz. letters 1811–1849 from J. van der Capellen-van Tuyll van Serooskerken (wife of G.A.G.Ph. van der Capellen) with a journal of her voyage to the Moluccas in 1824, and letters from P. Merkus and J.J. Rochussen (see also no. 208 e).

c. *D.C.A. van Hogendorp, September 23rd 1812 Rijssel – September 12th 1891 The Hague*

Son of C.S.W. van Hogendorp; 1838 secretary and fiscal of the Rechtbank van Ommegang (Court) of the Western District of Borneo; 1838 magistrate and fiscal in Amboina; 1839 member of the Court of Justice Semarang, 1840 member of the administration on the Gereformeerd Weeshuis (Calvinist Orphanage) there; 1840 assistant resident Buitenzorg; 1846 resident Kedu; 1850 resident Banjumas; 1853 on leave in the Netherlands; 1855 resident Japara; 1857 up to his departure to the Netherlands in 1862 resident Semarang; in the Netherlands member of the central committee of the Dutch Red Cross 1867–1882.

Originating from D.C.A. van Hogendorp are papers drawn up and/or received in his functions from 1838 with letters 1826–1879 and documentation about himself 1861–1891 (87 inventory items), and in addition letters 1838–1852 to his wife G.E. van Hogendorp-Hardy from J.J. Rochussen and the C.S.W. van Hogendorp family. For the names and subjects mentioned in the inventory see the index.

d. *M.A. van Hogendorp, March 31st 1864 The Hague – ?*
Daughter of D.C.A. van Hogendorp. No other facts known.

Originating from her are private papers 1811–1859 concerning Banka and various regions in Borneo and Java, written by various government officials and soldiers. For the names and subjects mentioned in the inventory see the index; 16 inventory items.

e. *Papers originating from various members of the Van Hogendorp family*
Letters 1816–1828 to W. van Hogendorp sr. from D. van Hogendorp sr. and C.S.W. van Hogendorp.

Entry: Inventory. In *VROA*, 1922, p. 177–214 and supplement inventory in *VROA*, 1927, p. 117–123.

Size: *C.* 1.70 running metres (272 inventory items).

**Van Hogendorp (third collection)** [208]

This collection consists for a large part of documentation (especially concerning G.K. van Hogendorp) and in addition papers concerning the Dutch East Indies originating from:

a. *G.K. van Hogendorp*
Letters to him from D. van Hogendorp sr. and C.S.W. van Hogendorp. See also no. 206 a.

b. *W. van Hogendorp sr., 1765–1835*
Letters from C.S.W. van Hogendorp.

c. *W. van Hogendorp jr.*
Letters to him from various persons, portrait of him with C.P.J. Elout as pangeran. See also no. 206 c.

d. *D. van Hogendorp jr., December 18th 1797 Amsterdam – March 18th 1845 The Hague*
Son of G.K. van Hogendorp; jurist; 1822 graduated at Leyden; 1830 substituut-officier (deputy prosecutor) district court Amsterdam; 1840 extraordinary member of the Second Chamber.

Originating from D. van Hogendorp jr. are regulations of the Weduwen- en Wezenfonds (Widows' and Orphans' Fund); papers concerning the government; list of requisites for an East Indies officer; decrees of the governor general 1839–1840; letters from various persons including D.J. de Eerens, C.S.W. van Hogendorp and D.C.A. van Hogendorp.

e. *C.S.W. van Hogendorp*
Diary 1837–1846 and letters from him to R. Fagel. See also no. 207 b.

Entry: Provisional inventory (on index cards).

Size: *C.* 7.50 running metres (1130 inventory items).

Permission is necessary for consultation of the papers.

See also J.H. Scheffer and D.O. Obreen. *Het geslacht Van Hogendorp van de 14de eeuw tot op heden;* naar onuitgegeven bronnen en oorspronkelijke stukken bewerkt. Rotterdam, 1867.

G.K. van Hogendorp. *Brieven en geschriften;* uitgegeven door H. van Hogendorp. 's-Gravenhage, 1866–1903.

D. van Hogendorp [sr.] *Mémoires;* publié par D.C.A. van Hogendorp. 's-Gravenhage, 1887.

H. van Hogendorp. *Willem van Hogendorp [jr.] in Nederlandsch-Indië, 1825–1830;* naar onuitgegeven bronnen bewerkt. 's-Gravenhage, 1913.

**Hora Siccama** [209]

This collection consists of papers originating from various members of the Hora Siccama family. Papers concerning the East Indies are mainly originating from:

*O.W. Hora Siccama, January 5th 1805 Groningen – 1879*
1821 official Ministry of Education, Industry and Colonial Affairs, accompanied A.R. Falck to London in 1823 for the negotiations about the East Indies; 1824 official Foreign Affairs; 1841 member of the General Audit Department; 1868 president of the General Audit Department.

Originating from him are records of service and letters 1822, 1843 from A.R. Falck, letters 1839–1874 from J. Schneither, letter (in Kawi) from Gusti Djilantik to Pangeran Sjarif Hassan, copies of letters from G.A.G.Ph. van der Capellen, C.Th. Elout, A.R. Falck and J. Schneither.

Besides papers originating from O.W. Hora Siccama (38 inventory items) this collection also includes letters 1800–1842 from A.R. Falck (originating from

H.H. Hora Siccama) and letters 1821–1850 from O.W. Hora Siccama (originating from H.H. and J.H. Hora Siccama).

Entry: Inventory. In *VROA*, 1909, p. 62–70 with supplement inventory.

Size: *C.* 2.80 running metres (63 inventory items).

**F.W.T. Hunger, April 3rd 1874 Amsterdam – June 3rd 1952 Leyden** [210]

Botanist; 1899–1905 employed at 's-Lands Plantentuin (National Botanical Gardens) at Buitenzorg; then in the Netherlands lecturer of botany University of Utrecht; 1906–1910 director Vereenigd Proefstation voor Bergcultures (Associated Experimental Station for Upland Cultures) in Java; from 1917 in the Netherlands, i.a. lecturer of the history of botany State University of Leyden and director of the Institute for the History of Medicines and Biology at Leyden.

The collection consists of papers collected by Hunger for a biography of the botanist F.W. Junghuhn, with correspondence.

Entry: Inventory.

Size: *C.* 0.25 running metre (5 inventory items).

See also no. 482.

**A.J. Immink, March 12th 1838 Doesburg – September 24th 1914 The Hague** [211]

Jurist; 1861 graduated at Utrecht, then training for East Indies government official at Delft; 1864 to the East Indies, deputy clerk Court of Justice Semarang; 1865 clerk with the circuit judge of the fourth district of Java; 1869 public prosecutor Council of Justice of Amboina; 1869 member of the Council of Justice of Surabaja; 1874 president of the Council of Justice of Makasar; 1876 president Landraad (District Council) Surabaja; 1879 public prosecutor Council of Justice Semarang; 1882 till his retirement in 1889 justice High Court of Justice of the Dutch East Indies; in the Netherlands he was appointed member of the Staatscommissie tot Herziening van het Indisch Privaatrecht (State Committee for the Revision of the East Indies Private Law) in 1892.

The collection consists of papers 1892–1902 drawn up and/or received by Immink when he was a member of the State Committee, concerning Dutch nationality and citizenship, equalization of Native and European, Native Christians, councils of priests, district councils, mixed marriages, widows' and orphans' fund, law of things and obligations, with letters from L.W.C. van den Berg, R.Z. Dannenbarg, J. van Gennep, F.H. Gerritzen, P.A. van der Lith and M.H.C. van Oosterzee.

Entry: *Inventarissen van de collecties A.J. Immink, F. Pinke, C. Pijnacker Hordijk;* [door F.G.P. Jaquet en W. Visser]. 's-Gravenhage, 1965.

Size: *C.* 0.10 running metre (40 inventory items).

**M.H. Jansen, August 11th 1817 Antwerp – September 9th 1893 The Hague** [212]

Training Koninklijk Instituut voor de Marine (Royal Institute for the Navy); 1835 naval cadet; 1836 to the Dutch East Indies, second lieutenant; 1841 as captain to the Netherlands; 1842 cartographic surveys of waterways Surabaja; 1848 in the Netherlands in the committee for the defence of South Netherland; 1850 employed at the Ministry of Colonial Affairs; 1851 to the West Indies for cartographic surveys; 1853–1857 activities in meteorology; 1857–1858 to Java; 1859 employed at the Ministry of Naval Affairs; 1861–1862 to Curacao; 1862–1864 travels through Europe; 1865 command of the training division in the North Atlantic Ocean; 1866 superintendence of the building of and later commander of the iron clad ship 'Prins Hendrik der Nederlanden'; 1868 retired; 1874 member of the Council of State; 1889 rear admiral, engaged in the foundation of the Koninklijk Nederlands Aardrijkskundig Genootschap (Royal Dutch Geographic Society).

The collection consists of papers 1819–1889, including reports 1819–1836 about military operations on the West Coast of Sumatra. statistics 1821 Bantam, shipping, trade to China and the Dutch East Indies, import and export 1848–1850, Suez Canal 1873–1889.

Entry: Provisional list.

Size: *C.* 0.30 running metre (30 items).

**J.W. Janssens, October 12th 1762 Nijmegen – May 23rd 1838 The Hague** [213]

1774 cadet; 1777 vaandrig (ensign); 1787 first lieutenant; 1788 captain; 1796 retired; 1797 commissioner for the French troops; 1800 secretary general of the Ministry of War; 1802 governor general of Cape of Good Hope; 1806 after the capitulation to the British to the Netherlands, where he became i.a. staatsraad (councillor of state) and secretary general of War; 1809 discharge; 1810 governor general of the East Indies, has to capitulate to the British in 1811; 1812 back in the Netherlands after having been prisoner of war; 1813 divisional general; 1814 commissioner of War in the Southern provinces, in 1815 till his discharge in the same year, commissioner of War in the Northern provinces.

The collection consists of papers 1797–1832, including records of service, diplomas and biographical data. Of this collection 11 inventory items were drawn up during or after his governor generalship of the Dutch East Indies, viz. registers of letters dispatched 1810–1812 with letters to Lord Minto and the Minister of Naval and Colonial Affairs [P. van der Heim] 1811 and a letter to A.R. Falck 1814; correspondence between Minto and S. Auchmuty 1811; correspondence with brigade general Jumel 1811; despatches from the Minister of Naval and Colonial Affairs 1810; memorandum H.M. de Kock about the capitulation of Java 1811; nominative record of officers in the East Indies army 1812; memorandum of D.F. van Alphen as commander at Surabaja 1807 and the political situation in Java 1814.

Entry: Inventory. In *VROA*, 1906, p. 84–86. Supplement in *VROA*, 1941, p. 19.
Size: *C.* 0.30 running metre (32 inventory items).

**A.M. Joekes jr., May 5th 1885 Bua – April 1st 1962 The Hague** [214]

Studies of law and political sciences at Leyden; 1910 secretary of the general agent of the Semarang-Joana and Semarang-Cheribon Steam Tram Companies, leaves for the East Indies in 1911; 1918 secretary of the board of directors of the Companies at The Hague; 1920 referendary, 1923 administrator at the Ministry of Labour; 1925 member of the Second Chamber, 1933 leader of the parliamentary party of the Vrijzinnig Democratische Bond (Progressive Democratic League), 1933–1940 member of the Vlootcommissie (Naval Committee); after his captivity during the second world war co-founder and member of the central committee of the Partij van de Arbeid (Labour Party); 1948–1952 minister of Social Affairs and Public Health, 1949 member of the Ronde Tafel Conferentie (Round Table Conference); 1952–1960 member of the Council of State; member of the committee of the Kartini Fund (1912–1954), of the Mr. C.Th. van Deventer Foundation and the Van Deventer-Maas Foundation (1915–1957) and of the Tjandi Foundation (1934–1959).

Papers concerning the Dutch East Indies were received and/or drawn up by Joekes when he was a functionary of the Steam Tram Companies (43 inventory items), expert of East Indies affairs in the Second Chamber (50 inventory items), member of the committee of the Kartini Fund, the Van Deventer Foundation and the Tjandi Foundation (26 inventory items) and as a private person. For the names and subjects mentioned in the inventory see the index.

Entry: Inventory.

Size: *C.* 5.25 running metres (643 inventory items).

**B.C. de Jonge, January 22nd 1875 The Hague – June 24th 1958 Zeist** [215]

Training for jurist Utrecht; 1900 lawyer at The Hague; 1901 official Ministry of Colonial Affairs; 1910 official Ministry of War; 1917 Minister of War; 1918 employed with the Bataafsche Petroleum Maatschappij (Batavian Petroleum Company), from 1919 temporary director, 1919 voyage to the Dutch East Indies, China, Japan and the United States, 1920 director, 1921–1930 at London, 1921 supervisory director; 1931–1936 governor general of the Dutch East Indies; 1936 retired in the Netherlands, 1939 member of the Board of Governors of the Fonds ten behoeve van de Indologische Studiën aan de Universiteit te Utrecht (Fund for the benefit of the Studies of Indology at the University of Utrecht).

The collection consists of diaries and memoirs, correspondence from 1902, papers drawn up and received in his functions from 1900 (especially when he was governor general), documentation, photographs, private papers from 1883, papers originating from his wife and son (J.A. de Jonge). For the names and subjects mentioned in the inventory see the index.

Entry: F.J.M. Otten. *Inventaris van de papieren van Jhr. Mr. B.C. de Jonge.* 's-Gravenhage, 1971.

Size: *C.* 3 running metres (180 inventory items).

Permission is necessary for consultation of the papers.

See also B.C. de Jonge. *Herinneringen;* met brieven uit zijn nalatenschap; uitgegeven door S.L. van der Wal. Groningen, 1968. (WHG; serie 5; no. 1).

**J.A. Jonkman, September 13th 1891 Utrecht – 1974 The Hague** [216]

Law studies Utrecht, Toulouse, Leyden, 1918 graduated; from 1919 official with the Judicature in Indonesia, (extraordinary) deputy prosecutor Councils of Justice of Meester Cornelis 1919, Semarang 1921, Batavia 1926; public prosecutor in Makasar 1931 and Semarang 1938; also from 1927 member of the Volksraad (People's Council), from 1939 president, member of the Stuw-groep (association); 1942–1945 interned by the Japanese; 1946–1948 minister of Overseas Territories; 1948 member of the First Chamber for the Partij van de Arbeid (Labour Party); 1951–1966 chairman of the First Chamber.

The collection consists of Hindia Poetra 1918–1919, papers People's Council 1927–1942, articles and speeches about colonial policy 1936–1953, letters H.F.L.K. Vredenburgh and L. Neher 1947, *De Stuw* 1930–1933, papers after 1949.

Entry: List drawn up at the transfer.

Size: *C.* 0.40 running metre.

Permission is necessary for consultation of the papers.

**J. van Kal** [217]

*C.* 1785 to Batavia; 1787–1795 official in Malaya and Riouw, was in Malaya secretary of police and collector of the public lands, owned there the trading house Jacob van Kal & Co.; 1795–1796 at Batavia; 1796–1811 retired in the Netherlands. (Supplements *RAvNI*).

The collection consists for the greater part of papers concerning Malaya and Riouw before 1796, 5 inventory items relating to the period after 1795, viz. papers concerning his residence in Batavia 1795–1796; papers 1797–1811 written in the Netherlands, including an "apologie aan het Comité tot de Zaken van de Oost-Indische Handel en Bezittingen" 1797 and memorandums about the situation in the East Indies 1797–1801.

Entry: Inventory. In *VROA,* 1895, p. 196–198.

Size: *C.* 0.40 running metre (30 inventory items).

Deposited at the First Section.

**C.J. van Kempen, March 18th 1872 Zierikzee – November 25th 1955 The Hague** [218]

1895 government official; 1899–1912 (aspirant) comptroller East Coast of Sumatra; 1914 assistant resident Timor; 1918 assistant resident East Coast Sumatra (Asahan); 1921 resident Southern and Eastern District of Borneo; 1924–1928 governor East Coast Sumatra; in the Netherlands member of the Second Chamber.

The collection consists of reports left by him for his successor with annexes 1917–1928 drawn up by Van Kempen when he was government official in South Timor, the Southern and Eastern District of Borneo and the East Coast of Sumatra; papers 1936 concerning socio-economic measures in South and East Borneo; papers 1945–1955 concerning the Dutch-Indonesian relations.

Entry: Inventory.

Size: *C.* 0.15 running metre (7 inventory items).

**A.H.W. de Kock, March 20th 1808 Surabaja – June 17th 1891 The Hague** [219]

Son of H.M. de Kock; 1824 second lieutenant, 1824–1828 adjutant lieutenant of the lieutenant governor general, 1825–1830 Java War; 1830 adjutant of the governor general; 1832–1836 captain adjutant, 1833 expedition on the West Coast of Sumatra; 1838 major general staff West Coast of Sumatra, 1841 expedition there; 1841 lieutenant colonel, temporary resident Palembang; 1846 on leave; 1848 government commander Bali, second commander Bali War, temporary resident Besuki, discharge from military service; 1848 resident Djokjakarta; 1851 discharge; 1852 temporary resident Madiun; 1854 resident Batavia; 1855 member of the Indies Council, 1855–1856 president Protestantsch Kerkbestuur (Protestant Clerical Board); 1857 on leave; 1859 till his pension in 1860 vice president of the Raad van Indië (Indies Council).

The collection consists of papers 1838–1849, viz. report left by him for his successor, report of the demang of Djambi Darpa Wiguna, notes about Native officials at Palembang, memorandum about the country of Nangulon, notes about the court of Djokjakarta, journal and letters about his voyage to Bali.

Entry: Inventory. In *VROA*, 1891, p. 22.

Size: *C.* 0.05 running metre.

See also no. 636.

**H.M. de Kock, May 25th 1779 Heusden – April 12th 1845 The Hague** [220]

1795 employed with the Comité van Algemeen Welzijn (Committee for General Welfare); 1797 at the Ministry of War; 1801 naval captain, chief clerk of the navy; 1803 fiscal and secretary of the navy; 1804 colonel, chief general staff; 1806 chief equipment officer of the navy east of Cape of Good Hope; 1807 military commander Eastern Part of Java, Madura and Grisee; 1808 adjutant general of the governor general; 1809 brigadier and commander Semarang; 1810 chief general staff of the navy; 1811 chief general staff of the army; 1811–1813 prisoner of war of the British; 1815 general major; 1818 governor of the Moluccas; 1819 commander of the East Indies army; 1821 lieutenant general, expedition Palembang; 1822 lieutenant governor of 's-Rijks Aziatische Bezittingen (Dutch Asian Possessions), member of the Raad van Indië (Indies Council); 1825–1830 commander of the army during the Java War; 1826 temporary governor general; 1826–1830 lieutenant governor general; 1831 commander of the forces in Zeeland; 1836 minister of Home Affairs; 1841 minister of State; 1842 member of the First Chamber. (A.W.P. Weitzel. *De oorlog op Java van 1825 tot 1830;* hoofdzakelijk bewerkt naar de nagelaten papieren van [H.] M. de Kock. Breda, 1852, 1853. 2 vol.)

The collection consists of papers 1807–1821 drawn up and received in his functions of that period (9 inventory items); papers 1825–1830 when he was lieutenant governor general and commander of the forces during the Java War, with preliminary papers concerning Djokjakarta from 1780 (158 inventory items); papers 1830–1836 after he had left the Dutch East Indies (12 inventory items); papers 1825–1830 originating from other persons, including letters G.A.G.Ph. van der Capellen, F.G. Valck, J.F.W. van Nes, Raden Ario Noto Diningrat, Mangon Kusumo, Siento Negoro and papers originating from Diponegoro (46 inventory items). For the names and subjects mentioned in the inventory see the index.

Entry: Inventory. In *VROA*, 1905, p. 57–77.

Size: *C.* 2.50 running metres (225 inventory items).

**P.J. Koets, 1901 Macon –** [221]

Study of classic languages; 1930–1937 teacher lyceum Batavia, member of De Stuw (association); 1938–1939 teacher Bandung; 1939–1942 official Department of Education; 1942–1945 prisoner of war; 1946–1950 director Department of Education; 1950–1951 teacher Baerleusgymnasium Amsterdam; 1951–1961 chief editor *Het Parool;* 1962–1970 alderman of education Amsterdam; 1970 retired.

The collection consists of papers from 1945, of which those concerning Asia mainly pertain to the Indonesian question, including papers from, by and/or about the following persons and subjects: I.J. Brugmans, Bep Vuyk, J. Greshof, Sukarno, J.W. Meyer Ranneft, Makasar, NEFIS, minutes republican cabinet, Komité Nasional Indonesia, Linggardjati, Malino, Pangkal Pinang, Den Pasar, East Indonesia, East Sumatra, West Borneo, Commissie-Generaal (General Committee), 1st and 2nd police action, Madura, West Java, Bijeenkomst Federaal Overleg (Assembly Federal Consultations), Commissie Goede Diensten (Committee of Good Offices), Renville agreement, Security Council, Pegangsaan, M. Cochran, communism, RTC, Asian-Arabic Conference New Delhi 1947.

Size: *C.* 1 running metre.

Permission is necessary for consultation of the papers.

**H. Kolkman** [222]

President of the Algemeene Vereeniging van Rubberplanters ter Ooskust van Sumatra (General Association of Rubber Growers on the East Coast of Sumatra).

The collection consists of reports of the AVROS 1946–1948, letters 1945 from K. Raadsheer, planters regulations by W.J. Haisma Rahder, papers coolie ordinance 1924, assistants regulations East Coast Sumatra, Atjeh, Tapanuli, Palembang, Lampong, Benkulen, West Coast Sumatra 1926.

Size: *C.* 0.10 running metre.

**L.H. de Langen** [223]

In *Kleian's adresboek* 1929 and 1930 De Langen is indicated as engineer with Stork & Co at Surabaja, in the *RAvNI* 1932–1934 as municipal councillor of Pasuruan.

The collection consists of papers when he was municipal councillor, including minutes 1931–1933 and budgets 1931–1934 and papers concerning the Vaderlandsche Club (Fatherland Club) including minutes meeting 1930–1932 department East Java, circulars 1931–1932 circle Surabaja, papers 1931–1933 circle Pasuruan.

Entry: List drawn up at the transfer of the collection.

Size: *C.* 0.20 running metre.

**Van Lansberge** [224]

Papers concerning Asia are originating from:

*J.W. van Lansberge, November 16th 1830 Santa Fé de Bogota – December 17th 1905 Menton*
Training for jurist Leyden; 1856 diplomatic service (1857 Madrid, 1858 St. Petersburg, 1860 Brussels, 1864 Paris); 1866 temporary secretary general of Foreign Affairs; 1874–1880 governor general, was i.a. concerned with the organization of the government in the Southern and Eastern District of Borneo, in Palembang, Benkulen and the West Coast of Sumatra, abolition of slavery in Bali, Ternate and Tidore, regulations concerning the monetary system and education.

The papers originating from him consist for a large part of papers concerning the Dutch East Indies, including letters from 1866, publications from 1854, papers collected and drawn up when he was governor general, necrologies, publications 1906–1929 collected by his wife. Moreover a number of letters and articles 1889–1921 originating from G. van Lansberge (employee of the BPM) and his wife E.V.A. van Lansberge-Stuur about Indonesia, including the Orphans' Court at Batavia. For the names and subjects mentioned in the inventory see the index.

Entry: Inventory.

Size: *C.* 2 running metres (322 inventory items).

Permission is necessary for consultation of the papers.

**J.P. van Limburg Stirum, February 2nd 1873 Zwolle – April 17th 1948 The Hague** [225]

Jurist; 1895 graduated at Leyden; up to 1915 diplomatic service, subsequently official Foreign Affairs, ambassador to Stockholm, Rome, Constantinople, Peking, Copenhagen, Christiania; 1916–1921 governor general; 1922–1924 ambassador to Cairo; 1927–1937 ambassador to Berlin; 1937–1939 ambassador to London.

The collection consists for the greater part (58 inventory items) of papers drawn up and received by Van Limburg Stirum when he was governor general, including papers concerning the Volksraad (People's Council), proceedings of the Raad van Indië (Indies Council) 1920, correspondence with (former) ministers of Colonial Affairs and with government officials, soldiers and private persons in the Dutch East Indies 1916–1921; moreover letters after he had left the East Indies 1924–1926 and a lecture given by him in 1923. For the names and subjects mentioned in the inventory see the index.

Entry: Inventory.

Size: *C.* 0.35 running metre (62 inventory items).

Permission is necessary for consultation of the papers.

N.B: In the Van Limburg Stirum collection (ARA no. 93), containing papers originating from other members of the Van Limburg Stirum family, are deposited letters 1840–1857 from P. Pluygers of the Militaire Geneeskundige Dienst (Military Medical Service) at Semarang.

**A.F.C. List, November 23rd 1813 Muiden – 1850** [226]

In 1844 he left for the Dutch East Indies as adjutant to F. von Gagern for study of the defence there, his fellow adjutants were J.H.W. le Clerq and W.F.K. George; 1846–1850 captain and adjutant to the governor general.

The collection consists of journals of his voyage to the Dutch East Indies and his stay in Java 1844–1847 with a journal of W.F.K. George.

Size: *C.* 0.10 running metre.

**J.H.A. Logemann, January 19th 1892 Rotterdam – November 12th 1969 Leyden** [227]

Up to 1912 study of indology at Leyden, graduates in 1923; 1912 aspirant comptroller Blora; 1914 comptroller Bodjonegara; 1915 official Dienst der Pestbestrijding (Service for the Prevention of the Plague) Surakarta; 1919 inspector housing improvement Semarang; 1922 on leave; 1924 inspector Generale Thesaurie (General Treasury) Batavia; 1924 professor East Indies constitutional law Rechtshogeschool (University of Law) Batavia; 1939 on leave; 1940–1944 hostage of the Germans; 1945 minister of Overseas Territories; 1946 member Second Chamber for the Partij van de Arbeid (Labour Party); 1947–1962 professor of constitutional law of overseas territories at Leyden; in Indonesia he was from 1924 forward also i.a. government secretary of the Javasche Bank (Java Bank), member of the Commissie van Toezicht op het Landsarchief (Visiting Committee for the National Archives), Commissie voor Herziening Stelsels van Verkiezing (Committee for the Revision of the Electoral Systems), Commissie Stadsvorming (Committee for Town Formation), Verzoeningsraad voor de Spoor- en Tramwegen op Java en Madoera (Conciliatory Council for the Rail- and Tramways in Java and Madoera), Bataviaasch Genootschap van Kunsten en Wetenschappen (Batavian Society for Arts and Sciences), Royal Academy of Sciences, De Stuw.

The collection consists mainly of papers when he was professor and politician i.a. concerning the recognition of the sovereignty of Indonesia, the Statuut van het Koninkrijk (Constitution of the Kingdom) and Surinam. For the names and subjects mentioned in the inventory see the index.

Entry: C.M. Lont. *Inventaris van de papieren van Prof. Dr. J.H.A. Logemann.* 's-Gravenhage, 1975.

Size: *C.* 2 running metres (195 inventory items).

Permission is necessary for consultation of the papers.

**J. de Loos-Haaxman, ? – 1975 Rotterdam** [228]

Training Hague academy of arts, study history of art at Leyden; 1922 to Padang, then Tulung Agung and Madiun; 1925–1939 keeper of the Lands verzameling (National collection) at Batavia, member Bataviasche Kunstkring (Batavian Art Society) and Bond van Kunstkringen (League of Art Societies); publications about pictorial art and other arts in the Dutch East Indies. (J. de Loos-Haaxman. *Dagwerk in Indië.* Franeker, 1972).

The collection consists of letters received and copies of letters 1926–1975, report of the inspection National collection of paintings with correspondence 1930–1934, report about lost paintings of governors general, correspondence with the KITLV about Johannes Rach, photographs, articles.

Entry: List drawn up at the transfer.

Size: *C.* 0.40 running metre.

**J. Loudon, June 8th 1824 The Hague – May 31st 1900 The Hague** [229]

Law studies at Leyden, 1846 dissertation; 1846 lawyer and solicitor at Batavia; 1854 referendary Algemene Secretarie (General Secretariat); 1859 secretary general Ministry of Colonial Affairs; 1861 minister of Colonial Affairs; 1862 King's Commissioner for South Holland; 1871–1875 governor general.

The collection consists of memoirs with annexed letters, letters from and to various persons. For the names and subjects mentioned in the list see the index.

Entry: List drawn up at the transfer.

Size: *C.* 0.30 running metre.

Permission is necessary for consultation of the papers.

**Mackay van Ophemert** [230]

This collection consists of the papers originating from various members of the Mackay family. Papers concerning the East Indies are mainly originating from:

*D.J. Mackay van Ophemert, December 22nd 1839 The Hague – July 31st 1921 Earlston*
Training for jurist at Leyden, dissertation about the government of Daendels in Java, then attaché to London; *c.* 1865 clerk at the Ministry of Colonial Affairs; *c.* 1870 deputy of the Dutch Red Cross; 1871 member of the Second Chamber, expert of colonial affairs and education; 1875 to Great Britain, where he is naturalized in 1877 and inherits the title of lord Reay; when Briton i.a. governor of Bombay from 1885–1890; 1894–1895 undersecretary of State for India; from 1891 on the staff of the University of London; 1892 vice president of the Royal Asiatic Society. (*The Dictionary of National Biography,* 1921).

Concerning the Dutch East Indies are originating from him papers 1865–1869 when he was clerk at Colonial Affairs and letters 1855–1917 from W.K. van Dedem, I.D. Fransen van de Putte, H.C. van der Houven van Oordt, O. van Rees, H.G. van de Sande Bakhuyzen, and papers when he was governor of Bombay.

Besides papers originating from D.J. Mackay (74 inventory items), concerning the Dutch East Indies this collection also contains letters 1816–1876 from I.D. Fransen van de Putte, W.R. van Hoëvell, the KITLV, J.J. Melvill van Carnbee (originating from A. Mackay) and papers 1840–1862 of the state committees for colonization of Europeans in the Dutch East Indies (also originating from A. Mackay), in addition letters from C. Pijnacker Hordijk, R. Steinmetz, M. Tydeman, J.D. Veegens and the Vereeniging van Christelijke Onderwijzers en Onderwijzeressen in Nederland en Overzeesche Bezittingen (Association of Christian Teachers and Woman Teachers in the Netherlands and Overseas Possessions), originating from A. Mackay jr. and J.J. Mackay.

Entry: G.J.W. de Jongh. *Beschrijving van een verzameling stukken afkomstig van het geslacht Mackay van Ophemert.* 's-Gravenhage, 1967.

Size: *C.* 30.50 running metres (1684 inventory items).

**Th.H. de Meester, December 16th 1851 Harderwijk – December 27th 1919 The Hague** [231]

Law studies Utrecht, 1875 graduated; 1875 official at the office of the clerk of the provincial states of Overijssel; 1885 secretary Groningen; 1892 administrator Ministry of Finance; 1898 vice president of the Raad van Indië (Indies Council) in the same period also i.a. president of the Central Committee of the Red Cross and member of the board of governors of the Koning Willem III school; 1905 Minister of Finance; 1909 member of the Second Chamber, 1912–1914 also chief editor of *Het Vaderland;* 1917 member of the Council of State.

The collection consists of photographic copies of papers from 1872, including letters 1898–1905 from i.a. A.W.F. Idenburg, C.B. Nederburgh, J. Rooseboom, R. van den Broek d'Obrenon; articles concerning De Meester including necrologies.

Entry: List of papers.

Size: *C.* 0.50 running metre (53 items).

**[P.] Melvill van Carnbee, May 20th 1816 The Hague – October 24th 1856 Batavia** [232]

1831 naval cadet; 1835 first voyage to the East Indies; 1837 second voyage to the East Indies; 1838 naval lieutenant; 1839 third time to the East Indies, official in the hydrographic service of the Department of Naval Affairs at Batavia to draw up marine charts; 1845 to the Netherlands; 1850 to the East Indies as adjutant to vice admiral E.B. van den Bosch; 1853 discharge from naval service, employed as cartographer by the East Indies sedentary maritime affairs; 1856 naval captain lieutenant, director of the naval yard in Onrust.

The collection consists of a journal 1837 by C.P.A. de Salis of the expedition to Padang under general major [F.D.] Cochius; reports about Amboina 1836–1839, Bali and Lombok 1843, an expedition for the NHM to Bali and Lombok 1839; letters 1850–1851 from E.B. van den Bosch to the governor general.

Entry: Inventory.

Size: *C.* 0.15 running metre (3 inventory items).

**S. Meijer, May 13th 1912 Zevenaar – March 17th 1967 Rijswijk** [233]

Study of indology at Leyden; 1936 to Indonesia, aspirant comptroller at the Department of BB (Interior Government) at Batavia; 1939 idem Kedu; 1940 idem Pekalongan; 1941 idem Riouw (Sambu); after his internment during the second world war, in 1946 comptroller for Chinese affairs in Riouw; 1947 idem West Borneo (Pontianak); 1949 assistant resident, to the Netherlands.

The collection consists completely of papers concerning Indonesia, of which a very small part concerns the period before 1942, viz. decrees of appointment etc.; report of special audit committee for the accounts of K. Huibregtse 1936; report of the first Japanese air raid on Singapore 1942, with notes etc. about his internment 1942–1946 (3 inventory items). The remaining part of the collection concerns his functions in Indonesia 1946–1949. For names and subjects mentioned in the inventory see the index.

Entry: Provisional inventory.

Size: *C.* 0.20 running metre (123 inventory items).

**J.W. Meyer Ranneft, October 31st 1887 Magelang – February 3rd 1968 The Hague** [234]

Grand examination for civil servants at Leyden; 1907 to the East Indies, aspirant comptroller Kedu; 1907–1913 comptroller residency of Pasuruan (1907 Wonosobo, 1908 Puspo, 1910 Kraksaan, 1912 Lawang, 1913 Turen); 1913 comptroller for the police Semarang; 1914 residency of Surabaja (Grisee); 1916 on leave in the Netherlands; 1918 assistant inspector for matters of agriculture at the Department of Home Affairs (from 1921 with the status of assistant resident), 1924–1925 also member of the Volksraad (People's Council); 1925 on leave in the Netherlands; 1926 assistant resident Semarang (Pati), 1927–1928 also member of the Volksraad; 1928–1929 resident at the disposal of Semarang; 1929 president Volksraad; 1933 up to his departure to the Netherlands in 1936 vice president of the Raad van Indië (Indies Council); in the Netherlands from 1939–1942 teacher tropical colonial economics Wageningse Landbouw Hogeschool (Agricultural University of Wageningen); during the war from 1942 interned in St. Michielsgestel; 1945–1958 member of the Council of State. Meyer Ranneft was a member of the following committees and boards: 1921–1923 secretary of the Kiesrechtcommissie (Electoral Committee); 1927–1933 vice president of the Commissie tot het Nagaan en Verzamelen van Gegevens Betrekking Hebbend op de Welvaart der Inlandsche Bevolking (Committee for Tracing and Collecting Data Concerning the Prosperity of the Native Population); 1928–1934 member and president of the Commissie voor de Volkstelling 1930 (Committee for the 1930 Population Census); 1928–1931 member of the Hollandsch-Inlandsch Onderwijs-Commissie (Dutch-Native Education Committee); 1929–1936 member (of the board) of the Koninklijk Bataviaasch Genootschap van Kunsten en Wetenschappen (Royal Batavian Society of Arts and Sciences) and president of the Afdeeling Staathuishoudkunde (Section Political Economics); 1931 president of the Commissie voor Regeeringspublicaties (Committee for Government Publications); 1932 president of the Tweede Begrootingscommissie (Second Budget Committee); 1936 corresponding member at Batavia of the Koninklijke Akademie van Wetenschappen (Royal Academy of Sciences); 1936–1938 member of the Commissie voor Economische Samenwerking Nederland en Nederlandsch-Indië (Committee for Economic Cooperation between the Netherlands and the Dutch East Indies); 1946 member of the Stichting Indië in Nood (Foundation for the East Indies in Distress) and the Nationaal Comité Handhaving Rijkseenheid (National Committee for the Preservation of the Unity of the Kingdom), in 1947 amalgamated and in 1950 liquidated; 1950–1962 member of the Stichting Rijksbehoud (Foundation for Preservation of the Kingdom).

The collection consists of papers drawn up and received in his functions from 1910. In addition to official papers there are papers concerning his memberships of the various committees and associations, correspondence, speeches, publications, documentation and numbers of several periodicals. For the names and subjects mentioned in the inventory see the index.

Entry: I.W.L.A. Caminada and F.J.M. Otten. *Inventaris van de papieren van Dr. J.W. Meyer Ranneft.* 's-Gravenhage, 1971.

Size: *C.* 12.40 running metres (873 inventory items).

Permission is necessary for consultation of the papers.

**H.J. van Mook, May 30th 1894 Semarang – May 10th 1965 l'Isle-sur-Sorgue** [235]

Study of chemistry at Amsterdam and Delft and East Indies law at Leyden; 1919 comptroller Semarang; 1921 comptroller for agricultural matters Djokjakarta; 1925 on leave in the Netherlands; 1928 referendary at the Bureau Bestuurshervorming (Bureau for Administrative Reform) at Batavia; 1931 member Volksraad (People's Council), publications in *De Stuw* and *Koloniale Studiën;* 1934 senior officer, from 1937 director Department of Economic Affairs at Batavia, member of several committees of the Algemeene Volkscredietbank (General People's Credit Bank), 1938, 1940–1941 economic negotiations with Japan; 1941 minister of Colonial Affairs, stays in the East Indies as attached lieutenant governor general to governor general A.W.L. Tjarda van Starkenborgh Stachouwer; 1942 minister of Colonial Affairs in the London Cabinet; 1944 as lieutenant governor general to Australia; 1945 returns to Indonesia; 1948 discharge as lieutenant governor general, professor at Berkeley; 1951 as functionary for government problems in developing countries employed by the United Nations; 1960 settles in France.

The collection consists of papers 1942–1964 of which the period 1942–1948 is the chief point, when he was subsequently minister of Colonial Affairs and lieutenant governor general. For the names and subjects mentioned in the inventory see the index.

Entry: M. Dutilh. *Inventaris van de papieren Dr. H.J. van Mook.* 's-Gravenhage, 1975.

Size: *C.* 5.50 running metres (337 inventory items).

**Nederburgh** [236]

The collection consists for the greater part (568 inventory items) of papers originating from S.C. Nederburgh. In addition some papers (9 inventory items) are originating from his father and his brothers.

a. *S.C. Nederburgh, March 7th 1762 The Hague – August 3rd 1811 's-Gravenzande*
1777 law studies at Leyden, 1782 graduated; 1783 schepen (sherif) of Kool; 1785 lawyer of the States General and the Council of State; 1787 second lawyer of the VOC, in the same year first lawyer; 1791 commissioner general, leaves end of 1791, 1792–1793 at Cape of Good Hope, arrives in Batavia end of 1793; 1799 discharge, returns to the Netherlands, up to 1802 retired; 1802 member of a committee for examination of the trade to the East Indies; 1804 member of the Council for the Asian Possessions and Establishments; 1806–1807 chief of the First Division of the Ministry of Trade and Colonial Affairs; 1809–1810 director general of the public treasury.

The papers originating from S.C. Nederburgh consist of papers drawn up and received when he was lawyer of the VOC with preliminary papers from 1606

(167 inventory items); papers when he was commissioner general, viz. decrees of appointment, journal, papers concerning Cape of Good Hope, meetings and "verbaal" (minutes) of the commissioners general, correspondence, papers concerning the High Government, the Indies Council, the navy, the defence of the East Indies, the landed forces, finance, Council of Justice, VOC, Batavia, Preanger and Batavian Uplands, Cheribon, North East Coast of Java, Timor, Amboina, Banda and Ternate, Makasar, Palembang, Riouw, Malaya, Bengal, Coromandel, Malabar, Surat, Ceylon, Japan and China (291 inventory items); papers after 1800 (110 inventory items). For the names and subjects mentioned in the inventory see the index.

b. *Other members of the family.*
Papers concerning the Dutch East Indies are originating from the following persons: H.A. Nederburgh as resident of Tegal, memorandums etc. 1798–1806 concerning rice (i.a. by . . . Hendriks), indigo and a report left by him for his successor A. Hartsinck; J.J. Nederburgh as secretary of the Orphans' Court of Batavia, viz. a report left by J.J. Vogelaar for his successor 1798.

c. *Supplement.*
This part (*c.* 5 running metres), without an inventory yet, contains papers of and concerning S.C. Nederburgh, S.C.H. Nederburgh, C.B. Nederburgh and others.

Entry: R. de Vries. *Inventaris van het archief Nederburgh.* 's-Gravenhage, 1968.

Size: *C.* 10 running metres (*c.* 5 running metres inventorised, 577 items).

The collection is deposited at the First Section.

**Pinke** [237]

This collection consists of papers of F. Pinke sr. and jr.

a. *F. Pinke sr., January 1st 1860 Hengelo – September 25th 1917 Utrecht*
1875 naval cadet; 1879 to the Dutch East Indies; 1880 second naval lieutenant; 1906 in the Netherlands naval captain; 1910 inspector torpedo service; 1912 commander of the navy, chief of the Department of Naval Affairs in the Dutch East Indies, rear admiral; 1914 vice admiral; 1916 returns to the Netherlands; 1917 retired.

Originating from him are papers drawn up and/or received from 1913, viz. diary 1914–1916 concerning the first world war in the East Indies; papers 1914–1916 concerning the defence and neutrality of the East Indies; correspondence 1913–1916 concerning military and political affairs; letters to his wife J.M.W. Pinke-Carpentier Alting in connection with his death. For the names and subjects mentioned in the inventory see the index. (27 inventory items).

b. *F. Pinke jr., February 25th 1890 Den Helder – April 7th 1939 Dutch East Indies*
Son of F. Pinke sr.; 1909 as naval cadet to the Dutch East Indies; 1914 second naval lieutenant; 1919 discharge; 1923 back to service; 1928 first naval lieutenant; 1932 first officer naval barracks Udjung; 1934 instructor meteorol-

ogy and astronomy Willemsoord; 1937 officer H.M. "De Ruijter" in the East Indies.

Originating from F. Pinke jr. are newspaper cuttings 1933 concerning affairs of the court martial of the East Indies Navy and letters 1938 to his mother (2 inventory items).

Entry: *Inventarissen van de collecties A.J. Immink, F. Pinke, C. Pijnacker Hordijk;* [door F.G.P. Jaquet en W. Visser]. 's-Gravenhage, 1965. With a list in typescript of supplements of papers originating from F. Pinke sr.

Size: *C.* 0.20 running metre (33 inventory items).

**A.S. Pinke, October 14th 1894 Heerde – July 8th 1966 Bussum** [238]

1915 third naval lieutenant, 1917 second, 1925 first; 1920–1935 in the submarine service, most of the time in the Dutch East Indies; 1937 naval captain lieutenant; 1938 commander of the division torpedo destroyers in the East Indies; 1940 returns to the Netherlands; 1942–1945 in German captivity; 1945 rear admiral; 1946 commander of the East Indies Navy, also head of the Department of Naval Affairs at Batavia; 1947 vice admiral; 1949 discharge and to the Netherlands; 1951–1954 head Bureau Toegepast Wetenschappelijk Onderzoek ten behoeve van de Rijksverdediging (Bureau Applied Scientific Research for the Defence of the Kingdom).

The collection consists for the greater part of papers drawn up and received when he was commander of the East Indies Navy and head of the Department of Naval Affairs. In addition the collection contains some papers 1950–1963 which are also concerning his activities 1946–1949 in Indonesia. For the names and subjects mentioned in the inventory see the index.

Entry: Inventory.

Size: *C.* 0.30 running metre (49 inventory items).

Permission is necessary for consultation of the papers.

**Ch.O. van der Plas, May 15th 1891 Buitenzorg – ?** [239]

Study of indology at Leyden 1907–1911; 1911–1915 (aspirant) comptroller Bandung, Sukabumi, Sampang; 1915 comptroller Surakarta for the prevention of the plague; 1918 on leave, study of economics, adatrecht (customary law), Islam, Arabic language at Leyden; 1921–1926 consul to Djeddah; 1926 member of the delegation International Sanitary Congress at Paris; 1927–1931 assistant resident, temporary advisor Inlandsche Zaken (Native Affairs); 1931 at the disposal of the director BB (Indies Civil Service), president of the Volkscredietbank (People's Credit Bank) at Blitar; 1932 resident of Cheribon; 1936 governor of East Java; 1941 member of the Raad van Indië (Indies Council); 1942 chief commissioner Netherlands Indies Commission in Australia, adviser of the Ministry of Colonial Affairs; 1945 adviser political affairs of the lieutenant governor general; 1946 adviser Islamic and Arabic Affairs; 1947 government commissioner East Java; 1949 commissioner for the Kingdom of the Netherlands at Surabaja; 1951 FAO expert in Saudi Arabia; from 1951 missions for the UNO to Gambia and Greece.

The collection consists of papers from 1943. For the names and subjects mentioned in the list see the index.

Entry: List drawn up at the transfer.

Size: *C.* 0.80 running metre.

Permission is necessary for consultation of the papers.

**Th.B. Pleyte, 1864 – March 25th 1926 The Hague** [240]

1892–1908 lawyer with the Council of Justice at Semarang, where he was also second lieutenant (1894–1897), first lieutenant (1897–1903) and captain (1903–1908) auditor of the civic militia; moreover he was (president) supervisory director of the Telefoon-Maatschappij (Telephone Company) of the Principalities, established at Semarang; 1910 chairman of the Raad van de Scheepvaart (Council of Shipping Trade) at Amsterdam; 1913–1918 minister of Colonial Affairs; 1922–1925 ambassador to Rio de Janeiro.

The collection contains the papers drawn up and/or received by Pleyte when minister of Colonial Affairs and ambassador to Rio de Janeiro, the greater part (6 inventory items) consisting of papers concerning the Dutch East Indies, viz. telegrams 1913 from A.W.F. Idenburg on his appointment as minister with newspaper cuttings; correspondence 1913–1918 with H. Colijn, C.Th. van Deventer, D. Fock, A.W.F. Idenburg and his wife, J.P. van Limburg Stirum and his wife, S. van der Plas, letters 1915 about the vacancy of the governor-generalship with a note about the defence of the Dutch East Indies by J.J. Rambonnet.

Entry: Inventory. In *IRAA*; vol. 1(1928), p. 113–114.

Size: *C.* 0.20 running metre (10 inventory items).

**M.J.M. van Poll, February 24th 1881 Roosendaal – September 25th 1948 Eindhoven** [241]

Journalist; 1929–1946 member of the Second Chamber for the Rooms Katholieke Staatspartij (Roman Catholic State Party), subsequently Katholieke Volkspartij (Catholic People's Party), specialist in East Indies matters, 1938 study trip to Indonesia together with F. Teulings, against the Sutardjo resolution; 1946–1947 to Indonesia as a member of the Commissie-Generaal voor Nederlands-Indië (Committee General for the Dutch East Indies); 1948 journalistic manager of the publishing company Neerlandia at Utrecht.

The collection consists mainly of papers about the Indonesian question 1945–1948. For the names and subjects mentioned in the inventory see the index.

Entry: Inventory.

Size: *C.* 0.20 running metre (54 inventory items).

Permission is necessary for consultation of the papers.

**C. Pijnacker Hordijk, April 13th 1847 Drumpt – September 3rd 1908 Haarlem** [242]

Training for jurist at Utrecht; 1873 lawyer Rotterdam; 1874 professor Amsterdam; 1881 idem Utrecht; 1882 minister of Home Affairs; 1882–1885 retired; 1885 King's Commissioner for Drenthe; 1888–1893 governor general of the Dutch East Indies (concerned with statute labour, opium lease, Native education, foundation Pasteur Institute, expeditions Atjeh, Toba area, Western District of Borneo, Flores); 1894–1902 member of the First Chamber, specialist in colonial matters.
The greater part (217 inventory items) of the collection consists of (copies of) letters received 1867–1908; in addition political papers drawn up and/or received during his functions from 1882; papers 1875–1905 received when he was a member of several committees, including president of the board of governors of the Indische Instelling (East Indies Institution) at Delft and member of a committee for collection money in the aid of the distressed population of Amboina and Ceram; notes concerning Dutch history. For the names and subjects mentioned in the inventory see the index.

Entry: *Inventarissen van de collecties A.J. Immink, F. Pinke, C. Pijnacker Hordijk;* [door F.G.P. Jaquet en W. Visser]. 's-Gravenhage, 1965.

Size: *C.* 1 running metre (239 inventory items).

**Th.S. Raffles, July 5th 1781 Port Morant – July 5th 1826 Highwood** [243]

1795 clerk East India Company; 1805 under secretary Penang; 1806 secretary; 1811 lieutenant governor of Java; 1816 to England; 1817 lieutenant governor of Benkulen; 1824 to England.

The collection consists of photographic copies of letters 1810-1814 to lord G.E. Minto. The copies are also present in the Koninklijk Instituut voor Taal-, Land- en Volkenkunde (Royal Institute of Linguistics and Anthropology), see no. 660.

Size: *C.* 0.15 running metre.

**Rembges** [244]

Part of this collection is probably originating from:

*K. Rembges*
In 1809 K. Rembges was third bookkeeper of the Koloniaal Etablissement (Colonial Establishment) at Rotterdam.

The collection consists of papers from 1782, the bulk from 1801, viz. papers 1801–1808 concerning the stores of the Asian Council/Ministry of Colonial Affairs at Delfshaven, papers 1805–1809 concerning sales of Rotterdam, papers 1808–1853 concerning the navy, papers *c.* 1802 concerning the West Indies and Levant trade.

Entry: Inventory. In *VROA*, 1913, p. 78–80.

Size: *C.* 0.10 running metre (21 inventory items).

**Reynst** [245]

This collection consists of the papers originating from the Reynst family. Papers concerning the East Indies are mainly originating from:

*J.C. Reynst, January 23rd 1798 Velp – October 11th 1871 The Hague*
1815 in East Indies service, 1816 to Batavia where he was subsequently clerk, second and first officer of excise at the Algemene Rekenkamer (General Audit Department); 1818 secretary at Cheribon; 1819 secretary of the chief inspector of Finance; 1819 temporary resident of Sambas; 1822 assistant of the director of import and export duties; 1823 assistant of the commissioner and resident of Palembang; 1826–1830 (temporary) director of 's-Lands Producten en Civiele Magazijnen (Produce and Civil Stores of the Country), in 1827 also temporary general secretary; 1830–1836 (temporary) director general of Finance, in 1833 also temporary director of 's-Lands Producten en Civiele Magazijnen; 1836–1842 member of the Raad van Indië (Indies Council), interrupted from 1839–1841 when he was on leave to the Netherlands, in 1838 also temporary director general of Finance; 1842–1850 vice president of the Indies Council, from 1844–1845 also temporary governor general; from 1850 retired, in 1858 founder of the Jongens-weezeninrichting (Boys' Orphanage) at Surabaja; 1860 to the Netherlands.

Originating from him are records of service; journals 1816 and 1839 of his voyages the Netherlands-Batavia and return; correspondence from 1823; letters from his son P.H. Reynst from 1839; diaries with correspondence 1839–1871; notes 1822–1823 concerning Palembang; papers 1848–1858 concerning the foundation of the Boys' Orphanage at Surabaja. In addition the collection contains: papers concerning the landed estate of J. van den Bosch: Pondok Gedeh; papers concerning the firms of Reynst & Vinju and Hoboken; legal cases 1880 at Batavia; poetry albums; genealogies; engravings by J. Wap; papers concerning other members of the family (62 inventory items). For the names and subjects mentioned in the inventory see the index.

Entry: Inventory.

Size: *C.* 1.30 running metres (121 inventory items).

**Von Römer** [246]

This collection consists of genealogical documentation concerning the Von Römer family at Malang and related families from the 17th up to the 20th century. Included are letters and photographs. Besides documentation concerning the family at Malang there are also data concerning i.a. the Javanese descendants of L.S.A.M. von Römer and the families of Bach, Beer, Bik, Caspersz, Gallois (i.a. E.M. Gallois), Mayor, Van der Plas (i.a. Ch.O. van der Plas), Reeb, De Rochemont, Satoor and others.

Size: *C.* 1.75 running metres (22 vol.).

**Rollin Couquerque** [247]

In this collection all Asian papers are originating from L.M. Rollin Couquerque, 9 inventory items.

*L.M. Rollin Couquerque, 1791 Nielles les Adres – March 16th 1855 Maastricht*
1822 East Indies service; 1823 first clerk Finance, 1824 officer of excise Comptabiliteit (Accounting Department); 1825 overseer Landelijke Inkomsten (National Revenue) at Cheribon; 1829 assistant resident Banjuwangi; during his leave member of a committee charged with the examination of silk in France; 1838 in the East Indies at the disposal of the Directie Cultures (Direction Cultures) in connection with the silk cultures; 1841 inspector silk cultures; 1845 discharge.

The papers consist of reports and notes 1829–1841 about Banjuwangi, buildings, sugar, roads, rice, land rent, with drafts of letters to the resident of Besuki and Banjuwangi, the president of the factory at Canton and . . . Bogaard.

Entry: Inventory.

Size: *C.* 0.50 running metre (59 inventory items).

**W. Roosegaarde Bisschop, December 31st 1866 Semarang – c. 1940** [248]

Training for jurist at Leyden; 1893–1899 at the India Office in London for search of archives; 1897–1913 secretary of the Nederlandse Kamer van Koophandel (Dutch Chamber of Commerce) there; 1898 lawyer of the Dutch legation at London; 1901 admitted to the English bar; 1907 member and secretary of the Commissie voor den Nederlandsch-Zuid-Afrikaansche Handel (Dutch-South African Trade); 1920–1923 founder and secretary of the Anglo Batavia Society London.

The collection consists of copies of papers of the India Office at London, concerning Java 1812–1830; Sumatra 1818–1830; Malaya 1826–1830; Singapore 1827–1830; Straits-Settlements 1818–1830; Prince of Wales' Island 1818–1830; Bengal 1818–1830.

Entry: Inventory on index cards.

Size: *C.* 1 running metre.

**C.C.F.M. le Roux, June 8th 1885 Assen – September 8th 1947 Amsterdam** [249]

1908 as lieutenant to the Dutch East Indies; 1913 at the disposal of Burgerlijke Openbare Werken (Civil Public Works) in connection with road construction, up to 1918 in Flores; *c.* 1918 discharge from military service; 1920 head Civil Public Works; 1921–1927 teacher Koningin Wilhelmina School at Batavia; 1929–1933 curator of the museum of the Koninklijk Bataviaasch Genootschap van Kunsten en Wetenschappen (Royal Batavian Society for Arts and Sciences); 1934 to the Netherlands; 1937–1947 curator and director of the Rijksmuseum voor Volkenkunde (State Museum for Anthropology) at Leyden.

The part with an inventory consists of reports of retiring officials and descriptions made by various persons concerning i.a. Flores, Timor and New Guinea 1907–1937 (for the names and subjects mentioned in the inventory see the index), the part without an inventory consists of journals etc. of scientific expeditions to North New Guinea 1914, 1915 and to Central New Guinea 1920 i.a. concerning the language and anthropology made by i.a. J. Kremer, H.J. Lam and H.J.T. Bijlmer and of reports left by L.J. Oerlemans for his successor in Samosir, *c.* 1915 and 1937.

Entry: Inventory (of the first 20 items).

Size: 0.50 running metre (43 items).

**P. Sanders, September 21st 1912 Leyden – ?** [250]

Lawyer; 1945 secretary general Department of General Warfare; 1946 secretary general Commissie-Generaal (General Committee), one of the staff of W. Schermerhorn; 1947 back in the legal profession; 1959 professor of constitutional law Rotterdam.

The collection consists of the papers drawn up and received when he was secretary general of the General Committee: diary, correspondence with A.D. Belinfante, E.H. van der Beugel, O. Damsté, R.W. van Diffelen, J. Drion, H.J. Friedericy, J.H.J. de Jong, J.M. Kiveron, M. Kohnstamm, J. Landré, H. Lenderink, J.H.A. Logemann, S.L. Mansholt, W.B. Meyboom, . . . de Meester, C.L. Patijn, S. Posthumus, H.M. van Randwijk, J.H. van Royen, I. Samkalden, H.W. Sandberg, J. de Visser, K. Vorrink, D. Vriesman, J.J. Vriesendorp, M. Weisglas, M. Wright, copies of letters from M. de Niet to J. Barents and J.A. Jonkman about the succession of H.J. van Mook and the Indonesian question, photographs, drawings by H. Ngantoeng, negotiations in Linggardjati.

Entry: Inventory.

Size: *C.* 0.20 running metre.

Permission is necessary for consultation of the papers.

**B.Chr. de Savornin Lohman, July 13th 1883 Groningen – March 17th 1946 Utrecht** [251]

Law study at Utrecht, 1910 graduated, 1910 lawyer and solicitor at The Hague; 1912 professor of constitutional and administrative law at Utrecht, in 1925 concerned in the foundation of the Indologische Faculteit (Faculty of Indology) at Utrecht, also from 1926 member of the First Chamber for the Christelijk Historische Unie (Christian Historical Union); supporter of the concept of Dutch-Indonesian unity, in 1928 visit for the CHU to Java and Sumatra; 1940–1944 taken as a hostage by the Germans; 1946 member of the Van Poll Committee, dies just before he was to leave for Indonesia.

The collection consists of diaries, travel records, letters, papers when he was professor, member of the CHU, member of the First Chamber, documentation;

and also papers originating from some members of his family. For the names and subjects mentioned in the inventory see the index.

Entry: A.M. Tempelaars. *Inventaris van de papieren van Prof. Jhr. Mr. B.C. de Savornin Lohman en van enige van zijn familieleden.* 's-Gravenhage, 1975.

Size: *C.* 2 running metres (226 inventory items).

Permission is necessary for consultation of the papers.

**J.Ph.D. Scheidius, 1787–1833** [252]

*C.* 1800 naval cadet; *c.* 1808 naval lieutenant; *c.* 1830 divisional commander.

The collection consists completely of journals, i.a. of his voyages to the Dutch East Indies and return to the Netherlands and of his voyages through the East Indies archipelago 1800–1832, and also a journal of his voyage Batavia-Palembang and return in 1821.

Entry: Inventory. In J. de Hullu. *De archieven der admiraliteitscolleges.* 's-Gravenhage, 1924 (p. 327–328).

Size: *C.* 0.15 running metre (7 inventory items).

Deposited at the First Section.

**W. Schermerhorn, December 17th 1894 Akersloot – ?** [253]

Student Polytechnical University Delft, 1918 civil engineer; up to 1926 assistant geodesy Polytechnical University; 1926–1946 professor of Delft Polytechnical University; 1936 adviser Koninklijke Nederlandsch-Indische Luchtvaartmaatschappij (Royal Dutch East Indies Airlines), 1937 adviser China aerial surveying; 1938–1940 president Eenheid Door Democratie (Unity through Democracy); 1945 prime minister Schermerhorn/Drees cabinet; 1946–1961 member of the board of the Partij van de Arbeid (Labour Party); 1946–1947 chairman of the Commissie-Generaal (General Committee); 1948 member of the Second Chamber; 1951–1963 member of the First Chamber; after 1967 no longer active in politics.

The collection consists mainly of papers after 1945, including correspondence with C. Smit, official notes about Indonesia, diaries about his voyage to Indonesia, papers General Committee, military activities, correspondence C. Gerretson, G.J. van Heuven Goedhart, P. Sanders, I. Samkalden, . . . Ruygers, P. Schmidt, . . . Wansink in East Indonesia, Indonesian federalists, F.J. Goedhart, . . . Posthumus, . . . Honig, papers New Guinea and Polytechnical University Bandung.

Entry: List drawn up at the transfer.

Size: *C.* 0.80 running metre.

Permission is necessary for consultation of the papers.

**W.J.M. van Schmid, November 1806 – August 11th 1903 The Hague** [254]

1829–1837 official Algemene Secretarie (General Secretariat), interrupted in 1834 when he was on leave, (1829 second clerk, 1832 first clerk); 1838 secretary residency of Padang, after his leave of 1837; 1840 assistant resident Moluccas (Saparua and Haruku); 1842 discharge; 1845 assistant resident Pasuruan (Malang, also auctioneer there); 1854 assistant resident Rembang (Tuban); 1856 on leave; 1858 member of the Algemene Rekenkamer (General Audit Department) at Batavia; 1860 discharge.

The collection consists completely of papers concerning the Dutch East Indies 1741–1855, mainly concerning agriculture and cultures (tobacco, rice) and land taxes, i.a. regulations, instructions, guides, statistics concerning Java, Malang, Tuban with papers by Th.S. Raffles, H.W. Muntinghe and F. Klein; moreover an index for the archives of the General Secretariat 1610–1832, regulations 1819–1820 concerning action at law and titles and ranks of regents in Java, notifications 1821–1838 at Padang by residents and chief factors of the West Coast of Sumatra with programmes of ceremonies 1834–1839 and a paper in Malay.

Entry: Inventory on index cards.

Size: *C.* 0.35 running metre (24 inventory items).

**G.J.Chr. Schneither, February 19th 1795 Leyden – March 16th 1877 Arnhem** [255]

Jurist; 1819 third class official in the Dutch East Indies; 1821 principal clerk Algemene Secretarie (General Secretariat); 1821–1826 private secretary of governor general G.A.G.Ph. van der Capellen; 1826 on leave to the Netherlands; 1828 up to his departure for the Netherlands in 1832 (temporary) member of the Hooggerechtshof (High Court of Justice) at Batavia; 1839 justice of the provincial court of justice of Gelderland.

In addition to a few papers this collection consists of papers concerning the Dutch East Indies, of which the period 1816–1828 is the chief point. The earliest papers date from 1788, the latest from 1831. The collection is subdivided as follows: letters etc. to the governor general; papers concerning European government, Native government, personnel, justice, military matters, trade, cultures and industry, finance, monetary system, taxes, slavery, Java and Madura general, Batavia, separate residencies in Java and Madura, outside districts, relations with China, Japan and Persia. For the names and subjects mentioned in the inventory see the index.

Entry: Inventory. In *VROA*, 1915, p. 342–355.

Size: *C.* 4.25 running metres (137 inventory items).

**P. Spies** [256]

Employed at the Java Bank subsequently from 1927 as cashier at Medan; 1928 Surabaja; 1934 substitute agent Malang; 1936 Surabaja; 1939–1942 agent at

Pontianak; 1946–1952 director and second substitute president at Batavia; after his discharge in 1952 adviser. *(Verslag van de president van de Javasche Bank, 1926–1952/Report for the financial year 1952–1956 [of] the Java Bank).*

The collection consists of papers 1946–1954, including papers concerning committees and presidents of the Java Bank 1946–1954, the Bank Negara Indonesia 1950, the Java Bank in Djokjakarta 1949–1950, politics and economics in Indonesia 1947–1950, communism 1947, politics in Makasar 1950, sultan Hamid 1952–1954, correspondence with i.a. R.E. Smits, J.C. van Waveren and H.J. Manschot.

Size: *C.* 0.80 running metre.

**S.H. Spoor, January 12th 1902 Amsterdam – May 25th 1949 Batavia** [257]

After his training at the KMA (Royal Military Academy) in 1923 as officer to the Dutch East Indies; 1929 training Hogere Krijgsschool (Army-staff College) at The Hague; 1932 drafted to the general staff at Bandung; 1934 tutor KMA Breda; 1938 officer general staff at Bandung; 1941 tutor Army-staff College at Bandung; 1942 taken refuge in Australia, charged with the organization of the Netherlands Eastern Forces Intelligence Service (NEFIS); 1945 as colonel to Indonesia; 1946 commander of the Indies land forces, also head of the Department of War at Batavia; 1949 general.

The collection consists mainly of papers 1946–1949 drawn up and received by Spoor as commander of the Indies land forces and head of the Department of War. Moreover the collection contains private correspondence 1948–1949. For the names and subjects mentioned in the inventory see the index.

Entry: Inventory of the S.H. Spoor and D.C. Buurman van Vreeden (see no. 168) collections.

Size: *C.* 0.40 running metre (93 inventory items).

Permission is necessary for consultation of the papers.

**G.L. Tichelman, January 31st 1893 Palembang – January 3rd 1962 Overveen** [258]

Training colonial agriculture at Wageningen and Deventer; 1913–1916 planter in South Sumatra and Tapanuli; 1916–1923 (aspirant) commander in Amboina (1917 Amboina, 1917 Ceram (Piru), 1918 Ceram (Amahai)); 1923–1929 commander Southern and Eastern District of Borneo (1923 South coast (Tanahbumbu), 1926 Hulu Sungei (Barabai)); 1931–1933 comptroller Atjeh (1931 Meureudu, 1932 Lho Seumawé); 1933–1937 comptroller East Coast of Sumatra (1933 Asahan (Batu Bara and Labuhan Batu), 1934 Simalungun and Karo area (Pemantang, Siantar)); 1936 assistant resident titular; in the Netherlands i.a. at the scientific staff of the Koninklijk Instituut voor de Tropen (Royal Institute for the Tropics), member of the board of the Vereniging Oost en West (Society East and West) section Haarlem, honorary member of the Vereniging van Indische Verlofgangers en Gerepatrieerden (Association of Persons on Leave and Repatriates from the East Indies), editor of the periodicals *Indië Roept* (1948) and *Wordend Indië* (1948).

The collection consists completely of diaries, memorandums and notes drawn up by Tichelman as government official 1918–1937 with some reports of others and some letters. For the names and subjects mentioned in the inventory see the index.

Entry: Inventory.

Size: *C.* 0.45 running metre (38 inventory items).

See also no. 655.

**A.W.L. Tjarda van Starkenborgh Stachouwer, March 7th 1888 Groningen – ?** [259]

Law studies at Groningen; 1912 official Ministry of Foreign Affairs; 1915–1925 in the diplomatic service in Washington, Paris, Berlin; 1925 Queen's Commissioner for Groningen, from 1928 also president of the board of governors of the university of Groningen; 1933 ambassador to Brussels; 1936 governor general; 1942–1945 interned by the Japanese; 1946–1948 ambassador to Paris; 1950 permanent representative with the Council of the North Atlantic Pact; 1956 Minister of State.

The collection consists mainly of papers from the time when he was governor general, including papers about the Japanese invasion, . . . Wavell, Volksraad (People's Council), J.M. Kiveron, . . . Tjalling Bakker, W. Schilling, C.W.A. Abbenhuis, W. van Boetzelaer, . . . Reinderhoff in Tarakan, H.J. Levelt, H. Hopkins, . . . Marshall, . . . King, Immamura, Ch.J.I.M. Welter, Queen Wilhelmina, J.A. Jonkman, L.J.M. Beel, . . . Six, . . . Snouck Hurgronje, P.S. Gerbrandy, P.J.A. Idenburg, H.J. van Mook, Mangku Negara VII, Visman Committee, Parlementaire Enquête Commissie (Parliamentary Committee of Inquiry), the NSB.

Entry: List drawn up at the transfer.

Size: *C.* 1 running metre.

Permission is necessary for consultation of the papers.

**Ver Huell** [260]

The collection consists almost completely of the papers originating from Chr.A. Ver Huell and his brother C.H. Ver Huell.

a. *Chr.A. Ver Huell, April 7th 1760 Doetinchem – March 14th 1832 Paris*
   1771 naval cadet; 1780 naval lieutenant; 1783 naval captain; 1788 to the East Indies as member of the military committee; 1793 returns to the Netherlands; 1795 discharge from naval service; 1802–1805 member of the Commissie van de Oost-Indische Zaken (Committee for the East Indies Affairs); 1806 member of the Staatsraad (State Council), chairman of the section for the navy; 1807 ambassador to Spain; 1811–1814 member of the Corps Legislatief (Legislative Corps) for the Department of the Upper IJssel. (Supplements *Nederland's adelsboek;* 1918. 16th vol.)

Besides a great number of papers 1788–1792 when he was a member of the military committee, there are some papers relating to his membership of the Committee for the East Indies Affairs and private letters 1804–1819 from G. Brantsen, F.S. van Bylandt Halt, T.F. van de Capellen, G.N. de Haan, J.H. van Kinsbergen, L.M.J. van Reede van Oudshoorn, H.A.W. van Rouwenoort, H. van Royen, H. van Stralen and E.A. Ver Huell (62 inventory items).

b. *C.H. Ver Huell, February 4th 1764 Doetinchem – October 25th 1845 Paris*
1775 cadet; 1779 naval cadet; 1781 naval lieutenant; 1793 captain lieutenant; 1795 naval captain, discharge; 1802 member of the municipal government of Doetinchem; 1803 commander of the Batavian fleet; 1805 Secretary of State for Naval Affairs, subsequently minister of Naval Affairs; 1806 marshal; 1807 ambassador to Petersburg; 1809 commander of the naval forces on the Meuse and the Zealand estuaries; 1810 commander of the squadron of Texel and inspector general of the coast of the North Sea; 1814 inspector general of the French navy; 1815 naturalized Frenchman.

The papers originating from C.H. Ver Huell consist of papers from 1779, of which a small portion concerns Asia, viz. letters 1807 from A.A. Buyskes concerning his mission to the East Indies; papers 1805–1806 concerning a squadron in the East Indies archipelago, with an instruction of C.A. Ver Huell for P. Hartsinck; letters 1794–1796 from S.H. Frijkenius at Batavia concerning A. Kuvel; survey 1806 by S.C. Nederburgh concerning the East Indies; memorandum 1806 by D. van Hogendorp concerning the East Indies; list of officials who accompanied C.Th. Elout and C.H. van Grasveld to the East Indies; papers 1807 concerning the project of P.C. Roth for the construction of a harbour in Banka; reports of the ministers of Naval Affairs, Foreign Affairs, Trade and Colonial Affairs concerning the trade after the conclusion of peace with England; private letters from i.a. Chr.A. Ver Huell 1791–1814, Q.M.R. Ver Huell 1804–1837, H.M. de Kock 1806–1833, C.H. van Grasveld 1806 and I. Titsingh 1802 (422 inventory items).

Entry: Inventory. In *VROA*, 1931, p. 94–134.

Size: *C.* 5.50 running metres (493 inventory items).

**P.F. Vermeulen Krieger, March 1782 – September 25th 1865 Etten** [261]

*C.* 1796 in military service; 1815 to the East Indies; 1817 expedition Saparua; 1818 captain, commander of the troops at Cheribon; 1819 major, commander of the troops in Malaya, expedition to Riouw; 1823 retired; 1829 returns to the East Indies; 1830 commander of a battalion of rifles; 1832 to Padang, lieutenant colonel, civil and military commander Padang Uplands, expedition against Padris, journey from Pisang to Agam; 1833 via Batavia to Djokjakarta, where he took over the command of the troops; 1834 retired.

The collection consists of a register of letters received and despatched 1817–1834; diaries of his travels through the Indies and of his voyages the Netherlands – the East Indies and return; narrative "het begin van den Bondjolschen opstand" and the "merkwaardigen terugtocht van Pisang op Agam" 1833; notes 1817 about a telescope captured at Saparua.

Entry: Inventory. In *VROA*, 1913, p. 58–59.

Size: *C.* 0.05 running metre (5 inventory items).

**E. de Waal, November 27th 1821 The Hague – July 1st 1905 The Hague** [262]

1837 to Batavia, clerk at the Algemene Secretarie (General Secretariat); 1838 clerk; 1848 referendary; 1850 government secretary; 1856 director of the Departement van Middelen en Domeinen (Department of Finance and Public Lands); 1858 to the Netherlands and in 1860 discharge; 1868–1870 minister of Colonial Affairs, i.a. agriculture act 1870 and sugar act 1870; publicist, in the Indies editor of *De Kopiist* 1842–1844 and *Indisch Magazijn* 1844–1846.

The collection consists of personal notes, appointments etc.; papers, notes and letters 1845–1875 referring to his publications and activities in the Indies and during his ministerial term; moreover papers originating from his father J.H. de Waal (1797–1858) and his son J. de Waal (1848–1922). For the names and subjects mentioned in the inventory see the index.

Entry: Inventory.

Size: *C.* 0.40 running metre (47 inventory items).

**J.Ph. van de Water, October 8th 1898 Rotterdam – ?** [263]

1920 employed with the Borneo Sumatra Handelmaatschappij (Borneo Sumatra Trading Company) at The Hague; 1932 sent out to Japan; 1941 interned and in 1942 exchanged, employed with the Netherlands Shipping and Trading Committee at London; 1943 to Australia; 1944 civil supply officer; 1945 Netherlands Indies Governmental Export Organization; 1947 Foreign Office, sent out to Japan; 1954 sub manager Borsumij; 1958 retired.

The collection consists of a biography 1898–1954 and papers 1940–1952, including papers Borsumij, NIGEO, Nederlandse Militaire Missie (Dutch Military Mission), Supreme Commander for the Allied Powers, trade with Japan 1946–1953, memorandum H. van Geuns 1944.

Entry: Provisional list.

Size: *C.* 0.10 running metre.

**A.W.P. Weitzel, January 6th 1816 The Hague – March 29th 1896 The Hague** [264]

1831 in military service; 1837 second lieutenant; 1853 as captain adjutant to F.V.H.A. de Stuers to the Dutch East Indies; 1856 expedition under J.A. Waleson to the Lampongs; 1860 returns to the Netherlands; 1873 general major; 1873–1875, 1883–1888 minister of War. Weitzel has published much about military matters in the Dutch East Indies.

The collection consists completely of his memoirs 1831–1888 with annexes. He wrote about the East Indies "Mijn verblijf in Indië van 1853 tot 1860", with 2 vol. of annexes viz. 1848–1860 and 1848–1887.

Entry: Inventory. In *VROA*, 1918, p. 504–506.

Size: *C.* 0.20 running metre (23 inventory items).

Permission is necessary for consultation of the papers.

**Ch.J.I.M. Welter, April 6th 1880 The Hague – ?** [265]

Training government officer East Indies Institute at Delft; 1902 third clerk at the Departement of Binnenlands Bestuur (the East Indies Civil Service) at Batavia; 1903 aspirant comptroller Kedu (Karanganjar); 1905–1908 (aspirant) comptroller Pekalongan (1905 Bumidjawa, 1906 comptroller Brebes, 1907 Bumiaju); 1908 official Algemene Secretarie (General Secretariat) at Batavia; 1911 on leave in the Netherlands, drafted to the Ministry of Colonial Affairs; 1915 official at the disposal of the General Secretariat at Batavia; 1917 government secretary at Batavia, and in 1918 also a member of the Commissie tot Herziening der Indische Staatsregeling (Committee for the Revision of the East Indies Form of Government); 1921 head of the General Secretariat; 1924 member of the Raad van Indië (Indies Council); 1925 minister of Colonial Affairs; 1926 member of the Indies Council; 1929 up to his departure for the Netherlands in 1931 vice president of the East Indies Council; 1931 president of the Staatscommissie tot Verlaging van de Rijksuitgaven (State Committee for the Reduction of the National Expenditure); 1933 president of the Ondernemersraad voor Nederlandsch-Indië (Employers' Council for the Dutch East Indies); 1937–1941 minister of Colonial Affairs; 1945–1946 member of the First Chamber; 1948–1963 member of the Second Chamber (up to 1955 for his own party the Katholieke Nationale Partij (Catholic National Party), supporting the preservation of the bonds between the Netherlands and Indonesia).

The collection consists of papers drawn up and received when he was minister of Colonial Affairs 1925–1926; member and vice president of the East Indies Council 1926–1931; president of the Employers' Council for the Dutch East Indies 1933–1937; minister of Colonial Affairs 1937–1941; member of the Chambers 1945–1963 (40 inventory items). For the names and subjects mentioned in the inventory see the index.

Entry: Inventory.

Size: *C.* 0.30 running metre (62 inventory items).

**S.I. Wiselius, February 4th 1769 Amsterdam – May 15th 1845 Amsterdam** [266]

Law student 1786 at Franeker and 1787 at Leyden, graduates in 1790; 1790 lawyer at the Hof van Holland (Court of Holland); 1793 in trade; 1795 secretary, subsequently president of the Provisioneele Representanten van het Volk van Nederland (Provisional Representatives of the People of the Netherlands); 1796–

1804 member of the Comité tot de Zaken van de Oost-Indische Handel en Bezittingen (Committee for the Affairs of East Indies Trade and Possessions), after 1800 member of the Raad der Aziatische Bezittingen en Etablissementen (Council for the Asian Possessions and Establishments); 1804 retired; 1813 chief inspector of police at Amsterdam; 1840 discharge.

The collection consists of papers from 1795, approximately one third of these papers concern his activities 1796–1804 with preliminary papers from 1602, and in addition letters 1789–1815 to Wiselius. For the names and subjects mentioned in the inventory see the index.

Entry: Inventory.

Size: *C.* 1.40 running metres (67 inventory items).

**Wolterbeek** [267]

This collection consists of papers originating from C.J. Wolterbeek, D.G. Muller and C.J.D. Wolterbeek Muller. Papers pertaining to the East Indies are originating from the former two. The larger portion of the collection consists of the papers of C.J. Wolterbeek.

*C.J. Wolterbeek, April 5th 1766 Steenderen – May 23rd 1845 The Hague*
1782 naval cadet; 1783 naval lieutenant; 1789 to the East Indies; 1791 expedition to Goram (south east of Ceram); 1794 second voyage to the East Indies; 1799 naval captain; 1808 eskaderbrigadier (squadron brigadier); 1810 discharge; 1813 back in the service; 1817 rear admiral, to the East Indies, appointed to take over Malaya from the British; 1819 expedition to Palembang; 1821 returns to the Netherlands; 1830 director general of the Ministry of Naval Affairs; 1831 vice admiral; 1844 retired.

Besides papers 1781–1815 concerning his period of service in the Netherlands the collection contains papers 1817–1821 (with letters 1822–1824) concerning his term of service in the East Indies, including journals and minutes of his voyages the Netherlands – the East Indies and return, Surabaja, east coast Java-Semarang, Malaya, Riouw and Lingga, Palembang with papers concerning his expeditions, and also miscellanies 1780–1838. For the names and subjects mentioned in the inventory see the index.

Originating from his son in law D.G. Muller are journals 1816–1819 of his voyages Vlissingen-Batavia and return and Batavia-Banda and return.

Entry: Inventory. In J. de Hullu. *De archieven der admiraliteitscolleges.* 's-Gravenhage, 1924 (p. 353–363).

Size: *C.* 0.70 running metre (88 inventory items).

Deposited at the First Section.

See also P.H. van der Kemp. De commissiën van C.J. Wolterbeek naar Malakka en Riouw in juli-december 1818 en februari-april 1820. In *BKI;* vol. 51 (1900), p. 1–101.

**Zoutman** [268]

Papers concerning the East Indies (18 inventory items) are originating from:

*J.A. Zoutman, 1797–1846*
1814 first naval cadet; 1819 second naval lieutenant; 1827 first naval lieutenant; 1839 naval captain lieutenant.

The collection consists of papers concerning his travels through Europe and the East and West Indies. Pertaining to the Dutch East Indies: journals of his voyages to Batavia 1815–1817 under the command of captain lieutenant d'Everdingen; 1818–1819 through the archipelago under the command of J. Groot; 1819 to the Netherlands; 1822–1826 to Batavia and through the archipelago; 1826 to the Netherlands; copies of reports 1825–1826 and a journal 1825 by J. Frijtag van Overstraten about the expedition against Boni.

Entry: Inventory. In J. de Hullu. *De archieven der admiraliteitscolleges.* 's-Gravenhage, 1924, p. 363–367.

Size: *C.* 1.50 running metres (25 inventory items).

Deposited at the First Section.

**Ph.J. van Zuylen van Nyevelt, April 21st 1785 Rotterdam – June 1st 1864 Utrecht** [269]

1802 cadet; 1806 second lieutenant; 1810 in French service; 1811 first lieutenant; 1813 in Dutch service; 1814 captain; 1819 transferred to the East Indies army; 1821 to the East Indies; 1822 aide-de-camp to governor general G.A.G.Ph. van der Capellen; 1823–1825 resident of Banjuwangi; 1825 lieutenant colonel; 1826 returns to the Netherlands; 1827 discharge from military service; 1844 extraordinary gentleman-in-waiting.

This collection consists completely of papers drawn up and received by Van Zuylen during his term of service in the East Indies, containing besides reports, memorandums, letters etc. 1821–1826 (i.a. concerning the army 1821–1825, expedition 1821 against Palembang, regencies in Padang and Minangkabau, residency Banjuwangi) also papers from 1760 up to 1830. For the names and subjects mentioned in the inventory see the index.

Entry: Inventory. In *VROA,* 1902, p. 46–47, no. 95–109.

Size: *C.* 0.20 running metre (15 inventory items).

See also no. 270.

**Van Zuylen van Nyevelt (collection 105)** [270]

This collection consists of papers originating from various members of the Van Zuylen family, of which a small portion (19 inventory items; no. 25–43) is originating from:

*Ph.J. van Zuylen van Nyevelt*
Only a small portion of these papers (1798–1858) concerns the Dutch East Indies, viz. papers drawn up when he was aide-de-camp to G.A.G.Ph. van der Capellen and resident of Banjuwangi; letters 1822–1825 from [G.A.G.Ph.] van der Capellen, . . . van Hogendorp, the resident of Surakarta I.J. van Sevenhoven, . . .Taets van Amerongen, . . . van der Veen and various members of the family; papers 1820 concerning his departure to the East Indies.

Entry: Inventory.

Size: *C.* 1.40 running metres (154 inventory items).

See also no. 269.

## Asian Papers in Dutch Collections [271]

By Dutch collections are understood the papers originating from persons, who, from 1796–1949, occupied no functions in Asia and no functions in organizations related with Asia.

a. **Asser**

Papers 1856–1913 originating from T.M.G. Asser, mainly concerning legal questions, including Soerabaja Droogdokmij. (Surabaja Dry-Dock Company) *c.* 1900, petition G. Rollin Jacquemijns for technical assistance for Siam, opium conference 1912, Costa Rica Packet 1896, Russo-Japanese War 1904–1905, government regulations 1911, Nederlandsch-Indische Handelsbank (Dutch East Indies Commercial Bank), Nederlandsch-Indische Zee- en Brand-Assurantiemij. (Dutch East Indies Marine and Fire Insurance Company) 1874–1876, New Guinea conference 1885 with a letter from G. Rosenthal, G.M. Campbell and the Koninklijke Siamese Spoorwegmij. (Royal Siamese Railway Company) 1897, Javasche Bank (Java Bank) 1904–1911, Japan and international law 1897, 1905, A.N. van Horn van der Hork and the Deli Mij. (Deli Company) 1895, Borneo tobacco, Borneo gold, legation in China, correspondence with J.J. Rochussen.

b. **Becker**

Letters from J.C. Becker about his voyage from the East Indies – the Netherlands 1902 and from J.W.C. Becker from Onrust, Siak and Riouw 1865.

c. **L.J.M. Beel**

Papers 1945–1950 about the Indonesian question, including letters from M. van der Goes van Naters, C.P.M. Romme, H. Hermans, H.J. van Mook, L. Neher, A.W.L. Tjarda van Starkenborgh Stachouwer, E.N. van Kleffens, H.F.L.K. van Vredenburch, E.M.J.A. Sassen, S.H. Spoor, with photographs.

Permission is necessary for consultation of the papers.

d. **D. Bos**

Printed articles by himself and by C.Th. van Deventer about secondary and higher education for Europeans and non Europeans, papers concerning the financial records of the government and recruitment in the East Indies for Surinam, with a report of J. van Vollenhoven, papers life insurance J. des Amorie van der Hoeven 1905–1915.

Permission is necessary for consultation of the papers.

e. **C.J. de Bruijn Kops**

Letters 1888–1900 from D. Fock, J.H. Gunning, S.C.H. Nederburgh and R.A. van Sandick.

f. **J.L. de Bruijn Kops**

Letters 1875–1888 from J. de Clercq, A. [J.L.] Couvreur, P.H. Holtzman, J. Jacobson, I. Jacquemart, J. Kappijne van de Coppello, R.A. Kerkhoven, M.J. de Lange, J.D. Leupen, . . . van der Minne, H. Muller, N.G. Pierson, E.N. Rahusen, J.C. Reepmaker, B. Reiger, Rueb & Ledeboer, W.A. Scholten, W. Schöffer, A. Uittenhoven, W.A.P. Verkerk Pistorius, J.C. Zimmerman and from De Bruijn Kops himself concerning sugar, tea and East Indies railways.

Permission is necessary for consultation of the papers.

g. **C.J.E. van Bylandt**

Papers 1856–1862 of the Nederlandsche Maatschappij ter Bevordering van de Afschaffing van de Slavernij (Netherlands Society for the Promotion of the Abolition of Slavery).

h. **S. Dassevael**

Papers *c.* 1795–*c.* 1810 concerning the Comité van de Oost-Indische Handel en Bezittingen (Committee for the Affairs of East Indies Trade and Possessions), the Raad der Aziatische Bezittingen en Etablissementen (Council for the Asian Possessions and Establishments) and the Ministry of Colonial Affairs.

i. **F.J. Domela Nieuwenhuis**

Letters from Singapore and Bangkok when he was a diplomat there 1889–1919.

j. **H.L. Drucker**

Letters 1903 from D. Bos concerning i.a. C. Pijnacker Hordijk and East Indies affairs, letters and papers 1905–1908 from C.Th. van Deventer concerning i.a. Atjeh, papers 1906–1911 concerning i.a. the East Indies budget, papers from C.B. Nederburgh about T. de Meester as vice president of the Raad van Indië (East Indies Council).

k. **Dumonceau**

Papers 1790–1820 originating from J.B. Dumonceau including letters from H.W. Daendels, A.R. Falck, P. van der Heim, D. van Hogendorp and J.W. Janssens.

l. **D. and T.G. van Eck**

Letters 1836–1888 from i.a. J. Heemskerk Azn, [J.] Loudon and O. van Rees, photographs of i.a. [I.D.] Fransen van de Putte, [W.] van Goltstein, O. van Rees, J. van Swieten, letters 1865–1926 from i.a. L. Couperus and P.J. Veth.

m. **H.F.Chr. Forstner van Dambenoy**

Rules and regulations 1851, 1854 of the Oost- en West-Indische Opvoedings-maatschappij (East and West Indies Society for Education), letters 1849–1866 from A.F.C. List, N.C. Mulder, J.J. Rochussen, F.V.H.A. de Stuers, papers of the NHM (Dutch Trading Company).

n. **Geisweit van der Netten**

Lists of East Indies arms in the possession of the generals Van der Capellen and Nahuys 1845, s.a.

o. **P.S. Gerbrandy**

Pertaining to Indonesia correspondence 1940–1942; papers after 1945 concerning the "rijkseenheid" and about the ARP (Calvinist Party) and Amboina. For the names and subjects mentioned in the inventory see the index.

Permission is necessary for consultation of the papers.

p. **G. Groen van Prinsterer**

Papers 1826 about the Maatschappij van Weldadigheid (Society of Charity) in the East Indies, letters 1815–1876 from J.C. Baud, C. Busken Huet, G.G. Clifford, L.W.C. Keuchenius, H.G. Nahuys van Burgst, O. van Rees, J.J. Rochussen, T. Roorda.

q. **Van Hall**

Originating from F.A. van Hall report 1845 of J.C. Baud about the boundaries in Borneo with a map and annexes, papers 1860–1867 about the Billiton Company.

r. **Van Hamel**

Papers originating from J.A. van Hamel, including diaries and letters of his travels from 1914 forward; papers 1945–1962 concerning the Indonesian question. For the names and subjects mentioned in the inventory see the index.

Permission is necessary for consultation of the papers.

s. **J. Heemskerk Azn**

Letters 1853–1882 from i.a. C. Busken Huet, L. Gericke, J.J. Hasselman, J.W. van Lansberge, J.J. Pompe van Meerdervoort and J.J. Rochussen, speech for the sixth Oriëntalisten Congres (Congress of Orientalists).

t. **Heldring**

Diaries E.J. Heldring of a voyage to East Asia and the Dutch East Indies 1897–1898, papers of E.J. Heldring (president of the NHM) about the Indonesian question 1945–1950.

u. **Prins Hendrik**

Despatches 1808, 1809 from the Minister of Naval and Colonial Affairs (P. van der Heim). Deposited at the First Section.

v. **J.C. van der Hoop**

Letters 1813–1824 from A.A. Buyskes, G.A.G.Ph. van der Capellen, A.R. Falck, J. Goldberg, P. van der Heim, J.W. Janssens, J.H. van Kinsbergen, J.J. Melvill van Carnbee, P. Melvill van Carnbee, C. von Römer, C.H. Ver Huell, Ph.J. van Zuylen van Nyevelt. Deposited at the First Section.

w. **S. van Houten**

Papers 1882–1892 concerning concessions Billiton, 1887–1890 coffee cultures in Java, articles P. Brooshooft, letter 1879 I.D. Fransen van de Putte.

x. **G.W.M. Huysmans**

Papers 1946–1948 about the economic system and the Indonesian question of and/or about the Commissie-Generaal (General Committee), Linggardjati, C.P.M. Romme, International Trade Organization, Foreign Affairs, J.A.A. Schagen van Leeuwen, first "politiële actie" (military action), firm of Philips, P.F.S. Otten, W. Bruynzeel, colonial debts, rehabilitation Dutch undertakings.

y. **C. de Jong van Rodenburg**

Report concerning the project of G.M. Roentgen about steam packet service between the Netherlands and the East Indies 1824.

z. **J.C. de Jonge**

Papers concerning the navy in Asia 1796–1803 i.a. about A.A. Buyskes, E. Lucas and S. Dekker, papers concerning the navy in the East and West Indies 1831–1842. Deposited at the First Section.

aa **Lenting/De Brauw**

First draft of a report of the Kamercommissie tot onderzoek van het reglement op drukwerken in Nederlands-Indië (Parliamentary Committee for examination of the regulations for printed matter in the Dutch East Indies) 1857.

ab **H.P. Marchant**

Notes about the Djambi question 1925, the East Indies and the NSB, colonial section of the Vrijzinnig Democratische Bond (Progressive Democratic League), papers concerning H. Colijn, notes about the colonial prosperity-cooperation-policy of C. Snouck Hurgronje *c.* 1920.

ac **Middelberg**

Papers originating from C.I. Idenburg, L.R.M. Vroezon and M. Vroezon-de Haan about Indonesia *c.* 1900.

ad **Van Pestel**

Letter from lieutenant F.L.F.C. van Pestel to mrs. Tobias with a description of the surprise attack by Dajaks on the ship "Onrust" near Bandjarmasin 1859.

ae **F.E. Posthuma**

Memorandums 1916–1918 to the Minister of Colonial Affairs T.B. Pleyte about the Petkovic affair, Djambi oil fields, foundation of a colonial council, papers 1939 of the Stichting Nederlandse Volksplanting (Foundation Dutch Settlement) in Sumatra about colonization in the Dairi area, correspondence 1921–1938 as a member of the Jan Pietersz. Coen Society, member of the board of the society Nederland in den Vreemde (The Netherlands in Foreign Countries), papers about the world exhibition at Antwerp 1930 (shipping and colonies) and Chicago 1933.

af **A.G.A. van Rappard**

Letters 1820–1862 from i.a. J.C. Baud, Ch. F. Pahud, C.G.C. Reinwardt, J.J. Rochussen, Ph.F. von Siebold, J.P. van der Hucht about emigration the Netherlands – the East Indies and P. Mijer about East Indies administrative matters.

ag **Rochussen**

Autobiography by J.J. Rochussen.

ah **Roëll**

Originating from W.F. Roëll papers 1825 concerning import and export duties with a letter from Van Ouwerkerk de Vries, letters 1825–1826 De Haan, memorandums 1826–1827 of C.A. van der Hoop about the NHM, papers 1819–1820 about tax on coffee and sugar; originating from J. Roëll papers 1892 about the Billiton Company, 1892–1893 about import and export, 1894–1895 about Lombok, 1902 about government coffee cultures, the submission of Panglima Polem in Atjeh, 1905 East Indies budget, 1894–1896 Costa Rica Packet, 1896 Li Hung Chang, grand chancellor of China.

ai **C.P.M. Romme**

Papers 1947–1949, including papers about his travel in Indonesia, Ch.J.I.M. Welter, Centraal Missie Bureau (Central Mission Bureau), L.J.M. Beel, . . . Teuling, Flores, Timor, youth movement, Catholic political organizations in Indonesia, Linggardjati, socio-economic problems, Catholic press, Catholic and Protestant mission, radio, film, Commissie van Negen (Committee of Nine), the Sundanese, New Guinea, E.M.J.A. Sassen, Van Royen-Rum agreement, Round Table Conference, with *c.* 3 running metres of correspondence 1945–1952, alphabetically arranged. Permission is necessary for consultation of the papers.

aj **A.F. de Savornin Lohman**

Papers 1902–1923 concerning education in the East Indies, Christelijke Ethische Partij (Christian Ethical Party), Nederlandsch Indische Vrijzinnigen Bond (Dutch East Indies Modernist League), Koninklijke Maatschappij tot Exploitatie van Petroleumbronnen (Royal Company for the Exploitation of Petroleum Wells), colonial policy, cholera problem, Lombok harvest festivals, J.J.B. Fanoy and A.W.F. Idenburg, memorandum 1908 by J.J. van Marle about Atjeh, letters 1875–1924 from H.C. van Capelle, P.J. Elout van Soeterwoude, A. Fanoy-Wilkens, D. Fock, L.W.C. Keuchenius, C. Pijnacker Hordijk, J.R. Steinmetz. Permission is necessary for consultation of the papers.

ak **J.R.H. van Schaik**

Letters 1917–1940 from H. Colijn, [A.C.D.] de Graeff, [Ch.J.I.M.] Welter, memorandum 1935 concerning the defence of the Dutch East Indies, papers 1946–1952 concerning the Indonesian question, papers C.P.M. Romme and . . . da Costa Gomez about the police operations 1949, papers 1955–1956 about New Guinea. Permission is necessary for consultation of the papers.

al **A. and A.W. Schimmelpenninck van der Oije**

Letters 1841–1916 from J.C. Baud, G.A.G.Ph. van der Capellen, P.J. Elout van Soeterwoude, I.D. Fransen van de Putte, W. Goltstein, J. Heemskerk, K. van der Heyden, E.H. s' Jacob, L.W.C. Keuchenius, J.W. van Lansberge, J.P. van Limburg Stirum, J. Loudon, P. Mijer, O. van Rees, J.P. Sprenger van Eyk.

am **H.J. Smidt**

Papers concerning L.W.C. Keuchenius and notes about the article 'Het conflict Snouck Hurgronje-Van Heutsz-Van Daalen'.

an **D.U. Stikker**

Under the heading general and international politics: letters and other papers 1945–1951 about the Indonesian question, report development Indonesia up to 1942, RTC 1949, transfer of sovereignty 1949, publications of Foreign Affairs about the discussion of the Indonesian question in the United Nations, data about the post war situation in Indonesia; under the heading miscellaneous: texts of speeches, i.a. about Indonesia. Permission is necessary for consultation of the papers. See also D.U. Stikker. *Memoires;* herinneringen uit de lange jaren waarin ik betrokken was bij de voortdurende wereldcrisis. Rotterdam etc., 1966.

ao **B.D.H. Tellegen**

Letters 1844–1885 from I.D. Fransen van de Putte and H.N. van der Tuuk; book review of W.R. van Hoëvell's *De emancipatie der slaven in Nederlandsch-Indië.*

ap **Tets van Goudriaan**

Originating from G.J. Tets a memorandum for the Second Chamber 1872/73 about the relations Atjeh-the Netherlands since 1824; originating from D.A.W. Tets papers 1879 about the settlement of the British in Borneo, subsidizing Java-China-Japan line 1902.

aq **Thorbecke**

Originating from J.R. Thorbecke letters 1810–1872 from i.a. . . . Anemaet, Bataviaasch Genootschap van Kunsten en Wetenschappen (Batavian Society for Arts and Sciences), J.C. Baud, C. Busken Huet, [W.] Dozy, A.J. Duymaer van Twist, A.R. Falck, I.D. Fransen van de Putte, J. Heemskerk Azn, W.R. van Hoëvell, L.W.C. Keuchenius, J. Kruseman, J. Loudon, J.F.W. van Nes, [Ch.F.] Pahud, O. van Rees, W.A. van Rees, J.J. Rochussen, T. Roorda, P.P. Roorda van Eysinga, D.C. Steyn Parvé, I.J. Stieltjes, J. van Swieten, D. Veegens, P.J. Veth, papers 1835–1872 mainly drawn up and received when he was a member of the Second Chamber and Minister for Home Affairs, i.a. concerning government regulations, finance (surplus, budgets, monetary system), education (training administrative officers, education Indonesians), agriculture, cultures, hereditary lease, waste lands, industry, railways (Dutch-East Indies Railway Company, line Semarang-Surakarta-Djokjakarta), shipping, harbours, mission work, regulations for the press, memorandums H. Dekking about free labour at Rembang, advices for the tobacco manufacturers Brilman and Barkey; originating from A. Thorbecke-Solger letters from W.R. van Hoëvell. Permission is necessary for consultation of the papers.

ar **H.W. Tilanus**

Papers 1948 when he was parliamentary observer in Indonesia, papers concerning the policy of the CHU (Christian Historical Union) in connection with the Indonesian question, chronological survey of the relations between the Netherlands-Indonesia 1945–1948, papers about New Guinea 1956–1957, letters Perhimpoenan Indonesia (R.A. Simatoepang Siboerain) 1945, L. Polhaupessie 1949 and Nikijuluw 1949. Permission is necessary for consultation of the papers.

as **D. Veegens**

Papers 1879–1880 East Indies budget.

at **Vreede/Van Marle**

Originating from G.W. Vreede papers about the funeral of P.F.C. Vreede at Batavia, letters 1865–1880 from G.L. Baud, E. Douwes Dekker, J.W. van Lansberge, D.J. Mackay [van Ophemert], [H.G.] Nahuys van Burgst, J.J. Rochussen, J. Schneither, papers 1845 of the Nederlandsche Maatschappij ter Bevordering van de Afschaffing der Slavernij (Dutch Society for Promoting the Abolition of Slavery).

au **Vrolik**

Papers 1800–1913 originating from G. Vrolik, W. Vrolik, A. Vrolik and Th.C. Gevers concerning C.G.C. Reinwardt, J.L. Schröder van der Kolk, biology (Javanese lory), F.W. van Eeden, Colonial Museum at Haarlem, steamship communications the Netherlands-the East Indies, Stoomvaart Maatschappij 'Nederland' (Steamship Company 'Nederland'), Dutch East Indies Escompto Company, Oost-Indische Zee- en Brand- en Assurantie-Mij. (East Indies Marine and Fire Insurance Company), NHM, Nederlandsch Bijbelgenootschap (Dutch Bible Society) and the Council of Justice at Surabaja.

av **J.J. Westendorp Boerma**

Papers *c.* 1950–1970 drawn up and received in consequence of his studies on J. van den Bosch and the NHM.

## Miscellaneous Papers [272–273]

The papers pertaining to Asia 1796–1949, mentioned below, are arranged according to the year and number of acquisition.

### Papers deposited at the First Section [272]

a. *1861 A III, col. acq. 243 a,* report P. van der Heim for Louis Napoleon about the sending of an expedition to Batavia, 1805.

b. *1861 A III, col. acq. 243 b,* 'register der wekelijkse gagementen' (register of the weekly rates of pay) of the Ministry of Naval and Colonial Affairs, 1808.

c. *1866 A X, col. acq. 62,* copy of a memorandum by J. van den Bosch about the West Coast of Sumatra, 1833.

d. *1876, col. acq. 236,* collection of copies 17th–19th century by J.K.J. de Jonge, including a report by J.J. de Keyser and W.H. van IJsseldijk for J.F. van Reede tot de Parkeler (governor of the North Eastern part of Java) 1799, a letter from H. van Lennep concerning A. Valckenier, 1876.

e. *1877 A XVI, col. acq. 179 b,* copy of a letter from J.E. de Sturler concerning 'aangelegenheden van het Nederlandsche gouvernement met het Japansche rijk' (affairs of the Dutch government with the Japanese empire), Batavia 1826.

f. *1878 A VIII, col. acq. 196 b,* history of the first settlement [at the West Coast of Sumatra] and the present situation; by J. Boelen, 1838.

g. *1878 VIII, col. acq. 196 b,* copy of a secret memorandum 'over den tegenwoordigen toestand van het rijk Atjeh' (about the present situation of the empire of Atjeh); by W.H.R. van Loon and . . . Ritter, 1837.

h. *1879 C XXV, col. acq. 197,* copy of a daily record of Batavia, 1807.

i. *1879 C XXV, col. acq. 198,* copies of extracts from the register of the resolutions of the governor general, May–July 1811.

j. *1879 C XXV, col. acq. 199,* copies of the general resolutions of the governor general and councillors, August–September 1811.

k. *1880 A III, col. acq. 202,* 'rang-, stam- en dienstboek' (record of rank and service and roll) of the officers of the Departments of Naval and Colonial Affairs in the East and West Indies, *c.* 1809.

l. *1880 A III, col. acq. 203,* lists and inventories of the Coloniaal Militair Magazijn (Colonial Military Magazine) at Amsterdam, 1809.

m. *1880 A III, col. acq. 204,* first draft of orders of the Ministry of Naval and Colonial Affairs, 1808–1810.

n. *1880 A IV, col. acq. 201,* papers concerning private ships taken on lease by the government for the transport of troops to the Dutch East Indies, 1815–1819.

o. *1881 A I, col. acq. 212 a 1,* copy of military capitulations of the VOC to the regiments of Luxemburg, Meuron, Waltner, Württemberg, Sternbach, 1781–1801.

p. *1881 A I, col. acq. 212 a 2,* papers concerning the regiments of Luxemburg, [Meuron and Württemberg], 1782–1799.

q. *1881 A VIII,* papers and correspondence relating to the assistance to be given to the OIC by means of the '20th penny', 1792–1796.

r. *1882 B XIV, col. acq. 256,* drafts of considerations of C. Brunings concerning the 'waterstaat van Batavia en Ommelanden met betrekking tot deszelver ongezonde luchtgesteldheid' (water management of Batavia and Surrounding Districts with regard to the unhealthy state of the atmosphere) with letters from W. Six and S.C. Nederburgh, 1803–1804.

s. *1884 A VI, col. acq. 214,* papers 1791–1831 of and concerning R.G. van Polanen when a member of the Council of Justice at Batavia and of the East Indies Council, with letters from S.C. Nederburgh, D. van Hogendorp, H.W. Daendels, pamphlets and libels, and also papers about his period in Cape of Good Hope and the United States.

See also Briefwisseling van R.G. van Polanen [met onder meer J. van den Bosch], 1828–1832; medegedeeld door J.J. Westendorp Boerma. In: *BMHG;* vol. 68 (1953), p. 89–122. (See also no. 273 c).

t. *1884 A VIII, col. acq. 223 c,* letters etc. to G.A.G.Ph. van der Capellen from the committees charged with the auction offices at Surabaja and Semarang, the director in chief of finance [J.C. Goldman] and the resident of Surabaja [A.M.Th. de Salis], 1820–1821.

u. *1884 A VIII, col. acq. 223 f,* letters from J.P. Barto to H.M. de Kock, from J.W. Janssens and H.M. de Kock to [G.] von Lutzow, 1811.

v. *1884 A VIII, col. acq. 223 and 1886 A X, col. acq. 262,* copies of memorandums of J. van den Bosch for the East Indies Council about the West Coast of Sumatra, 1831.

w. *Col. acq. 7,* 'reflectiën over het bestuur van [H.W.] Daendels' (reflections about the government of [H.W.] Daendels), s.a.

x. *Col. acq. 241 a,* copy of a letter from the governor of Celebes A. van der Hart to the governor general concerning Pare-Pare and the Palos Bay, 1854.

y. *Col. acq. 241 c,* copies of political reports 1860–1862 of the governors of Celebes [A.J.F. Jansen and W.E. Kroesen].

z. *Col. acq. 241 d,* note of the assistant resident at the disposal of the governor of Celebes concerning the ownership of land by the population, s.a.

aa *1896 C,* despatches from the chief factor of the Dutch trade in Japan to the governor general at Batavia, 1820.

ab *1897 L III 3,* including 'ordonnantie bevattende de nodige orders in de burgerlijke zamenleving der Chinezen in en buiten Batavia' (ordinance containing the necessary instructions for the civil society of the Chinese in and outside of Batavia), 1802.

ac *1905 LIX 1,* pictures of a Japanese banquet, [*c.* 1800].

ad *1905 LIX 2,* Chinese papers, [*c.* 1800].

ae *1920 XXXIV 1–9 and 1935 VI 1–7,* copies from the archives of the Gereformeerde Kerk (Calvinist Church) in Malaya, *c.* 1640–1890.

af *1920 XLV 4*, papers 1696–1869 pertaining to Makasar.

ag *1948 III*, correspondence and notes of J.E. Heeres concerning the Dutch rights over Miangas [and Las Palmas], 1918–1919.

ah *1965 I 1*, papers originating from C.Ph.C.E. Steinmetz concerning the VOC, military expeditions 1829–1833 in Java, reports 1828–1833 about the Lampongs, Lampong letters 1833–1834, letters from and to the governor of the West Coast of Sumatra [A.V. Michiels], 1840–1844.

Entry: Inventory on index cards.

**Papers deposited at the Second Section** [273]

a. *1887 A XL 5 b*, journal by J. Boelen of his voyages Ghent – the East Indies, 1820–1825.

b. *1891 24 cc*, marching orders of H.W. Daendels, 1808.

c. *1895 LXV 2*, letters and notes 1817, 1832 of R.G. van Polanen. (See also no. 272 s).

d. *1897 XL*, 'algemeene rekening aantoonende de hoe-grootheid der Oost-Indische remises. . .' (general account showing the size of the East Indies remittances. . .), 1844–1848.

e. *1915 XI 4*, address by H. Coerman, member of the Comité tot de Zaken van de Oost-Indische Handel en Bezittingen (Committee for the Affairs of East Indies Trade and Possessions), to his fellow-members, 1797.

f. *1923 III 1*, nominal rolls of prisoners, hostages, State prisoners, male and female slaves taken into custody at Batavia, handed over to the Council of Justice, 1807.

g. *1923 XIV 79*, instrument of procuration of W.A.L. Romswinkel (clerk with the Orphans' Court at Batavia) for J. van Wijk at Hattem to receive money for the benefit of his mother, 1824.

h. *1931 XI 2*, memorandum about a new organization of administration and commercial policy for the former possessions of the VOC, *c.* 1806.

i. *1940 VIII*, register of copies of letters received by the Minister of Colonial Affairs G.H. Uhlenbeck, 1862–1863.

j. *1956 VI*, biography of J. Ilcken (inland waterway engineer in the East Indies), 1856–1896.

k. *1958 XII*, diary of J. van Hinsbergen as soldier in Java, 1826–1830.

l. *1959 XIX 2*, circular letters from the Minister of Naval and Colonial Affairs P. van der Heim, 1808.

m. *1961 (without a number)*, papers concerning pensions of officials in the East Indies, mainly in Pasuruan, 19th century.

n. *1963 XI,* papers 1817–1865 concerning the government in the Dutch East Indies, viz. report 1817 concerning agriculture and cultures; 'Javaansche wetten' (Javanese laws), 1818; 'algemeen verslag van het middel op het zout over 1824' (general report of the revenue of salt for 1824); note of C.M. Baumhauer about the public revenue, 1827; budget residency of South and East Borneo, 1831; 'verslag over het landbouwetablissement te Tjilangkap' (report about the agricultural establishment at Tjilangkap), 1834; 'algemeen overzicht der Indische administratie over 1837' (general survey of the East Indies accounts for 1837); extract daily record of M. Lange, comptroller at Kotta, 1843; statistics 1835–1859 Cheribon; marriages 1829–1862 Amboina; report increase of salaries of Native chiefs, 1865; report 1865 for the King; report withdrawal culture percentages and extension of the BB (East Indies Civil Service), 1865; report statute labour, 1865.

o. *1964 III,* memoirs of D. Fock (1858–1941).

p. *1968 319,* photographic copies of letters 1873 by J.M. Wap about the Atjeh expedition and the East Indies press.

q. *1968 XVIII 1 and 2,* correspondence of B. ter Haar, 1936–1937; copies of *Asia Raya,* 1945.

r. *1974 IV,* papers of the auditor in the court martial of the West Coast of Sumatra at Padang 1889.

Entry: Chronological list of miscellaneous acquisitions.

## Maps and Marine Charts [274]

The collection (*c.* 55,000 pieces) consists of maps and atlases originating from the various archives and collections deposited at the ARA ('archiefkaarten' (archival maps)) and of maps and atlases added to the collection as miscellaneous acquisitions. The maps belonging to the archives of the Ministry of Colonial Affairs are still partly without an inventory. The collection has the following entries:

a. [P.A. Leupe]. *Inventaris der verzameling kaarten berustende in het Rijks-Archief.* 's-Gravenhage, 1867. With a register in typescript. Of the *c.* 700 maps and charts concerning Asia and Oceania *c.* 30 were published after 1796.

b. [S.P. Honoré Naber]. *Inventaris der verzameling kaarten berustende in het Algemeen Rijksarchief;* eerste supplement [op de inventaris door P.A. Leupe, 1867]. 's-Gravenhage, 1914. With registers. Including *c.* 200 maps concerning the Indonesian archipelago, subdivided into the following headings: Sumatra, Bintang, Riouw, Banka, Billiton, Borneo, Strait of Gaspar, Strait of Sunda, Java, Bantam, Batavia, Preanger, Krawang, Cheribon, Tegal, Pekalongan, Central Java, Eastern Java, Madura, Timor, Lesser Sunda Islands, Celebes, Moluccas, Amboina, Ceram, Banda, New Guinea; *c.* 50 maps concerning Asia and Oceania, subdivided into: Malabar, Ceylon, Coromandel, Bengal, Arakan, Pegu, Malaya, Siam, Cambodia.

c. [P.A. Leupe]. *Catalogus der verzameling van kaarten van het Ministerie van Marine*. 's-Gravenhage, 1872. Register is forthcoming. Included are *c.* 400 marine charts concerning the Indonesian archipelago, subdivided into: Sumatra, Riouw, Strait of Banka, Strait of Gaspar, Strait of Sunda, Java Sea, Java, Borneo, Celebes, Strait of Makasar, Amboina, New Guinea; *c.* 750 marine charts concerning Asia and Oceania, subdivided into: Indian Ocean, Persian Gulf, Malabar, Maldives, Lakedives, Ceylon, Bengal, Andaman and Nicobar Islands, St. Paulus, Amsterdam, Cocos Islands, Philippines, Sulu Archipelago, Strait of Malaya, Singapore, Durian, Tambalan, Anambas, Natuna Islands, Condor Islands, Chinese Sea, Siam, Cochin China, Cambodia, China, Japan, Korea, Tartar Gulf, Australia, Bass Strait, Van Diemensland, Torres Strait, Coral Sea, New Zealand, Pacific, New Caledonia, Loyalty Islands, New Hebrids, Louisiade, New Britain, Solomon Islands, Carolina Islands, Maria Islands, Tonga, Samoa, Hawạii Archipelago, Viti Archipelago, Christmas Islands, Marquesas Islands, Galapagos Islands, Gambier Group, Auckland Group.

d. Catalogue on index cards (alphabetically to names of places), mentioning *c.* 220 maps concerning the Indonesian archipelago and *c.* 80 concerning Asia and Oceania. For the names of the places see the index.

# CHAPTER II

# EERSTE KAMER DER STATEN-GENERAAL (FIRST CHAMBER OF THE STATES GENERAL)

## Eerste Kamer der Staten-Generaal

21–23 Binnenhof, 2513 AA The Hague.
Admission by appointment.
Research finished in 1972.

**Archief van de Eerste Kamer (Archives of the First Chamber) 1814, 1815–** [275]

The right of legislation is vested in the King and the Staten-Generaal (States General). According to the constitution of 1814 the States General were composed of 55 members, elected by the Provinciale Staten (Provincial States), their principal right being the voting of the annual State budget. By the constitution of 1815 two chambers were established (see also no. 6). The First Chamber was composed of 40–60 members, appointed by the King for life. According to the constitution of 1840 after the secession of Belgium this number became 20–30. The constitution of 1848 provided that the 39 members should be elected by the Provincial States from the highest rate payers for direct tax, for the period of 9 years. One third would retire once every 3 years. The rights of the First Chamber included the right of interpellation. Their meetings became public. By the constitution of 1887 the number was fixed at 50, and besides the highest rate payers now also those persons who had occupied higher public offices in the past or present could become a member. They were also granted the right of instituting an inquiry. By the constitution of 1917 the requirements for eligibility were made similar to those for the Second Chamber. The constitution of 1922 provided the principle of proportional representation also for the First Chamber, the session became 6 years, every 3 years one half of the members retired. From 1940–1945 the States General did not meet. In 1956 the number of the members was fixed at 75. The First Chamber may only pass or reject a resolution which has been voted (and in some cases amended) by the Second Chamber. (G.W. Bannier. *Grondwetten van Nederland*. Zwolle, 1936. W.F. Wijthoff. *De Staatsinrichting van Nederland*. Haarlem, 1957).

The archives consist of a series of '(registers) notulen' ((registers) minutes) from 1814, which are 'minuut-notulen' (fair copies of minutes) copied out, comparable to the proceedings. There is also a series of 'minuut-notulen' from 1814, containing a kind of 'guidance' to be used by the president of the

Assembly, and the papers received by the Chamber, whether in print or not. The larger portion of these are printed as annexes to the proceedings, but the series of 'minuut-notulen' contains also non published papers such as letters from the members and petitions. The series 1815–1830 are in Dutch and French. The papers up to 1848 were not published, with the exception of the papers of 1814, when the States General consisted of only one Chamber, and a 'kort verslag van het verhandelde in de Eerste Kamer' (brief report of the proceedings in the First Chamber) 1827/28–1845/46. The larger portion of the papers after 1848 was printed in the *Verslag der handelingen van de Staten-Generaal.* (see no. 6).

Entry: Register to the printed proceedings.

Size: '(Registers) notulen' 1814–1950 *c.* 16 running metres, 'minuut-notulen' 1814–1950 *c.* 32 running metres.

Permission is necessary for consultation of the papers.

# CHAPTER III

# MINISTRIES

## Ministerie van Binnenlandse Zaken (Ministry of Home Affairs)

1 Plein, 2511 CR The Hague.
The archives of Colonial Affairs are deposited at 125 Leeghwaterstraat, The Hague.
Permission in writing of the Ministry is necessary for access.
Photocopying equipment available.
Research finished in 1971.

**Ministerie van Koloniën (Ministry of Colonial Affairs), 1901–1959, 1959–1963** [276]

This part of the archives is the continuation of the part which is deposited at the ARA (see no. 25, 26). From 1945 the Ministry of Colonial Affairs was called Ministerie van Overzeese Gebiedsdelen (Ministry of Overseas Territories), from 1949 Ministerie voor Uniezaken en Overzeese Rijksdelen (Ministry of Union Affairs and Overseas Dominions), from 1952 Ministerie voor Overzeese Rijksdelen (Ministry of Overseas Dominions) and from 1957 Ministerie voor Zaken Overzee (Ministry for Overseas Affairs). From 1959 up to 1963 New Guinea belonged to the Ministry of Home Affairs; Surinam and the Netherlands Antilles were added to the Cabinet of the vice prime minister. The archives fall into two parts, viz. the periods 1901–1940 and 1940–1963.

A. *Period 1901–1940*

This part is divided into:

a. Non secret "verbaal" (minutes) with postbooks, alphabetic and systematic indexes (from 1922 only alphabetic and systematic indexes of departments).
b. Secret/cabinet "verbaal" (minutes) with postbooks and alphabetic indexes.
c. Mail reports with lists of mail reports (from 1914 divided into a secret and non secret series).
d. Archives of the Commissariaat voor Indische/Indonesische Zaken 1927–1940 (Commissariat for East Indies/Indonesian Affairs 1927–1940). These archives consist mainly of papers concerning the sending and pay of officials on leave and of papers concerning the purchase of goods for the Indies. The archives are arranged according to the series of minutes and have postbooks serving for entries (not accessible to subject because of the absence of indexes).

B. *Period 1940–1963*

This part is divided into:

a. "Londens archief" (London archives). These archives were formed during the second world war by the Ministry of Colonial Affairs of the Dutch government in exile at London. They consist mainly of bundles of papers 1940–1945, arranged according to heading, concerning the policy of the government respecting the colonies, with postbooks and an inventory list to subject serving for entries. The papers drawn up in the Netherlands during this period are of little importance for the conduct of affairs relating to the colonies.

b. Non secret "verbaal" (minutes) 1940–*c.*1950 with postbooks, systematic and alphabetic indexes, non secret file *c.*1950–1963 with postbooks and file inventory. From 1950 department by department adopted the file system. From that time forward the classification system was no longer based on the chronology of date and number of the resolution on the matter dealt with, but on the subject in hand.

c. Secret/cabinet "verbaal" (minutes) 1940–1954 with postbooks, systematic and alphabetic indexes, secret file 1955–1963 with postbooks and file inventory.

d. Mail reports up to 1953 with mail report lists.

e. Archives of the Commissariat for East Indies/Indonesian Affairs up to 1949 with postbooks.

Size: *C.* 2700 running metres. However, the archives are in course of preparation so that a final statement about the size is impossible.

For consultation of papers more recent than 50 years permission in writing is necessary.

For the remaining archives belonging to the Ministry of Colonial Affairs see no. 18–37, 277.

**Overzeese Pensioenen (Overseas Pensions), c. 1830–** [277]

This collection consists of records of service, card index systems, personal files etc. drawn up in connection with payment of pensions to KNIL soldiers. After 1949 the papers were sent over to the Netherlands.

Entry: Alphabetic indexes.

Size: It is impossible to give an exact statement because of the dissimilarity of the documents, probably several hundred running metres.

See also no. 27.

# Ministerie van Buitenlandse Zaken (Ministry of Foreign Affairs)

23 Plein, 2511 CS The Hague. Central administration old archives 16 Casuariestraat, 2511 VB The Hague. "Statisch archief" (old archives) Fluwelen Burgwal, The Hague.
Admission by appointment.
Photocopying equipment available.
Research finished in 1972.

## Introduction to the archives [278]

The archives, which are deposited at the Ministry of Foreign Affairs, consist of papers from 1871 up to the present. They are the continuation of that part of the archives which are deposited at the ARA (see no. 38–68). The papers up to 1918 are gradually being transferred to the ARA. For this survey they are still included among the archives which are deposited at the Ministry of Foreign Affairs.

The archives are divided into:

The archives of the Ministry (see no. 279–282).

The archives of the diplomatic posts, sent over by the Dutch representations in foreign countries to the Ministry (see no. 283–321).

The archives of committees (see no. 322–325).

The collections originating from various persons (see no. 326–335).

Entry: *Overzicht van de archieven van het Ministerie van Buitenlandse Zaken en de overgezonden archieven van de Nederlandse vertegenwoordigingen in het buitenland.* The central administration of the old archives contains all the inventories which have been made for the various parts of the archives.

Permission in writing is necessary for consultation of papers less than 30 years old.

See also C. Smit. *Bescheiden betreffende de buitenlandse politiek van Nederland;* uitgegeven door de Rijks Geschiedkundige Publicatiën; grote serie. (RGP no. 100: 1899–1903; no. 102: 1903–1907; no. 106: 1907–1914; no. 109: 1914–1917; no. 116, 117: 1917–1919; no. 128: 1899–1914 foreign sources; no. 137: 1914–1917 foreign sources).

J. Woltring. *Bescheiden betreffende de buitenlandse politiek van Nederland;* uitgegeven door de Rijks Geschiedkundige Publicatiën; grote serie. (RGP no. 107: 1871–1874; no. 118: 1874–1880; no. 122: 1881–1885; no. 126: 1886–1890; no. 132: 1891–1894; no. 138: 1895–1898).

## Ministrial archives [279–282]

These archives consist of the "statisch archief" (old archives) 1871–1945 (see no. 279–281) and the "semi static" and "dynamic" archives 1945 up to the present (no. 282). From the old archives the most important archives for the history of Asia and Oceania are the kabinetsarchief (archives of the cabinet) (no. 279), the groot archief (larger archives) (no. 280) and the Londens archief (London archives) (no. 281).

Entry: In addition to the inventories to the various archives (see there), there are also postbooks, alphabetic indexes and card indexes to all the old archives 1871–1940.

### Kabinetsarchief (Archives of the Cabinet), 1871–1940 [279]

These archives consist of two parts:

a. Political reports 1871–1940 of Dutch diplomatic representatives in foreign countries, with the political information of consular officials in countries where no embassy was established. For the names and subjects mentioned in the inventory see the index.

   Entry: Inventory.

   Size: *C.* 35 running metres (812 inventory items).

b. Files of the Cabinet section 1871–1940, classed under lettered and unlettered headings. The lettered headings include "koninklijke reizen en bezoeken" (royal travels and visits) of i.a. Asian princes to the Netherlands and Indonesia, "buitenlandse gezantschapsaangelegenheden" (foreign embassy affairs) i.a. China, Japan, Persia, Siam, "koloniale aangelegenheden van vreemde natiën" (colonial affairs of foreign nations), "varia" (miscellanies) i.a. Persia, "rapporten over Chinese zaken" (reports about Chinese affairs), "marinezaken" (naval affairs), "oorlog" (war) i.a. China and Japan, "aangelegenheden betreffende Nederlands-Indië" (matters concerning the Dutch East Indies), running ashore of the Nisero 1884, Sumatra treaty 1871. The unlettered headings include letters of credence China, treaty of friendship the Netherlands-Afghanistan.

   Entry: List of headings.

   Size: *C.* 65 running metres.

### Groot archief (Larger archives), 1871–1940 [280]

These archives consist of three parts:

a. A-files, 1871–1940, containing legal, political and diplomatic papers. The A-files up to 1918 have been transferred to the ARA (see no. 39). For the names and subjects mentioned in the inventory see the index.

   Entry: Inventory. New inventory in preparation.

   Size from 1871: *C.* 134 running metres (281 file nos.)

b. B-files, 1871–1940, containing papers about consular and commercial affairs. For the names and subjects mentioned in the inventory see the index.

   Entry: Inventory.

   Size: *C.* 84 running metres (275 file nos.)

c. DEZ-files 1919–1940 originating from the Direction of Economic Affairs, up to 1921 arranged according to "zaken-dossiers" (a case file system), from 1921 arranged according to a heading system. For the names and subjects mentioned in the inventory see the index.

   Entries: Inventories 1919–1920, 1921–1923 and 1924–1940.

   Size: *C.* 90 running metres (respectively 4071 case files, 330 and 339 headings).

**Londens archief (London archives), 1940–1945** [281]

These archives originated in London during the second world war, where the Ministry of Foreign Affairs of the Dutch government in exile had its headquarters.

These archives consist of two parts:

a. Ordinary or larger archives, subdivided into diplomatic affairs, political affairs, bureau Germany, economic affairs, general affairs, legal affairs, cabinet, general secretariat and the committee archives of the Comité Interallié pour l'Etude de l'Armistice. For the names and subjects mentioned in the inventory see the index.

   Entry: Inventory.

b. Secret archives, subdivided into diplomatic affairs, legal affairs, consular and commercial affairs, general secretariat, safe files. For the names and subjects mentioned in the inventory see the index.

   Entry: Inventory.

Total size of the London archives: *C.* 105 running metres.

**Semi-statisch en dynamisch archief (Semi static and dynamic archives), 1945–** [282]

These archives have been arranged according to the file system. The files are filed away according to a decimal classification. The ten main classes are: 0 "algemeen" (general); 1 "organisme" (organism); 2 "protocol" (protocol); 3 "juridische aangelegenheden" (legal affairs); 4 "aangelegenheden van openbare orde, rust, veiligheid, zedelijkheid, volksgezondheid" (matters of public order, peace, safety, morality, public health); 5 "waterstaat, verkeer en vervoer, verbindingen en electronica" (water management, traffic and transport, communications and electronics); 6 "economie en arbeid" (economics and labour); 7 "maatschappelijke steun en voorzorg" (social assistance and precaution); 8 "culturele en wetenschappelijke aangelegenheden, voorlichting" (cultural and scientific matters, information); 9 "staatkunde, landsverdediging, internationale organisaties" (politics, national defence, international organizations). These main classes are again subdivided on a decimal basis. Each subject is marked by a code number. For each subject an inventory of the included files is kept. All the archives are subdivided into units of 10 years, so: 1945–1954, 1955–1964 etc.

Entries: Archives code of the filing plan; inventories according to code number; moreover the usual entries such as postbooks and card indexes.

Size: *C.* 5165 running metres (3431 semi static).

Permission in writing is necessary for consultation of papers less than 30 years old.

## Legations and consulates [283–321]

Included are the archives of the Dutch representations in foreign countries.

Present are the archives of posts in Argentina, Australia, Australian New Guinea, Belgium, Belgian Congo, Bolivia, Brazil, Bulgaria, Burma, Canada, Ceylon, Chili, China, Columbia, Costa Rica, Cuba, Denmark, Germany, Ecuador, Egypt, Ethiopia, Finland, Formosa, France, French territories in Africa, French territories in the Pacific, Greece, Hungary, Ireland, India, Indonesia, Iran, Israel, Italy, Japan, Libanon, Liberia, Luxemburg, Morocco, Mexico, NATO and OEEC, New Guinea, New Zealand, Norway, Austria, Pakistan, Peru, Philippines, Poland, Portugal, Rumania, Russia, Saudi Arabia, Singapore, Spain, Thailand, Trinidad and Tobago, Czechoslovakia, Turkey, Uruguay, Vatican, Venezuela, United Kingdom, British territories in Asia, British territories in Africa, United Nations, United States, Vietnam, South Africa, Yugoslavia, Sweden and Switzerland.

Many papers (especially those of a general political importance) may also be found in the ministrial archives. Papers of local importance, however, are often only to be found in the "postenarchieven" (archives of the legations and consulates). The archives most relevant for the history of Asia and Oceania up to 1949 are mentioned below.

### Argentina, 1911–1954 [283]

These archives consist of two parts: archives of the legation 1911–1944 and archives of the legation and embassy 1945–1954 to Buenos Aires. The first part, with table of contents, includes concerning Asia: tea seed for Buitenzorg 1922–1925; East and West Indies export to Argentina 1934; Japanese emigration Paraguay 1935; trade of the Dutch East Indies with Argentina 1939–1941; Japanese fishermen in the Indonesian archipelago 1937, 1938; defence of the Dutch East Indies 1940; war with Japan 1941–1942; Japanese prisoners of war 1943–1944; East Indies officials and medical service 1942–1944.

Entry: Table of contents 1911–1944.

Size: 1911–1944 *c.* 6.50 running metres, 1945–1954 *c.* 11 running metres.

**Australia** [284]

Relevant for the history of Asia and Oceania up to 1949 are:

a. *Canberra (legation), 1940–1946*
Before 1940 no archives, for the period 1940–1946 papers concerning consulates at Adelaîde, Melbourne and Perth, . . . Hobart, personnel, shipping, second world war, Manilla, hadji's, Djeddah, Javanese coolies in New Caledonia, Red Cross, education, New Zealand, Linggardjati, Indonesian question 1945–1949, New Guinea, Papua, Malino conference, communism, agriculture, fishing industry, stockbreeding industry Anzac Pact, military affairs, Netherland-Indies Women's Army.

Entry: Table of contents.

Size: *C.* 2.50 running metres.

b. *Brisbane (consulate), 1940–1959*
Most of the papers concern the period after 1949, papers of the preceding period include letters [J.] Donaldson 1943–1948, N.S. Pixley 1944–1958, accountancy papers, minutes 1940–1945, Dutch association in Brisbane, 1943–1946.

Entry: Summary table of contents.

Size: *C.* 1.50 running metres.

c. *Melbourne (consulate), 1924–1945; (consulate general), 1945–1957*
The large portion of the archives consists of papers before 1949. For the names and subjects mentioned in the tables of contents see the index.

Entries: Tables of contents 1924–1945 and 1945–1957.

Size: 1924–1945 *c.* 0.50 running metre, 1945–1957 *c.* 0.20 running metre.

d. *Sydney (consulate general), 1930–1954*
For the names and subjects mentioned in the table of contents see the index.

Entry: Table of contents.

Size: *C.* 11 running metres.

**Belgium** [285]

For the history of Asia and Oceania up to 1949 only the archives of the legation/embassy to Brussels 1875–1954 are relevant. For the names and subjects mentioned in the inventories see the index.

Entries: Postbooks, table of contents 1875–1923, inventories 1920–1940, 1945–1954, secret archives 1945–1954.

Size: 1875–1923 *c.* 20 running metres, 1920–1954 *c.* 43.50 running metres.

See also no. 40 a, 41.

**Brazil** [286]

Some papers in the archives of the legation/embassy to Rio de Janeiro 1910–1955, viz. trade with the Dutch East Indies 1934–1940, Indonesia 1946–1954, Iran 1946–1954.

Entry: Table of contents 1910–1945, 1945–1955.

Size: 1910–1945 *c.* 6 running metres, 1945–1955 *c.* 2 running metres.

**Burma** [287]

The archives of the legation/embassy to Rangoon (1945–1969) mainly consist of papers after 1949. From the period before 1949 papers concerning the establishment of the legation, protocol etc., in the secret archives also papers Attlee-agreement, Russia, Pakistan, New Guinea, Indonesia, ICAO, Far Eastern Commission, ECOSOC, UN, Asian Relations Conference, Boguio Conference, economic affairs, communism, Indo China, RTC, Fox contract.

Entries: Tables of contents non secret and secret.

Size: *C.* 13 running metres.

**Canada** [288]

The following archives are relevant:

a. *Ottawa (legation/embassy), 1939–1954*
Including papers Indonesia during the second world war, prisoners of war Japan 1943–1944, Indonesia in the Pacific 1943, Java Bank 1943, Colombo project, Indonesian question 1945–1949, Korea, Commonwealth.

Entries: Tables of contents 1939–1945, 1945–1954.

Size: 1939–1945 *c.* 5 running metres, 1945–1954 *c.* 7 running metres.

b. *Montreal (consulate general), 1924–1946*
Including papers Dutch East Indies 1929–1940, Sinologisch Instituut (Institute for Sinology) 1936, China 1929, Japan 1928–1939, New Zealand 1939, Institute for Pacific Relations 1942–1943, annual fair Bandung 1932–1938, women's force KNIL 1944–1945, Japan 1933–1941, personal files.

Entry: Tables of contents 1924–1946 and personal files.

Size: *C.* 14 running metres.

c. *Toronto (consulate), 1929–1944*
Including papers women's force KNIL 1944.

Entry: Table of contents.

Size: *C.* 1 running metre.

d. *Vancouver (consulate general), 1924–1945*
Including papers women's force KNIL 1943.

Entry: Table of contents.

Size: *C.* 0.30 running metre.

**Ceylon** [289]

The archives of the consulate of Colombo consist of two parts, the periods 1940–1945 and 1945–1953. For the names and subjects mentioned in the tables of contents see the index.

Entries: Tables of contents 1940–1945 and 1945–1953.

Size: *C.* 0.50 running metre.

**China** [290]

These archives consist of the following parts:

a. *Peking (legation/embassy), 1877–1954*
In 1941 the legation was removed to Chungking, in 1943 erected into an embassy. This embassy was transferred to Nangking in 1946, and from 1950 re-established in Peking.

These archives also contain papers concerning the protection safeguarding of the interests of Germany, Hungary and Austria 1917–1923. For the names and subjects mentioned in the inventory see the index.

Entry: Inventory.

Size: *C.* 62 running metres.

b. *Amoy (consulate), 1874–1894, 1928–1949*
The inventory for these archives only covers the first period, including correspondence 1874–1894, correspondence with Peking 1890–1891, with the governor general of the Dutch East Indies 1890–1891, with consular officials of the contracting powers in China 1890–1891, with trading houses and private persons 1890–1891, instructions about the admission of Chinese and foreign Orientals into the Dutch East Indies 1891–1893. In 1895 Amoy came under the control of Hongkong (see no. 315).

Entry: Inventory 1874–1894.

Size: 1874–1894 *c.* 0.70 running metre (20 inventory items), 1928–1949 *c.* 0.40 running metre.

c. *Canton (consulate), 1929–1941, 1943–1951*
The consulate was closed in 1941 as a result of the Japanese advance, in 1943 re-opened in Kunming, then again in Canton and finally abolished in 1951. For the names and subjects mentioned in the inventory see the index.

Entry: Inventory.

Size: *C.* 2.50 running metres.

d. *Foochow (consulate), 1948–1949*
This small amount of archives consists of two files concerning visa.

Entry: Table of contents.

Size: *C.* 0.10 running metre.

e. *Peking (office of the doyen), 1891–1943*
For the names and subjects mentioned in the table of contents see the index.

Entry: Table of contents.

Size: *C.* 16 running metres.

f. *Shanghai (consulate general), 1874–1941, 1945–1951*
In 1941 the consulate was closed as a result of the second world war, the Swiss consulate protected the Dutch interests then, in 1945 the consulate was re-opened, in 1951 abolished. For the names and subjects mentioned in the inventory see the index.

Entry: Inventory.

Size: *C.* 56.50 running metres.

g. *Swatow (consulate), 1947–1949*
This small amount of archives consists of letters received and despatched 1947–1949.

Size: *C.* 0.10 running metre.

h. *Tientsin (consulate), 1905–1951*
These archives consist of legal and notarial papers and papers pertaining to various persons, companies and subjects.

Entry: Table of contents.

Size: *C.* 0.75 running metre.

See also no. 63.

**Egypt** [291]

These archives consist of the following parts:

a. *Alexandria (consulate general), 1861–1885; (consulate), 1885–1945*
In 1884 the consulate general was removed to Cairo. A honorary consul continued the activities in Alexandria. The archives 1885–1914 are destroyed. For the names and subjects mentioned in the inventory see the index.

Size: 1861–1885 *c.* 1 running metre, 1915–1945 *c.* 5 running metres.

b. *Cairo (consulate), 1850–1949; (consulate general), 1876–1921; (legation), 1922–1955*
From 1884 diplomatic affairs were managed by the station in Cairo. Besides the archives mentioned above there is also a small amount of archives kept

separately, of the protection of the Austrian and Hungarian interests. For the names and subjects mentioned in the inventory see the index.

Size: *C.* 25 running metres.

c. *Ismailia (vice consulate), 1879–1889*
The vice consulate was founded here in connection with the establishment of the Suez Canal Company, and abolished in 1889 when the Company was removed to Port Said. The small amount of archives consists of correspondence with various consulates.

Size: *C.* 0.10 running metre.

Entry to the total amount of the archives: Inventory.

**Formosa** [292]

The archives of the consulate of Taihoku 1924–1941 consist mainly of correspondence, papers concerning visa, instructions etc.

Size: *C.* 1 running metre.

**France** [293]

Relevant for the history of Asia and Oceania up to 1949 are:

a. *Paris (legation), 1870–1931*
For the names and subjects mentioned in the inventory see the index.

Entry: Inventory.

Size: *C.* 32 running metres.

b. *Paris (legation), 1887–1940*
Included are papers concerning Chinese debts 1931–1939, pensions Indo China 1937, consulates Saigon 1888–1940, Hanoi 1934–1935, Sydney, New Hebrides, New Caledonia, Noumea 1932–1939, reports on the Dutch East Indies by Titayna 1934–1935, Russians in the Dutch East Indies, communism in the Dutch East Indies, Association France-Java.

Entry: Table of contents.

Size: *C.* 3 running metres.

c. *Paris/Algiers (legation), 1942–1945*
Included are reports of the Dutch representation from Chungking 1943–1944, papers Indo China 1944.

Entry: Table of contents.

Size: *C.* 0.30 running metre.

d. *Paris (legation secret archives), 1945–1951*
For the names and subjects mentioned in the table of contents see the index.

Entry: Table of contents.

Size: *C.* 3.50 running metres.

e. *Paris (legation), 1945–1954*
For the names and subjects mentioned in the table of contents see the index.

Entry: Table of contents.

Size: *C.* 9 running metres.

See also no. 40 a, 44.

**French territories in the Pacific** [294]

This part consists of the archives of the vice consulate in Noumea (1934–1952). In 1952 the consulate was abolished and its tasks were performed by the consulate general at Sydney. For the names and subjects mentioned in the inventory see the index.

Entry: Inventory.

Size: *C.* 1 running metre (88 inventory items).

**Greece** [295]

The archives of the legation of Athens 1834–1954 contain some papers concerning the Dutch East Indies, viz. consular convention 1896, 1905, 1908 and papers concerning the Bandung quinine factory, 1915–1920.

Entry: Table of contents.

Size: *C.* 12 running metres.

**India** [296]

This part consists of:

a. *New Delhi (legation), 1901–1955*
For the names and subjects mentioned in the table of contents see the index.

Entry: Table of contents 1922–1956.

Size: *C.* 14 running metres.

b. *Bombay (consulate general), 1940–1961*
This small amount of archives consists for the period up to 1949 of secret files 1940–1949 of the consul general at Simla, and also a list Turkish archives handed over to Afghan consulate at Bombay on 25–5–1931.

Entry: Summary table of contents 1940–1950, table of contents 1951–1961.

Size: 1940–1950 *c.* 0.10 running metre, 1951–1961 *c.* 0.10 running metre.

**Indonesia** [297]

For the period up to 1949 are only relevant the archives of the Directie Verre Oosten (Direction Far East) (DIRVO, 1946–1952) at Djakarta. For the names and subjects mentioned in the table of contents see the index.

Entry: Table of contents.

Size: *C.* 30 running metres.

**Iran** [298]

This part consists of:

a. *Teheran (legation/embassy), 1900–1956*
For the names and subjects mentioned in the tables of contents see the index.

Entries: Tables of contents 1900–1945 and 1945–1956. Inventory is forthcoming.

Size: 1900–1945 *c.* 5.50 running metres, 1945–1956 *c.* 3.50 running metres.

b. *Buchire (consulate), 1911–1934*

Size: *C.* 0.20 running metre.

See also no. 48.

**Israel** [299]

The archives of the consulate general at Jerusalem (1932–1945) contain some papers concerning the Dutch East Indies (banknotes, companies, pensions, KNIL) and concerning Iran.

Entry: Table of contents.

Size: *C.* 2 running metres.

**Italy** [300]

Relevant for the history of Asia and Oceania up to 1949 are the archives of the legation in Rome (1911–1940). For the names and subjects mentioned in the table of contents see the index.

Entry: Table of contents.

Size: *C.* 10 running metres.

**Japan** [301]

These archives consist of:

a. *Tokio (legation/embassy), 1923–1954*
The archives of the legation 1890–1922 have been destroyed by fire. For the names and subjects mentioned in the tables of contents see the index.

Entries: Tables of contents 1923–1941 and 1945–1954.

Size: *C.* 13 running metres (1923–1941) and 2.50 running metres (1945–1954).

b. *Kobe (consulate general), 1918–1941, 1873–1966*
These archives consist of two parts. For the names and subjects mentioned in the inventory 1918–1941 see the index. The part 1873–1966 includes sentences

Osaka-Hiogo 1876–1879, notarial deeds 1923–1933, correspondence 1874–1897, alphabetic indexes, passports, postbooks.

Entries: Inventory 1918–1941, table of contents 1873–1966.

Size: 1918–1941 *c.* 2 running metres (160 inventory items), 1873–1966 *c.* 2 running metres.

c. *Tokio (Dutch military mission), 1945–1954*
For the names and subjects mentioned in the table of contents see the index.

Entry: Table of contents.

Size: *C.* 12 running metres.

d. *Tokio, Reparatie en Restitutie (Reparation and Restitution), 1945–1952*
Included are papers India, personnel, . . . van Vreeswijk, . . . Kroese, economic records Japan, NHM, oil industry, steel works, whale fishery, shipping, banking, war damages, Java Bank, Nederlandsch-Indische Escompto Mij. (Dutch East Indies Discount Company), tin, rubber, opium, diamond, RTAC, RAC.

Entry: Table of contents.

Size: *C.* 6 running metres.

See also no. 47, 62, 67.

**Mexico** [302]

The archives of the legation/embassy to Mexico 1925–1956 include treaty Mexico-Japan 1926, disturbances in Java 1928, extra territoriality China 1936, politics Dutch East Indies 1938, files Indonesia 1941–1956 "politiële acties" (military actions), communism, military affairs.

Entries: Tables of contents 1925–1939 and 1941–1956.

Size: 1925–1939 *c.* 1.50 running metres, 1941–1956 *c.* 1.50 running metres.

**New Zealand** [303]

This part consists of the legation archives Wellington, subdivided into the periods 1914–1944 and 1945–1954. For the names and subjects mentioned in the tables of contents see the index.

Entries: Tables of contents 1914–1944 and 1945–1954.

Size: 1914–1944 *c.* 1.50 running metres, 1945–1954 *c.* 4.50 running metres.

**Norway** [304]

The legation/embassy archives Oslo 1917–1954 include papers Norwegian interests Amoy 1930–1931, disturbances in Java 1926–1930, contraband trade of arms to Asia 1927–1928, voyage A. Mohr to the Dutch East Indies 1934–1936, Norway-China 1949–1954, Norway-Indonesia 1949–1953.

Entries: Tables of contents 1917–1940 and 1945–1954.

Size: 1917–1940 *c.* 1.50 running metres, 1945–1954 *c.* 3.50 running metres.

See also no. 40 c, 40 p, 42, 55, 66, 320.

**Pakistan** [305]

This part consists of the legation archives of Karachi 1945–1954. The archives of the consulate before 1945 have been destroyed. For the names and subjects mentioned in the table of contents see the index.

Entry: Table of contents.

Size: *C.* 2.50 running metres.

**Philippines** [306]

This part consists of the archives of the legation/embassy Manilla 1939, 1945–1961. For the names and subjects mentioned in the table of contents see the index.

Entry: Table of contents.

Size: *C.* 2 running metres.

**Portugal** [307]

This part consists of:

a. *Lisbon (legation), 1895–1944*
   Included are papers King of Siam 1897, Portugal-China 1899, Timor 1899–1944, slavery in the Portuguese colonies 1909–1925, rebellion Timor 1908–1919, Macao 1926–1927, transportation Portuguese revolutionaries to Timor 1928–1932, sale Portuguese colonies 1932–1937.

   Entry: Table of contents.

   Size: *C.* 3 running metres.

b. *Lisbon (government committee for the Dutch refugees in Portugal), 1940–1945*
   These archives consist of personal files, including those of Dutch refugees to the Dutch East Indies.

   Entry: Table of contents.

   Size: *C.* 4 running metres.

See also no. 40 j, 49, 60.

**Russia** [308]

The archives of the embassy to Moscow 1943–1955 include papers shipping China 1945 and peace treaty Japan 1951.

Entry: Table of contents.

Size: *C.* 10 running metres.

See also no. 40 l, 51.

**Saudi Arabia** [309]

This part consists of:

a. *Djeddah (consulate general), 1873–1950*
For the names and subjects mentioned in the inventory see the index.

Entry: Inventory to the archives of Djeddah and Mecca.

Size: *C.* 13.50 running metres (432 inventory items).

b. *Mecca (vice consulate), 1939–1950*
Included are papers Raden Ma'mun al Rasjid Kusumadilaga 1948–1950, Amanulah Sutaatmadjo 1947–1950, H. Mohamad Rasjad Tjitroatmodjo 1946–1948, shipments of money from Indonesia 1947, Raden Adjeng Fatmah 1948–1949, H. Junus bin Umar Sjeich 1939–1948, J.F. Everaerts 1947–1948.

Entry: Inventory of the archives of Djeddah and Mecca.

Size: *C.* 0.30 running metre (24 inventory items).

**Singapore (consulate general), 1946–1954** [310]

Included are, for the period up to 1949, survey of organizations of Javanese in South East Asia, interrogation reports NEFIS, Press Intelligence reports, trade, rubber, fishery, shipping, rice, coal, India, aviation, taxes, economics, palm oil, personnel files.

Entry: Table of contents.

Size: *C.* 3 running metres.

See also no. 65.

**Spain** [311]

In the archives of the legation/embassy to Madrid papers 1893–1954. The archives 1930–1940 have been destroyed. For the names and subjects mentioned in the inventories see the index.

Entries: Inventories 1893–1930, 1940–1954.

Size: 1893–1930 *c.* 5 running metres, 1940–1954 *c.* 10.50 running metres.

**Thailand** [312]

This part consists of the archives of the consulate general in Bangkok (1860–1942, 1945–1950). These archives are partially destroyed. For the names and subjects mentioned in the inventory and the table of contents see the index.

Entry: Inventory 1860–1942, table of contents 1945–1950.

Size: 1860–1942 *c.* 4 running metres (245 inventory items), 1945–1950 *c.* 7 running metres.

**Turkey** [313]

For the history of Asia and Oceania up to 1949 are relevant: (NB excluded are the archives of the consulate general Istanbul 1845–1900 (30 running metres), without an entry)

a. *Istanbul (legation), 1808–1920*
Included are papers pilgrims 1908–1915, Islam 1905–1917, Turkey-Atjeh 1868–1872, 1893, Dutch East Indies 1884–1920, Bombay-Djeddah, sultan Djambi 1858, consulate Singapore 1866–1874, Turkey-Persia 1870–1872.

Entry: Table of contents.

Size: *C.* 15 running metres.

b. *Ankara (legation/embassy), 1921–1954*
Included are papers Turkey-Germany-Japan 1945, press reviews governor general 1927–1932, cabinet Abe Nobujuki 1939, A.S. Blok on internment policy in the East Indies 1941, Turkey-Persia 1923–1924, assumption of power in Persia by Resa Shah Pahlevy 1925–1926, China 1942, emigration Chinese 1943, Rosadi 1941, Indonesian military affairs 1942–1945, trade with the East Indies 1941, opium supply to Batavia 1939–1942, P.C. Westerling and son R.P.P. Westerling 1945–1949, Korea, India, Indonesia 1945–1949, Linggardjati, Singapore, G. de Jonghe d'Ardoye, Turkey-Persia 1945–1949.

Entries: Inventory 1921–1945, table of contents 1945–1954.

Size: 1921–1945 *c.* 1.50 running metres, 1945–1954 *c.* 2.50 running metres.

See also no. 40 n, 53.

**United Kingdom** [314]

Important are the posts in London, as is apparent from the separate heading colonies in most of the archives. The archives are:

a. *London (legation), 1914–1937*
For the names and subjects mentioned in the inventory see the index.

Entry: Inventory.

Size: *C.* 19 running metres.

b. *London (legation/embassy), 1937–1945*
For the names and subjects mentioned in the table of contents see the index.

Entry: Table of contents.

Size: *C.* 6 running metres.

c. *London (legation/embassy, secret archives), 1938–1945*
For the names and subjects mentioned in the inventory see the index.

Entry: Inventory.

Size: *C.* 2.50 running metres.

d. *Postbooks, 1914–1945*
Size: *C.* 2.50 running metres.

e. *London (embassy), 1945–1954*
For the names and subjects mentioned in the table of contents see the index.

Entry: Table of contents.

Size: *C.* 15 running metres.

f. *London (embassy, secret archives), 1946–1954*
For the names and subjects mentioned in the table of contents see the index.

Entry: Table of contents.

Size: *C.* 7 running metres.

g. *London (consulate general), 1823–1944*
For the names and subjects mentioned in the table of contents see the index.

Entry: Table of contents.

Size: *C.* 36 running metres.

See also no. 45, 61.

**British territories in Asia** [315]

These include Hongkong. Of the archives of the consulate general there (1895–1957) the archives 1920–1941 were destroyed. For the names and subjects mentioned in the inventory and table of contents see the index.

Entries: Inventory 1895–1920, table of contents 1945–1954.

Size: 1895–1920 *c.* 7 running metres (196 inventory items), 1945–1954 *c.* 1.50 running metres.

**United Nations** [316]

Relevant are the archives of the Permanent Representation in New York (1946–1954). For the names and subjects mentioned in the table of contents see the index.

Entry: Table of contents.

Size: *C.* 12 running metres.

**United States** [317]

Relevant are:

a. *Washington (legation), 1910–1940*
For the names and subjects mentioned in the table of contents see the index.

Entry: Table of contents.

Size: *C.* 16 running metres.

b. *Washington (embassy), 1940–1952*
For the names and subjects mentioned in the table of contents see the index.

Entry: Table of contents.

Size: *C.* 26 running metres.

c. *Washington (Dutch delegates to the Far Eastern Commission), 1946–1951*
Included are papers Japan, Indonesia (Batavia), India, Pakistan, Dutch Military Mission in Japan, RTAC, Turhill & Co., trade, aluminium, gold, silver, platinum, wood, oil, rubber, sugar, tin, shipping, railways, iron, banking in Japan, with postbooks.

Entry: Table of contents.

Size: *C.* 2.50 running metres.

d. *Los Angeles (consulate), 1940–1959*
Included are papers commander Zeemacht Nederlands-Indië (Dutch East Indies Naval Forces) 1940, East Indies pensions 1943–1946, women's corps KNIL 1944, Indonesian question 1949.

Entry: Table of contents.

Size: *C.* 3.50 running metres.

See also no. 40 i, 54.

**Vietnam** [318]

The archives of the Dutch representation in Saigon consist of four parts: (vice) consulate 1926–1939 without table of contents; consulate general 1944–1950; consulate general 1946–1956; consulate general/legation/embassy 1947–1967 (including only a few accounting papers of before 1949). For the names and subjects mentioned in the tables of contents 1944–1950 and 1946–1956 see the index.

Entries: Tables of contents 1944–1950, 1946–1956, 1947–1967.

Size: 1926–1939 *c.* 0.30 running metre, 1944–1950 *c.* 0.70 running metre, 1946–1956 *c.* 1.20 running metres, 1947–1967 *c.* 6.50 running metres.

**South Africa** [319]

Relevant are:

a. *Pretoria (legation/embassy), 1888–1960*
These archives consist of two parts: the period 1888–1953 and 1930–1960. For the names and subjects mentioned in the tables of contents see the index.

Entries: Tables of contents 1888–1953, 1930–1960.

Size: 1888–1953 *c.* 20 running metres, 1930–1960 *c.* 9 running metres.

b. *Cape Town (consulate general), 1916–1945*
Included are papers Indonesia 1942–1945, evacuations to Australia and New Zealand, J. Fabius in Tokio 1938, trade Dutch East Indies 1933–1941, East

Indies pensions 1940–1947, W.F.M. Mansvelt 1933, F.D. Holleman, Dutch East Indies 1928–1934 (i.a. Bali and Java).

Entry: Table of contents.

Size: *C.* 4.50 running metres.

**Sweden** [320]

These archives consist of:

a. *Stockholm (legation and consulate general), 1910–1946*
Included are many papers concerning the second world war originating from the legation. For the names and subjects mentioned in the inventory see the index.

Entry: Inventory.

Size: *C.* 19 running metres.

b. *Stockholm (legation), 1945–1954*
Included are only a few papers concerning Asia up to 1949, viz. Indonesia 1945–1949, Verzetster Oost-Azië, Hankow-Tientsin import company.

Entry: Table of contents.

Size: *C.* 4.50 running metres.

See also no. 40 c, 40 p, 42, 55, 66, 304.

**Switzerland** [321]

Relevant are the archives of the legation/embassy to Bern 1916–1954. These archives are divided into three periods: 1916–1939 (East Indies papers are to be found by means of the alphabetic indexes to the postbooks, but the very summary table of contents does not mention these papers); 1939–1945 (second world war Indonesia and Japan); 1945–1954 (papers India, Indonesia, Raden Bagus Bhimo Jodjana, Chinese People's Republic, Philippines, Japan, Indonesia 1945–1949, Japanese prisoners in the Dutch East Indies in 1942, Persia, Portuguese Timor, L.E.J. Brouwer, Hongkong, Red Cross, Thailand, R. Schwenke, E.W. Hablützel).

Entries: Table of contents 1916–1939, tables of contents 1939–1945, 1945–1954.

Size: 1916–1939 *c.* 17 running metres, 1939–1945 *c.* 6.50 running metres, 1945–1954 *c.* 18 running metres.

See also no. 56, 58, 59, 68.

## Committees [322–325]

These are the archives of committees established by or closely connected with Foreign Affairs. Four of these contain papers concerning the history of Asia and Oceania up to 1949.

**Missie "Steenberghe" (Mission "Steenberghe"), 1940–1945** [322]

This small amount of archives consists of cabinet papers, including a report "price levels before and during Japanese occupation of the Netherlands East-Indies", papers Naval Intelligence Service about desertion of Javanese sailors, memorandum concerning terms of employment of Indonesian crew 1943.

Entry: Alphabetic list.

Size: *C.* 0.10 running metre.

The complete archives of the Missie "Steenberghe" are deposited at the Ministry of Economic Affairs.

**United Nations War Crimes Commission UNWCC** [323]

These archives are composed of papers after 1945, including circulars, lists of i.a. Japanese war criminals, papers concerning British soldiers in Japanese captivity.

Size: *C.* 1 running metre.

**Commissie "Verzetster Oost Azië, 1942–1945" (Committee "Resistance Star East Asia, 1942–1945")** [324]

These archives consist of 376 personal files, minutes, correspondence etc. 1946–1957 concerning the awarding of distinctions to Dutchmen on account of their part in the second world war in Asia.

Entry: Alphabetic index to personal names.

Size: *C.* 2.50 running metres.

**Security Committee Indonesia** [325]

These archives consist of papers 1948 concerning the situation in Indonesia.

Size: *C.* 1 running metre.

## Private persons [326–335]

Included are all the collections originating from functionaries of the Ministry of Foreign Affairs.

**F. Beelaerts van Blokland, January 21st 1872 The Hague – March 27th 1956 The Hague** [326]

Law studies at Leyden up to 1895; 1897 official management PTT; 1903 commies griffier (clerk) Second Chamber; 1909–1919 envoy extraordinary and plenipotentiary to Peking, then employed at Foreign Affairs; 1927 envoy to Brussels; 1927–1933 minister of Foreign Affairs; 1933 vice president of the Raad van State (Council of State); 1942–1944 president of the extraordinary advisory council at London; 1944–1946 member of the committee of control of the League of Nations. (*Keesings Historisch Archief*; 1955/56).

The collection (1942–1952) contains papers 1942–1945 concerning the Institute of Pacific Relations.

Size: *C.* 0.10 running metre.

**N.S. Blom, March 22nd 1899 Deventer–May 30th 1972** [327]

Law studies at Leyden up to 1923; 1924 deputy registrar High Court of Justice Batavia; 1925 official Department of Justice Batavia; 1928 member of the Council of Justice Makasar; 1931 member of the Council of Justice Padang; 1934 president of the district council in Magelang; 1934 senior officer Department of Justice Batavia; 1941–1946 director of Justice, 1942–1945 in England and Australia, 1944 also deputy of the lieutenant governor general; 1946 member of the Dutch delegation of the UN Assembly; 1947 member of the UN Special Committee on Palestine; 1947–1950 legal advisor Foreign Affairs, 1949 member of the Dutch delegation RTC; 1950 secretary of state Foreign Affairs; 1952–1964 government commissioner for Indonesian affairs and general advisor of Foreign Affairs; 1964–1966 director general Indonesian Affairs.

The collection has not been arranged yet. The larger portion consists probably of papers drawn up and/or received by Blom when he was government commissioner in Indonesia from 1952.

Size: *C.* 4 running metres.

**W.F.L. van Bylandt, December 31st 1896 The Hague–?** [328]

Law studies at Leyden up to 1921; 1923 in the diplomatic service, i.a. Cairo 1937; 1940 temporary secretary general Foreign Affairs in London; 1946 political adviser to the lieutenant governor general at Batavia; 1947 envoy extraordinary and plenipotentiary to Rome; 1953 high commissioner of the Netherlands in Djakarta; 1958 ambassador to Rome and Tunis (1960).

This collection (1940–1945) includes correspondence 1940–1945 with E.N. van Kleffens, A.H.J. Lovink, J.C. Pabst and others.

Entry: Table of contents.

Size: *C.* 0.20 running metre.

**H.A. van Karnebeek, March 21st 1874 The Hague – March 29th 1942 The Hague** [329]

Law studies at Utrecht up to 1900; 1902 employed at the Ministry of Colonial Affairs; 1911 burgomaster of The Hague; 1918–1927 minister of Foreign Affairs; 1927 minister of State; 1928 Queen's Commissioner for South Holland; his functions included the presidency of the assembly of the League of Nations in 1922. (*Nederland's Adelsboek*; 41st vol., 1943/48).

The collection (1919–1927) includes correspondence with H. Colijn, J.T. Cremer, A.C.D. de Graeff, J.P. van Limburg Stirum, C. van Vollenhoven, papers concerning the Anglo-Batavian Society 1921, memorandum of A.D.A. de Kat Angelino concerning proposals of China about the Far East 1920.

Entry: Inventory.

Size: *C.* 0.80 running metre.

**E.N. van Kleffens, November 17th 1894 Heerenveen–?** [330]

Law studies at Leyden up to 1918; 1919 employed at the secretariat of the League of Nations; 1921 at the secretariat of the Royal Shell at London; 1922 official at various departments of Foreign Affairs; 1939 envoy to Bern; 1939–1946 minister of Foreign Affairs; 1946 minister without portfolio with the Dutch representation on the Security Council and the economic and social council of the United Nations; 1947 ambassador to Washington; 1950 envoy to Portugal; 1954–1955 president of the general assembly of the United Nations; 1956–1958 permanent representative of the Netherlands to the NATO and the Organization for European Economic Co-operation; 1958–1967 chief representative in the United Kingdom of the High Authority of the European Coal and Steel Community.

This collection (1940–1950) includes correspondence with Raden Muhamad Musa Surianata Djumena, S.K. Hornbeek, papers 1945 concerning transport of troops to Indonesia, attitude of Australia towards the Dutch East Indies 1942, memorandum of Indonesian societies in the United States for the United Nations concerning the independence of Indonesia 1945, rubber and tin purchases in the Dutch East Indies, German internees in the Dutch East Indies, papers war in the Pacific, Japan and the Dutch East Indies in the second world war (including statements Matsuoka and Ishii, incidents Niyo, Hashida, propaganda Yoshisumi among the Chinese, rubber supply to Japan, New Guinea), papers Indo China, Hongkong, Singapore, Bangkok, Portuguese Timor, Thailand, China, Dutch East Indies and the United States, oil.

Entry: Table of contents.

Size: *C.* 2 running metres.

**A. Loudon, June 5th 1892 The Hague – February 4th 1953 The Hague** [331]

Law studies at Leyden; 1920 attaché London; 1922 second secretary to the legation Buenos Aires, subsequently Washington, 1926 first secretary to the legation Madrid, 1932 legation counseller and chargé d'affaires; 1934 envoy Bern and representative of the Netherlands to the League of Nations; 1938 envoy Washington, 1942 ambassador; 1947 member of the Raad van State (Council of State); 1952 secretary general of the Permanent Court of Arbitration. (*Keesings Historisch Archief*; 1953/54).

This collection (1940–1950) includes papers Far Eastern Lecture series, i.a. concerning the Dutch East Indies 1941, Institute of Pacific Relations 1942, "Indonesië gruweldossier" (Indonesian horror file) with American outporings about the Dutch actions in the Dutch East Indies 1947, report H.J. van Mook concerning i.a. the Dutch East Indies 1947, papers "terugkeer en hersteld gezag" (return and restored authority) of the Netherlands in Indonesia 1945, reports for A.W.L. Tjarda van Starkenborgh Stachouwer 1940–1941, report H.A. Cossten about Indonesia March 1942, underground activities against Japan in Java and Madura.

Entry: Table of contents.

Size: *C.* 1 running metre.

**J. Loudon, March 18th 1866 The Hague – November 11th 1955 Wassenaar** [332]

Law studies at Leyden up to 1890; from 1894 diplomatic service London, Rome; 1900 chief cabinet Foreign Affairs; 1901 chargé d'affaires Peking; 1903 legation counseller London; 1904 Paris; 1906 envoy Tokio; 1908 Washington; 1913–1918 minister of Foreign Affairs; 1919–1940 envoy Paris, i.a. deputy of the Netherlands to the League of Nations. (*Keesings Historisch Archief*; 1955/56).

This collection (1887–1940) includes letters W.J. Oudendijk from Peking 1908, papers about Japan by L. van der Polder 1912, papers about Lombok 1895, "machtsuitbreiding Japan" (enlargement of power Japan) 1916, granting of asylum to general Tsjang Shun in the Dutch legation in Peking 1917, papers concerning Indonesian students in the Netherlands and Moh. Hatta 1926–1927.

Size: *C.* 0.60 running metre.

**E.F.M.J. Michiels van Verduynen, December 2nd 1885 The Hague – May 13th 1952 London** [333]

Law studies at Leyden up to 1911; from 1915 employed at Foreign Affairs, 1920–1923 in Prague, from 1923 back at the Ministry, 1939 till his death (extraordinary and plenipotentiary) ambassador to London. (*Keesings Historisch Archief*; 1951/52).

This collection (1940–1945) includes papers concerning the war in the Pacific and the Japanese attack on the Dutch East Indies 1942.

Entry: Table of contents.

Size: *C.* 0.10 running metre.

**W. Riphagen, January 1st 1919 Batavia–?** [334]

Law studies at Amsterdam up to 1940; 1942–1947 official at the Ministry of Home Affairs; 1947–1949 member of the Dutch negotiating delegation with Indonesia; 1949–1952 assistant legal adviser of Foreign Affairs; 1954 council adviser in general service of Foreign Affairs.

The collection has not been classified yet. The larger portion probably consists of papers drawn up and/or received by Riphagen when he was legal adviser of Foreign Affairs from 1949 forward.

Size: *C.* 1.50 running metres.

**C.J.M. Schaepman, August 16th 1896 Arnhem – December 28th 1963** [335]

Law studies at Leyden; from 1921 employed at Foreign Affairs where his last function from 1950 forward was council adviser in general service (up to 1961).

The collection (1940–1945) includes Japanese press reviews 1941, statement E.N. van Kleffens concerning the Dutch East Indies Civil Service 1940, papers 1940 concerning Germans in the Dutch East and West Indies.

Size: *C.* 1 running metre.

# Ministerie van Defensie (Ministry of Defence)

4 Plein, 2511 CR The Hague.
Navy: 17 Koningin Marialaan, 2595 GA The Hague.
Bureau Maritieme Historie (Bureau Naval History): 112 Jan van Nassaustraat, 2596 BW The Hague.
Centraal Archievendepôt Ministerie van Defensie (Central Archive Repository Ministry of Defence): 19 2e Van der Kunstraat, 2521 BD The Hague.
Sectie Krijgsgeschiedenis KL (Military History Royal Land Forces): Frederikkazerne, 31 v.d. Burchlaan, 2597 PC The Hague.
Admission by appointment.
Photocopying equipment available.
Research finished in 1973.

**Introduction to the archives** [336]

The archives kept by the Ministry of Defence cover the period *c.* 1900 up to the present.
They are the continuation of that part of the archives which are deposited at the ARA (see no. 69, 70).
The archives are subdivided into:

The archives of the former Ministry of Naval Affairs, with the papers transferred from Indonesia (see no. 337–339).

The archives deposited at the Bureau Maritieme Historie (Bureau Naval History) (see no. 340–362).

The archives of the former Ministry of War, with the services in the Netherlands and England and the archives of KL and KNIL transferred from Asia (see no. 363–385).

The archives deposited with the Section Military History (see no. 386–391).

Permission of the Ministry is necessary for consultation of all the archives deposited at the Ministry of Defence. For the period 1945–1950 see also: "Nota betreffende het archievenonderzoek naar gegevens omtrent excessen in Indonesië begaan door Nederlandse militairen in de periode 1945–1950". In *Verslag der handelingen van de Staten-Generaal zitting 1968–1969*, 10008 no. 3. This memorandum includes a survey of archives containing papers about the subject mentioned above.

For the remaining archives belonging to Defence see no. 69–72.

## Navy: Archives [337–339]

### Ministerie van Marine (Ministry of Naval Affairs), 1928– [337]

The archives are the continuation of the part deposited at the ARA (see no. 70). The archives up to 1950 consist of "verbalen" (minutes), after 1950 of files.

They are subdivided into:

a. Non secret and secret archives.

   Entries: Postbooks, systematic and alphabetic indexes, archives code.

   Size: *C.* 800 running metres, of which *c.* 120 running metres secret archives.

   Archives non secret 1901–1928 and secret 1845–1928 have been destroyed by fire.

   For the period 1940–1945 see also London archives (no. 337 b).

b. London archives, consisting of non secret and secret archival documents originating from the Ministry of Naval Affairs of the government in exile at London.

   Entry: Card index system.

   Size: *C.* 50 running metres.

Permission of the Ministry of Defence is necessary for consultation of all the archives deposited at Naval Affairs.

## Archives transferred from Indonesia [338–339]

### Marine Commandant Australië (Naval Commander Australia), 1942–1947 [338]

These archives consist mainly of papers concerning the Navy during the second world war.

Entry: Index cards (alphabetically to persons).

Size: *C.* 5 running metres.

**Commando Zeemacht Oosten/Commando Zeemacht Nederlands-Indië (Command Naval Forces East/Command Naval Forces Dutch East Indies), CZMO/CZMNI 1942, 1945–1949** [339]

These command archives contain mainly papers concerning the Navy in the Indonesian archipelago. At first the Command was called Zeemacht Oosten, later Zeemacht Nederlands-Indië, after 1949 Commando Zeemacht Nieuw-Guinea (Command Naval Forces New Guinea) (CZMNG). These archives include data about personnel from 1942.

Entry: Inventory is forthcoming.

Size: *C.* 200 running metres.

## Bureau Maritieme Historie [340–362]

**Library** [340]

3000 books; 150 periodicals.

Particulary directed to the naval history of the Netherlands.

**Publications** [341]

*Jaarboek van de Koninklijke Marine.* 1875/76–1937/38, 1952– . 1938–1952 not published.

*Bijdragen tot de geschiedenis van het zeewezen;* vol. 1 (1969)– .

### Archives [342–360]

**Introduction** [342]

The larger portion of the archives are pertaining to the second world war.

In many cases it was impossible to trace the origine of the archives. If it appeared that they had been sent over from Indonesia to the Netherlands, this is stated with the description.

Copies of almost all regulations of the Royal Navy, now repealed, are present at the Bureau.

Permission of the Ministry of Defence is necessary for consultation of all the archives deposited at the Bureau.

**Marinestaf (Naval Staff), *c.* 1940–1945** [343]

These archives consist of correspondence and reports originating from the naval staff in London and Ceylon. They are subdivided into: very secret papers, secret papers, general affairs, merchant shipping and administration.

Size: *C.* 8.25 running metres.

**Nederland meidagen 1940 (The Netherlands May 1940), A-archives, 1938–1947** [344]

These archives contain mainly papers about the invasion of the Netherlands by Germany May 1940. They consist of two parts: commandementen in Nederland (commands in the Netherlands) (A a) and Krijgsverrichtingen (Military operations) (A b). These two parts consist of *c.* 50 headings. The headings A a 4, 10, 25, A b I 4 and II 4 contain papers concerning Asia. For the names and subjects mentioned in the shelflist see the index.

Entry: Shelflist.

Size: *C.* 4.75 running metres.

**Engeland; Oost- en West-Indië; Zuid-West Pacific (England; East and West Indies; South West Pacific), B-archives, 1938–1961** [345]

These archives contain mainly papers about the second world war. They consist of three parts: England (B a, subdivided into 5 headings), East and West Indies, general (B b, 16 headings), East Indies and South West Pacific, krijgsverrichtingen (military operations) (B c, 29 headings), B b and B c contain papers concerning Asia. In addition there are a number of papers 1940–1943 (numbered A–G) originating from legations and consulates of Sydney, Wellington, Manila, Tokio, Thailand, Washington and Honolulu. For the names and subjects mentioned in the shelflist see the index.

Entry: Shelflist.

Size: *C.* 2.25 running metres.

**Buitenlandse gegevens (Foreign data)/Combined Chiefs of Staff Washington, C-archives, 1940–1961** [346]

These archives consist of papers 1940–1961 mainly pertaining to the second world war. They are subdivided into: Japanese data (C a), American data (C b), English data (C c), German data (C d) and Italian data (C e). The archives contain also the archives 1941–1949 of the Combined Chiefs of Staff at Washington (C b-100), of which a separate archives list has been made. For the names and subjects mentioned in the shelflists see the index.

Entries: Shelflists.

Size: *C.* 9.50 running metres.

**Schepen en diensten (Ships and services), D-archives, 1937–1960** [347]

These archives consist of papers 1937–1960, for the greater part pertaining to actions of naval craft in Europe and Asia during the second world war. The archives are subdivided into: artillery ships (D a), torpedo destroyers (D b), Mine Service, mine layers and sweepers (D c), MTB service and boats (D d), sundries (Government Navy, merchant shipping, convoys, hospital ships, tankers, auxiliary ships, Naval Coastguard Station, diving business), ME, trade protection (D e) and submarine service and boats (D f). For the names and subjects mentioned in the shelflist see the index.

Entry: Shelflist.

Size: *C.* 3 running metres.

**Chronologie en dagboeken (Chronology and diaries), E-archives, 1939–1954** [348]

The larger portion of these archives consists of chronological reports about the second world war and of diaries of several divisions and functionaries. In addition there are some reports about Indonesia after 1945. The archives are subdivided into: England (E a), East Indies (E b) and America (E c). The heading E b contains papers concerning Asia. For the names and subjects mentioned in the shelflist see the index.

Entry: Shelflist.

Size: *C.* 0.90 running metre.

**Supreme Allied Command South East-Asia, SACSEA, F-archives, 1942–1946** [349]

SACSEA had its headquarters in New Delhi, after the war in Singapore. Commander (1943–1946) was L.F.A.V.N. Mountbatten. The archives, consisting completely of papers 1942–1946 concerning the second world war in the Pacific and Asia, are arranged chronologically in 8 subdivisions (F a-h). For the names and subjects mentioned in the shelflist see the index.

Entry: Shelflist.

Size: *C.* 5 running metres.

**Netherlands Purchasing Commission New York, NPC, G-archives, 1939–1946** [350]

The larger portion of the archives of this commission consists of papers 1939–1946 concerning purchases for and supply of naval ships and concerning protection of trade. For the names and subjects mentioned in the shelflist see the index.

Entry: Shelflist.

Size: *C.* 4.50 running metres.

**Archief Australië (Archives Australia), H-archives, 1940–1946** [351]

These archives consist of papers 1940–1946 originating from various functionaries and organizations in Australia. They are subdivided into 5 headings (H a–e), i.a. concerning NEFIS, naval craft, several places and districts in Indonesia, NICA, staff matters etc. The first impression is that the papers are of a secret nature, in contrast to the papers "Gewoon archief Australië" (Ordinary archives Australia) (see no. 354). For the names and subjects mentioned in the shelflist see the index.

Entry: Shelflist.

Size: *C.* 2 running metres.

**Marine Attaché Washington (Naval Attaché Washington), MAW, J-archives, 1938–1944** [352]

These archives consist of papers 1938–1944 originating from the naval attaché at Washington. They are chronologically arranged in 7 subdivisions (J a–g). A large portion of the papers concerns the second world war in Europe, Asia and the Pacific. For the names and subjects mentioned in the shelflist see the index.

Entry: Shelflist.

Size: *C.* 8.25 running metres.

**Handelsbescherming (Protection of trade), K-,V-,W-,X-,Y-,Z-archives, 1941–1946** [353]

These archives consist of papers 1941–1946 concerning the protection of ships in the European and Asian waters during the second world war. These archives include the papers originating from the handelsbeschermingsofficieren (officers for the protection of trade) (HBO) at Halifax (V-archives), San Francisco (W-archives), Colon (X-archives), New York (Y-archives) and New Orleans (Z-archives). For the names and subjects mentioned in the shelflists see the index.

Entries: Shelflists K-archives, V-archives, W-archives, X-archives, Y-archives, Z-archives.

Size: *C.* 5 running metres.

**Gewoon archief Australië (Ordinary archives Australia), S-archives, 1942–1947** [354]

These archives consist of papers 1942–1947 originating from various functionaries and organizations in Australia, i.a. concerning several naval craft and districts in Indonesia. The first impression is that the papers are of a non secret nature, in contrast to the papers "Archief Australië" (see no. 351). For the names and subjects mentioned in the shelflist see the index.

Entry: Shelflist.

Size: *C.* 1.25 running metres.

**Bevelhebber Strijdkrachten Oosten in Ceylon (Commander Eastern Forces in Ceylon), BSO, 1940–1945** [355]

These archives consist of reports and correspondence of the Commander Eastern Fighting Forces concerning the second world war. They were transferred from Indonesia in 1947. For the names and subjects mentioned in the shelflist see the index.

Entry: Shelflist.

Size: *C.* 3.50 running metres.

**Marine Luchtvaart Dienst (Naval Air Service), MLD, 1939–1962** [356]

These archives consist of reports and documentation 1939–1962 about the second world war. They are subdivided into chronologically arranged headings: Europe (A), Asia (B), USA and Canada (C), MAC ships (D), individual operations of allied units (E), general (F), post war period (G). For the names and subjects mentioned in the shelflist see the index.

Entry: Shelflist.

Size: *C.* 2.75 running metres.

**Mariniersbrigade (Marine Brigade), 1945–1950** [357]

These archives consist, as far as can be traced, of papers 1945–1950 transferred from Indonesia.

Size: *C.* 50 running metres.

**Intelligence Asia** [358]

This collection consists of reports of intelligence services in Asia, probably for the period 1940–1950.

Size: *C.* 0.50 running metre.

**Scheepsjournalen (Ship's logbooks), from 1940** [359]

This collection contains logbooks of ships which sailed in Asia. The collection is the continuation of the series of logbooks in the archives of the Ministry of Naval Affairs, present at the General State Archives (see no. 70). The collection is alphabetically arranged, according to the name of ship.

Size: *C.* 13 running metres.

**Nieuw Guinea (New Guinea), 1949–1962** [360]

These non arranged archives are mentioned for the purpose of completeness, because they may contain papers concerning the period 1796–1949.

Size: *C.* 3 running metres.

**Marine charts** [361]

The collection consists of *c.* 5000 marine charts. It is the aim to form a complete collection, such as is set out in the *Catalogus van Nederlandse zeekaarten en boekwerken*, 1964, and in the *Catalogue of admiralty charts*, 1970.

Entry: Catalogue on index cards.

**Photographs and drawings** [362]

This collection consists of:

a. *C.* 150,000 photographs, subdivided into the headings: ships (alphabetically to name), persons, harbours, bases.

b. *C.* 6000 aerial photographs (132 reels).

c. Marine Voorlichting (Navy Information) (MARVO) photographic archives, 1945–1950.

d. *C.* 800 16 mm films, including instructional and documentary films.

e. Archives of drawings of a historical-documentary nature. Being collected and drawn are i.a. pictures of old and modern uniforms, emblems, decorations, arms.

f. Archives of drawings of ships and objects of the Royal Navy from the second half of the 19th century and from the 20th century. *C.* 1500 cylinders.

## Centraal Archievendepôt [363–385]

**Ministerie van Oorlog (Ministry of War), 1914–1950** [363]

These archives are the continuation of the part deposited at the ARA (see no. 72).

The archives consist of "verbalen" (minutes).

They are subdivided into:

a. Non secret (general) archives.

Entries: Postbooks, indexes to subject, alphabetic indexes to name, card index systems to name and authority.

Size: *C.* 325 running metres.

"Verbalen" (minutes) 1915 – August 12th 1915, 1916–1917, 1934 – February 18th 1945 largely destroyed by fire, as were the postbooks 1934–1939, indexes 1924–1939 and card index systems 1934–1939.

b. Secret archives (1945–1949).

Entries: Postbooks, systematic indexes, alphabetic indexes to name and authority, card index systems.

Size: *C.* 2.75 running metres.

"Verbalen" (minutes) 1845–1945 destroyed by fire, as were the postbooks and systematic indexes 1845–1939 and card index systems 1919–1939.

These archives are deposited 36 Bagijnestraat, The Hague.

c. Very secret archives (1945–1949).

Entries: Postbooks and alphabetic indexes.

Size: *C.* 0.75 running metre.

These archives are deposited 36 Bagijnestraat, The Hague.

d. London archives (1940–1945).

These archives consist of non secret and secret "verbalen" (minutes) originating from the Ministry of War of the government in exile at London.

Entries: Alphabetic indexes to name and authority and card index systems to authority, name and subject.

Size: *C.* 30 running metres (non secret) and 11 running metres (secret).

e. Cabinet of the Minister of War (1945–1950).

Especially the period 1947–1949 is of importance for the history of Indonesia.

Entry: Alphabetic index and card index systems to authority and subject.

Size: *C.* 5 running metres.

Permission of the Ministry of Defence is necessary for consultation of all archives deposited at the Centraal Archievendepôt.

**Algemeen Hoofdkwartier; Volkenbond (General Headquarters; League of Nations), 1919–1935** [364]

The archives of the section League of Nations of the General Headquarters in the Netherlands include papers concerning the entry of Australia to the treaty for control of international trade in arms 1931, Manchuria-Japan question 1932 and the international congress of the Red Cross in Tokio 1934.

Entry: Inventory.

Size: *C.* 1 running metre.

**Volkenbond (League of Nations), 1926–1937** [365]

This collection of miscellaneous papers includes papers of the Koloniaal Comité (Colonial Committee).

Entry: Inventory.

Size: *C.* 2 running metres (18 inventory items).

**Nederlandsche Militaire Missie in Engeland/Militair Attaché te London (Dutch Military Mission in England/Military Attaché to London), 1940–1949** [366]

These archives consist of a secret and a "general" part. In addition the papers originating from the Military Mission and the Military Attaché to London there are also some papers of the Mission to Washington, i.a. concerning Japan and mrs. Chiang Kai Shek. For the names and subjects mentioned in the inventory see the index.

Entry: Inventory.

Size: *C.* 1.80 running metres.

**Hoofdkwartier van de Generale Staf in Nederland (Headquarters of the General Staff in the Netherlands), 1945–1950** [367]

The inventory of these chronologically arranged archives is forthcoming. From the already arranged part (5 running metres) appears that papers about the RTC and correspondence about the army in Indonesia are present.

Size: *C.* 32.50 running metres.

**Hoofdkwartier Nederlandse Troepen in Siam/Militaire Attaché Bangkok (Headquarters Dutch Forces in Siam/Military Attaché Bangkok), 1945–1950** [368]

There is no inventory yet to these in files arranged archives.

Size: *C.* 1 running metre.

**Staf Depôt Nederlands-Indië/Infanterie Depôt Indië met de kampen Hulten, Vilheide, Wanroy en De Rips (Staff Depot Dutch East Indies/Infantry Depot East Indies with the camps Hulten, Vilheide, Wanroy and De Rips), 1945–1948** [369]

In 1945/1946 the Division A included the 1st Aanvullingsdepôt (Supplementary Depot), consisting of the camps Vilheide, Wanroy and De Rips, and the 2nd Koninklijk Nederlands Infanterie Depôt (Royal Dutch Infantry Depot).

In 1946 the Division A was abolished and from its staff the Staf Depôt Nederlands-Indië was constituted. The commander of the KNID was stationed at Hulten to settle affairs. Hulten, being subordinate to the Staff Depot Dutch East Indies, acquired the command of the three camps (Vilhéide, Wanroy and De Rips) of the 1st Aanvullingscommando and was called Infanterie Depôt Indië (IDI). From this place troops were sent to the East Indies.

The archives consist of a secret and a "general" part, chronologically arranged according to unit.

For the names and subjects mentioned in the inventory see the index.

Entry: Inventory.

Size: *C.* 1.50 running metres.

**Leger Voorlichtingsdienst (Army Information Service), 1945–1951** [370]

This service was set up in 1945 by N.D. van Goethem. Its tasks included to keep informed in Indonesia and so it was closely connected with the Dienst voor Legercontacten (Service for Army Contacts) at Batavia (for the archives of this service see no. 291).

The small amount of archives is arranged chronologically according to subject. For the names and subjects mentioned in the inventory see the index.

Entry: Inventory.

Size: *C.* 0.25 running metre (15 inventory items).

**"Nationale Inspanning Welzijnsverzorging Indië" (National Effort Welfare Work East Indies), NIWIN, 1946–1952** [371]

This organization was established in 1946 at The Hague. There were sections in Indonesia, which, together with the head office in the Netherlands, was in charge of the welfare work for soldiers in Indonesia. The NIWIN was abolished in 1951.

The archives are arranged chronologically according to subject. For the names and subjects mentioned in the inventory see the index.

Entry: Inventory.

Size: *C.* 2.30 running metres.

See also no. 424.

**Archives transferred from Indonesia** [372–385]

**Legercommandant in Australië (Commanding Officer in Australia), 1942–1945** [372]

At the time of the capitulation in 1942 part of the KNIL succeeded in making their escape from Indonesia to Australia. Here the military forces became subordinate to the commanding officer Australia, who was subordinate to the commander in chief of the Dutch military forces in the East.

Important sections in Australia were the NEFIS and the Militaire Luchtvaart (Military Aviation). After the capitulation of Japan the commanding officer left for Indonesia (for this see no. 373–377).

This small and incomplete amount of archives consists of three subdivisions: information, aviation and sundries. For the names and subjects mentioned in the inventory see the index.

Entry: Inventory.

Size: *C.* 0.50 running metre (18 inventory items).

**Kabinet van de Legercommandant in Indonesië (Cabinet of the Commanding Officer in Indonesia), 1945–1950** [373]

After the capitulation of Japan in 1945 the commanding officer left Australia (see also no. 372) and settled at Batavia. Commanding officer was L.H. van Oyen, succeeded by S.H. Spoor in 1946. After a reorganization the army direction was composed of the following main groups:

The Algemeen Hoofdkwartier (General Headquarters) consisting of the Kabinet van de Legercommandant (Cabinet of the Commanding Officer), the headquarters of the Generale Staf (General Staff) (see no. 374–376) and the Centrale Opleidingsapparaat (Central Training Organ) of the army.

The Hoofdkwartier van de Adjudant-Generaal (Headquarters of the Adjutant General) (see no. 377).

The Hoofdkwartier van de Kwartiermeester-Generaal (Headquarters of the Quartermaster General).

After the transfer of sovereignty the title of Legercommandant (Commanding Officer) was changed into Commandant van de Nederlandse strijdkrachten in Indonesië (Commander of the Dutch military forces in Indonesia). After July 26th 1950 the Afwikkelingscommando (Liquidation Commando) (see no. 378) proceeded to settle the affairs.

The incomplete archives have been arranged according to subject. They are subdivided into two main groups: department affairs (consisting of political matters, accounting matters, army contacts, courses of conduct and decorations) and army affairs (consisting of operational, personnel, materials, medical and aviation matters). For the names and subjects mentioned in the inventory and shelflist see the index.

Entries: Inventory, shelflist.

Size: *C.* 2.50 running metres (293 inventory items).

**Hoofdkwartier van de Generale Staf in Indonesië (Headquarters of the General Staff in Indonesia), 1945–1950** [374]

With the reorganization of the army direction the Headquarters of the General Staff were established side by side with the Cabinet of the Commanding Officer (see no. 373). The Headquarters ended their activities July 26th 1950.

The incomplete archives consist of a very secret and a secret part. The papers have been arranged chronologically according to subject. As appears from the systematic survey there are papers concerning military and non military Dutch organizations, military and non military Indonesian organizations, international organizations and treaties. For the names and subjects mentioned in the inventory see the index.

Entry: Chronological and systematic inventory.

Size: *C.* 13.50 running metres (96, 1044 inventory items).

**Hoofdkwartier van de Generale Staf in Indonesië; Krijgsgeschiedkundige Sectie (Headquarters of the General Staff in Indonesia; Section Military History), 1945–1949** [375]

These archives consist of reports originating from all KNIL and KL divisions. The papers have been arranged chronologically according to region in Indonesia. For the names and subjects mentioned in the inventory see the index.

Entry: Inventory.

Size: *C.* 5.20 running metres (41 inventory items).

**Inlichtingendienst van het Hoofdkwartier Militaire Luchtvaart in Indonesië (Intelligence service of the Headquarters Military Aviation in Indonesia), 1942–1949** [376]

These incomplete archives consist of folders, reports, memorandums originating from various national and international aviation services. For the names and subjects mentioned in the inventory see the index.

Entry: Inventory.

Size: *C.* 4 running metres (33 inventory items).

**Assistent Adjudant-Generaal KL; Afd. IIIA en B (Assistant Adjutant General Royal Land Forces; Section IIIA and B), 1945–1951** [377]

With the reorganization of the army in Indonesia after the Japanese capitulation in 1945 the Headquarters of the Adjutant General were established side by side with the General Headquarters (see also no. 374). Their task was to relieve the Commanding Officer of duties connected with personnel affairs. Subordinate to the Adjutant General were the Assistants Adjutant General I (KNIL), II (KNIL/KL) and III (KL), from which these archives originate. Subordinate to the Assistant Adjutant General III were the section A (personnel), B (military administration), C (legal affairs) and D (field post). Papers have been preserved only of section A, with a small, not arranged part of section B. The archives of section A have been chronologically arranged according to subject. For the names and subjects mentioned in the inventory see the index.

Entry: Inventory of the archives section A.

Size: *C.* 25.30 running metres.

**Afwikkelings Commando van het Leger in Indonesië (Liquidation Commando of the Army in Indonesia), 1950–1951**

**Nederlandse Militaire Missie Indonesië (Dutch Military Mission Indonesia), 1949–1955**

**"Landmacht Nieuw-Guinea" (Land Forces New Guinea), 1949–1954** [378]

These three archives, being outside the period 1796–1949, are mentioned for completeness' sake as they contain papers connected with the period before 1950.

Entry: Inventories of the three archives.

Size: Respectively *c.* 0.25, 5.50 and 21.80 running metres.

**Netherlands Forces Intelligence Service/Centrale Militaire Inlichtingendienst (Central Military Intelligence Service), NEFIS/CMI, 1942–1949** [379]

The NEFIS was established in Australia during the second world war. After the Japanese capitulation the service continued its activities in Indonesia and changed its name into Centrale Militaire Inlichtingendienst after November 1948.

The archives consist of reports originating from various divisions of the army and other organizations and of daily, weekly and monthly reports and several kinds of surveys of the NEFIS itself. For the names and subjects mentioned in the inventory see the index.

Entry: Inventory.

Size: *C.* 3.70 running metres (30 inventory items).

**Militaire Politie in Indonesië (Military Police in Indonesia), 1948–1950** [380]

These archives consist of files and statements partly arranged according to name or case.

Size: *C.* 2.70 running metres.

**Eerste Divisie '7 december' (First Division '7th of December'), 1946–1950** [381]

The '7 december divisie' was established in 1946 and was intended for sending out to Indonesia. The division took part in the first and second police action. In 1949/50 the troops were repatriated.

The archives consist of papers originating from the several subdivisions of the division. They have been arranged chronologically according to subdivision in a secret and "general" volume. The papers of the staff are not included. For the names and subjects mentioned in the inventory see the index.

Entry: Inventory.

Size: *C.* 9.25 running metres.

**T-, U-, V-, W-, X-, Y- en Z-Brigade (T-, U-, V-, W-, X-, Y- and Z-Brigade), 1945–1950** [382]

These brigades were established in 1945 and repatriated in 1949/50. All of them took part in the police actions. The T-brigade was based in Semarang, subsequently Salatiga, operational area was Central Java, the U-brigade in Padang and Sungai Patani conducted its operations in Central Sumatra, the V-brigade carried out its activities near Tanggerang, Buitenzorg, Banjumas, Tegal, Cheribon, Pekalongan and Djokjakarta, the W-brigade operated in West and Central Java, the X-brigade in East Java, the Y-brigade in Bali, Palembang and South Sumatra and the Z-brigade in North Sumatra, based in Medan.

Of the archives of these brigades only the very secret and secret papers of the staffs and also of a subdivision of the W-brigade (staf II-13 RI) have been preserved. Of the X-brigade only corps histories are present, as well as of the Y-brigade, and only a few centimeters of archives. Of the other brigades more papers have been preserved with their corps histories (for these corps histories see also 'Nederlands-Indië 1945–1950' (The Dutch East Indies 1945–1950), which collection is present at the Sectie Krijgsgeschiedenis KL (Section Military History Royal Land Forces), no. 388). The archives have been arranged chronologically according to brigade. For the names and subjects mentioned in the inventory see the index.

Entry: Inventory.

Size: *C.* 3.70 running metres (32 inventory items).

**Aan- en Afvoertroepen (Supply and Transport Troops), AAT, 1945–1951** [383]

These archives consist of the papers originating from the several subdivisions of the AAT. They have been arranged chronologically according to subdivision.

Size: *C.* 19 running metres.

**Subsistenten-Bataljon KL in Indonesië (Subsistence Battalion Royal Land Forces in Indonesia), 1946–1951** [384]

To this battalion were admitted sick and recovering soldiers and soldiers waiting for a destination to be determined.

The archives consist mainly of personal notes of soldiers who stayed in this "refuge". The papers have been arranged numerically.

Size: *C.* 2 running metres.

**"Verspreide archivalia" (Scattered archives), 1945–1949** [385]

This collection consists of papers originating from subdivisions in East Java, viz. Mariniersbrigade (Marine Brigade), A-division, B-division, 4e Infanterie Brigade (4th Infantry Brigade), Territoriaal Troepencommando (Territorial Commando of Troops) East Java, Garderegiment Jagers (Guards Regiment of the Rifles), Inlichtingendienst (Intelligence Service) East Java and subdivisions in the Malang sector, Eastern Part of Java, Madiun sector, Madura sector, Modjokerto sector, Surabaja and Ponorogo sector. For the names and subjects mentioned in the inventory see the index.

Entry: Inventory.

Size: *C.* 6.80 running metres (350 inventory items).

## Sectie Krijgsgeschiedenis [386–391]

### Introductory data [386]

The archival material concerning Asia and Oceania consists for the greater part of a collection of papers from 1942, transferred from Indonesia. The papers from the period before 1942 were destroyed by bombing during the second world war. The collection is subdivided into three parts: 1. Dutch East Indies contra Japan; 2. Dutch East Indies 1945–1950; 3. the Warners collection. (The papers concerning New Guinea from 1950 are left out here).

Entries: Chronological catalogue on index cards. Alphabetic/systematic catalogue on index cards. The headings are alphabetically arranged, of which concerning Asia the most important are: the Dutch East Indies contra Japan and the Dutch East Indies 1945–1950. These headings are subdivided according to subject and name of place.

Alphabetic card index system to names of persons, including both the names from the archives and those from the collection of photographs (see no. 391).

Permission of the director of the Sectie Krijgsgeschiedenis is necessary for consultation of all the archives deposited at the section.

**Archival material** [387–389]

**Nederlands-Indië contra Japan (the Dutch East Indies contra Japan)** [387]

The main parts of this collection are: a collection of fighting reports, reports and papers concerning captivity; photocopies of Japanese studies of the second world war; reports of "NEFIS-parties" from Australia to New Guinea and Java; reports of the meetings of the Raad van Departementshoofden (Council of Heads of Departments) at Brisbane; reports Conica's and Sonica New Guinea; reports of expeditions of the Corps Insulinde from Ceylon to Burma; miscellaneous papers of J.M.R. Sandbergh concerning the Pacific war; miscellaneous papers of G.L. Reinderhoff concerning Indonesia during the second world war.

Entry: The alphabetic/systematic catalogue on index cards mentioned sub no. 386. For the subheadings (names of places and subjects) see the index.

Size: *C.* 6.50 running metres.

**Nederlands-Indië 1945–1950 (The Dutch East Indies 1945–1950)** [388]

The main parts of this collection are: reports, daily reports, situation reports etc. of the General Headquarters of the General Staff; sitreps GOB and sitreps TC-NG; reports Conica's and Sonica NG; papers of the Cabinet of the Commanding Officer (S.H. Spoor); papers originating from the various divisions and brigades (divided according to territories); papers concerning the shock troops in Malaya, Menado, Bali, Central Java and South Sumatra; corps histories KL and KNIL; draft staff work KNIL (draft description of the military history 1945–1950 by E.L.F. Couvreur); weekly military surveys NEFIS/CMI (also present in the archives of the Ministry of Defence); press news; papers concerning Menadonese and Amboynese in the 'Makasar-affaire' (Makasar affair) 1950.

Entry: The alphabetic/systematic catalogue on index cards mentioned sub no. 386. For the subheadings (names of places and subjects) see the index.

Size: *C.* 26 running metres.

**C.J. Warners, August 20th 1900 Leersum – ?** [389]

1925 reserve second lieutenant of the field artillery; 1926–1936 engineer with the PTT in the Dutch East Indies; 1927 reserve first lieutenant of the field artillery KNIL; 1936 head engineer PTT, mission to Siam; 1938 reserve engineer captain of the KNIL; 1940–1942 director Telegraaf-, Telefoon- en Radiodienst (Telegraph, Telephone and Radio Service); 1941 reserve lieutenant colonel, charged with the military supervision over the PTT, the broadcasting service and organization; 1942 liaison officer with the Commander in Chief of the ABDAC (general Wavell); 1942 council adviser of the Minister of Algemene Oorlogvoering (General Warfare) (P.S. Gerbrandy); 1943 reserve colonel, attached to the Dutch representation Combined Chiefs of Staff and liaison officer with the Office of Strategic Services Washington; 1944 as Chief of the Netherlands East-Indies Civil Affairs Section attached to the Staff Supreme Allied Commander South East-Asia Command Kandy (Ceylon); 1945 director Verkeer en Waterstaat (Transport and Water Management), also temporary Director of Economic Affairs in the Dutch East Indies government at Brisbane (Australia), October 1945 up to February 1948 at Batavia; 1947 leaves the active military service, 1948 secretary of state, head of the Department of Verkeer, Energie en Mijnwezen (Transport, Power and Mining Matters) of the provisional federal government at Batavia; 1949 to the Netherlands; 1951–1965 general manager KNMI at De Bilt, since then retired.

The collection consists of papers drawn up and/or received by him when he was council adviser of the Minister of General Warfare, liaison officer with the Combined Chiefs of Staff, representative of the Dutch government with the Supreme Commander SEAC, director of Transport and Water Management, secretary of state and head of the Department of Transport, Power and Mining Matters. For the names and subjects mentioned in the inventory see the index.

Entry: Inventory.

Size: *C.* 6 running metres (416 inventory items).

**Maps** [390]

This collection consists of *c.* 700 ordnance survey maps 1930–1945, geographically subdivided.

**Photographs** [391]

This collection consists of *c.* 200,000 photographs concerning the military history of the Netherlands and its colonies. *C.* 4500 of these are pertaining to Asia and Oceania. The main headings are: military operations in the Dutch East Indies *c.* 1825–1949 (*c.* 4000 photographs); warfare in the Pacific (*c.* 50 photographs); KNIL (*c.* 100 photographs); prisoners of war/prisoners of war camps (camps for women *c.* 30, camps for men *c.* 30, Burma/Siam *c.* 30 photographs); uniforms (KL and KNIL *c.* 500, British India 18, China 2, Japan 2 photographs); besides a collection of portraits of soldiers of the KL and KNIL *c.* 4000 photographs.

Entries: The photographs have been arranged according to headings. Alphabetic/ systematic catalogue on index cards.

Alphabetic card index system for names of persons (see no. 386).

## Ministerie van Economische Zaken (Ministry of Economic Affairs)

30 Bezuidenhoutseweg, 2594 AV The Hague.
One should apply in writing to the secretary general for permission to consult the archives.
Photocopying equipment available.
Research finished in 1978.

**Archives, 1917–** [392]

Although that part of the archives of the Ministry of Economic Affairs that has been deposited at the General State Archives ARA, is not included in this "Guide", we will mention the part that is still present in the archives of the Ministry, because we have an impression that during the twentieth century and especially during and after the second world war many affairs were dealt with without the knowledge of the Ministry of Colonial Affairs. For this see no. 80 and 81 where the archives of the NSTC and the Directie Indische Zaken (Direction East Indies Affairs) are described.

The Ministry belonged successively to: Home Affairs, 1813–1818; Education, National Industry and Colonial Affairs, 1818–1824; National Industry and Colonial Affairs, 1824–1825; Home Affairs, Education and Water Management, 1825–1829; National Industry, Water Management and Colonial Affairs, 1830–1831; National Industry and Colonial Affairs, 1831–1833; Foreign Affairs, 1834–1841; the ministries of Home Affairs, Justice, Finance and Foreign Affairs, 1841–1877; Water Management, Trade and Industry, 1877–1905; Agriculture, Trade and Industry, 1905–1922; Labour, Trade and Industry, 1922–1931; Economic Affairs and Labour, 1931–1933; Economic Affairs, 1933–1935; Trade, Industry and Shipping, 1935–1937; from 1937 the Ministry was independent.

Relevant for the history of Asia and Oceania are the papers originating from the Direction for Trade and Industry. The papers are mainly pertaining to trade relations with the countries there. For the names and subjects mentioned in the inventory see the index.

Entry: Inventory.

Size: *C.* 10 running kilometres, of which 17.50 running metres of "Asian papers" originating from the Direction for Trade and Industry.

Permission is necessary for consultation of the papers.

**Londens archief (London archives), 1940–1945** [393]

These archives were formed during the second world war by the Ministry of Economic Affairs of the Dutch government in exile in London. For the history of Asia and Oceania are relevant the headings:

Commercial policy; monetary and financial matters; limited liability companies; provisions; judicial matters and war time; documentation. Included are papers about the following countries: Dutch East Indies, China, Afghanistan, Japan, New Zealand, Australia, Siam, British India, Philippines.

Entry: Inventory.

Size: *C.* 40 running metres.

Permission is necessary for consultation of the papers.

# CHAPTER IV

# STATE ARCHIVES

## Rijksarchief in Drenthe (State Archives in Drenthe)

4 Brink, 9401 HS Assen.
Admission free. Lending to institutions.
Photocopying equipment available.
Research finished in 1977.

**Oosting** [394]

Papers concerning the history of Indonesia are originating from H.J. Oosting, 70 inventory items.

*H.J. Oosting, 1842–1915*
1865 official Landelijke Inkomsten en Cultures (Public Revenue and Cultures) Pangkul; 1869 comptroller, on leave; 1871 returns to Pangkul; 1875 official for the study of the Sunda language; 1881 to the Netherlands, where he published his Sunda dictionaries; 1889–1913 municipal councillor of Assen; 1904–1915 registrar of births, deaths and marriages at Assen.

The archives consist of papers concerning his publications, including his dictionaries, financial papers, papers concerning his official career: diaries, notes and correspondence when he was comptroller with the Landelijke Inkomsten en Cultures and official for the Sunda language, papers concerning his membership of the Bataviaasch Genootschap van Kunsten en Wetenschappen (Batavia Society of Arts and Sciences) and the Koninklijk Instituut voor de Taal-, Land- en Volkenkunde van Nederlandsch-Indië (Royal Institute of the Linguistics and Anthropology of the Dutch East Indies).

Entry: J.E. Ennik. *Inventaris van het archief van de familie Oosting en de daaraan verwante geslachten.* Assen, 1976.

Size: *C.* 5 running metres (576 inventory items).

## Rijksarchief in Overijssel (State Archives in Overijssel)

20 Eikenstraat, 8021 WX Zwolle.
Admission free. Lending to institutions.
Photocopying equipment available.
Research finished in 1977.

**Van Bevervoorden** [395]

There are only a few papers concerning the history of Asia, partly originating from various persons of the Engelbert family in Ceylon and Negapatnam in the 18th century.

In addition there are letters from W.F. Engelbert van Bevervoorden 1890 and papers originating from E.E.H.P.K. van Bevervoorden tot Oldemeule (May 26th 1824 Groningen – October 29th 1887 Buitenzorg) as functionary of the Pondok Gedeh company.

Entry: A.J. Gevers. *Inventaris van de papieren Van Bevervoorden.* Zwolle, 1974.

Size: *C.* 1.50 running metres (507 inventory items).

## Rijksarchief in Gelderland (State Archives in Gelderland)

1 Markt, 6811 CG Arnhem.
Admission free. Lending to institutions.
Photocopying equipment available.
Research finished in 1977.

**Papierfabriek Padalarang (Paper Factory Padalarang), 1921 – *c.* 1950** [396]

The factory was established in 1921 at Padalarang.

The object was to run the factory and to make, process and sell paper.

The archives consist of papers up to *c.* 1950 of the factories Padalarang and Letjes at Probolinggo, including minutes and correspondence of shareholders and board of supervisory directors, articles of association, annual accounts, accounting, papers, monthly reports, correspondence, personnel records, technical reports.

Size: *C.* 15 running metres.

Permission is necessary for consultation of the papers.

**Van Goltstein** [397]

Relevant for the history of Asia are the papers originating from W. van Goltstein, 97 inventory items.

*W. van Goltstein, May 13th 1831 Hamburg – September 10th 1901 Putten*
Jurist, 1854–1861 official at the Ministry of Foreign Affairs; 1864–1871 member of the Second Chamber; 1872–1874 member of the First Chamber; 1874–1876, 1879–1882 Minister of Colonial Affairs; up to 1894 retired; 1894–1899 envoy to London; in 1890 co-founder of the Maatschappij tot Bevordering van Natuurkundig Onderzoek der Nederlandsche Koloniën (Society for the Promotion of Physical Research of the Dutch Colonies).

These archives include autobiographic notes, letters from F.E.M. van Alphen 1879–1886, F. s' Jacob 1880–1892, J.W. van Lansberge 1874–1892, J.J. Smits and

B. de Geer about the departure of Smits to the Dutch East Indies 1873, A. Vinkhuyzen about the departure of Prince Willem to the Dutch East Indies 1880–1881, papers when he was a member of the First and Second Chamber and minister of Colonial Affairs, letters from G.J. Mulder mainly about the opiumregie (monopoly of opium) 1875–1879, papers when he was a member of the Staatscommissie voor de Indische revenuen (State Committee for the East Indies revenue) 1868–1869.

Entry: Inventory.

Size: *C.* 8.50 running metres (464 inventory items).

## Asian papers in Dutch collections [398]

For a definition see no. 271.

a. **Brantsen**

Papers Th.L. Brantsen concerning The Mount Gray Mine Syndicate in Australia 1889, report about an expedition to Japan and the opening of the Strait of Shimonoseki 1864 with maps; papers J. Mossel after 1796 concerning his inheritance 1797, the Jacob Mossel Fund 1761–1899, payment of the building of the Lutheran Church at Batavia 1747, 1817.

b. **Van Heeckeren van Waliën**

Letter of appreciation from the King of the action of E.L. van Heeckeren van Waliën in sending of the steamship Soembang to Japan 1855.

c. **Van den Heuvel**

Certificate of the promotion of W.H. van den Heuvel Rijnders as second lieutenant in the Moluccas.

d. **Rink**

Articles concerning A. Rink, programme 1934 of the Rotary national meeting at Buitenzorg, correspondence of E. van Galen at Bandung *c.* 1850.

e. **Ver Huell**

Papers originating from Q.M.R. Ver Huell, including *Herinneringen van eene reis naar de Oost-Indiën, 1815–1819* (edited in 1835, 1836), letters 1803–1860, records of service, journal 1816–1818, autobiography.

f. **Ziegenhirt van Rosenthal**

Papers originating from W.G.A. Ziegenhirt van Rosenthal concerning the firing at the Medusa in the Strait of Shimonoseki by Japanese 1863.

g. **Huis Keppel (House of Keppel)**

Papers about the stay of A.J. Schimmelpenninck van der Oye at Batavia 1807.

h. **Van der Capellen**

Papers of R.J. van der Capellen, shares Amphioen-sociëteit (Opium Society) at Batavia, J.C. Smissaert, G.A.G.Ph. van der Capellen and R.L.J. van der Capellen.

## Rijksarchief in Utrecht (State Archives in Utrecht)

201 Alexander Numankade, 3572 KW Utrecht.
Admission free. Lending to institutions.
Photocopying equipment available.
Research finished in 1977.

**Van Boetzelaer** [399]

Relevant for the history of Indonesia are the papers originating from C.W.Th. van Boetzelaer van Asperen en Dubbeldam, 20 inventory items.

*C.W.Th. van Boetzelaer van Asperen en Dubbeldam, July 5th 1873 De Bilt – 1956* Theologian; 1906–1919 protestant missionary consul at Batavia; 1922–1937 member of the Second Chamber for the CHU.

Originating from him are letters and reports of his journey, together with J.W. Gunning, round protestant mission stations in the Dutch East Indies, description of a feast of the death at Galela, notes about the Kakihan alliance in Ceram, papers about religious education at the protestant missionary schools, papers when he was a member of the Chamber about colonial policy, texts of causeries, photographs, i.a. of the Christelijke Persvereeniging (Christian Press Association) in the Dutch East Indies 1913.

Entry: Inventory with supplements.

Size: *C.* 10 running metres (920 inventory items).

**Asian papers in Dutch collections** [400]

For a definition see no. 271.

a. **Aardspriesters te Utrecht (Worldly Priests at Utrecht)**

Papers originating from bishop Wijkersloot about the catholic mission in Indonesia, *c.* 1840.

b. **Zeister Zendingsgenootschap (Protestant Missionary Society of Zeist)**

Papers about protestant mission in Tranquebar, Tibet, Ceylon, China and Australia, *c.* 1750 and *c.* 1850.

c. **Van Vollenhoven**

Papers originating from E.J. van Vollenhoven about the Salatiga protestant mission, 19th and 20th century.

## Rijksarchief in Noord-Holland (State Archives in North Holland)

12 Ceciliasteeg, 2011 RJ Haarlem.
Admission free. Lending to institutions.
Photocopying equipment available.
Research finished in 1977.

### Asian papers in Dutch collections [401]

For a definition see no. 271.

a. **Nagtglas**

Papers originating from H.Ch.A. Harloff, H.W.J.A. Gaster, W.L.A. Harloff and H. Knotterus concerning the coffee estate Kemoening, 1876–1898; report of a revolt of a Chinese crew 1783, notes of captain J. Ingerman of the voyage Amsterdam-Batavia first half 19th century, journal of L.W. Mondt of a voyage to Batavia, 1825.

b. **Semeyns de Vries van Doesburg**

Papers originating from A.A. Buyskes.

# CHAPTER V

# MUNICIPAL ARCHIVES

## Gemeentelijke Archiefdienst Amsterdam (Municipal Archives of Amsterdam)

67 Amsteldijk, 1074 HZ Amsterdam.
Admission free. Lending to institutions.
Photocopying equipment available.
Research finished in 1977.

### Private organizations in the economic field [402–419]

**Louis Bienfait & Soon** [402]

The firm was engaged in plantations, shipping insurance and trading business and factories.

The archives consist of papers 16th up to 20th century; relevant for the history of Asia are papers concerning shipping, insurances and trade, including papers P. Zunderdorp at Tjilatjap 1838–1840, papers A.J.F. Tijdeman 1853–1854, N. Biorn 1854, papers about the Philippines 1845–1865, China, Canton 1846–1848, P. van Vlissingen 1857, Oost-Indische Paketvaart Mij (East Indies Packet Boat Company) 1857, papers L.A. Bienfait 1852–1865, sugar factory Krian 1852–1865, Nederlandsch-Indische Katoen Mij (Dutch East Indies Cotton Company) 1864–1868, tobacco estates Maeng and Ajer Mendidi 1895–1896, sarongs and other textile fabrics at Makasar 1859, Van Hemert & Co at Batavia and Bicker & Co at Surabaja 1869, tea plantations 1875, sugar 1871, H.J.C. Hoogeveen and the firm of Gijselman & Van Rinkhuijzen about concessions in Pulau Sibesi and Sibuka 1873–1875, trade in Makasar 1855.

Entry: Inventory.

Size: *C.* 3.25 running metres (numbered PA 646, 336 inventory items).

**Boissevain & Co, 1777–1924** [403]

The firm was established in 1812, composed of merchants, shipowners and insurers.

The archives consist of papers 1777–1924, viz. papers administration, business (general, shipping industry, insurances, financial records), shipping and shipping industry of a special nature, papers trade and papers family affairs. Relevant for the history of Asia and Oceania are papers concerning trade and shipping to the Dutch East Indies (Batavia, Surabaja, Makasar, Padang) 1841–1881, Japan (Deshima, Nagasaki) 1825–1869, China 1854–1857, Australia 1850–1867, papers NHM 1865–1876, Jan Pietersz Coen 1856, expedition to Bali 1849, papers average at Tjilatjap 1869, insurance Singapore, Pacific Steamship Company 1866, papers J. Boissevain concerning the steamship service to the Dutch East Indies, papers W. Boissevain concerning his official journey in Java 1898, Stoomvaart Mij Nederland (Steamship Company the Netherlands) 1898–1910, war China-Japan 1895, inheritance H.A. von Henrici 1838.

Entry: Inventory.

Size: *C.* 8 running metres (numbered PA 647, 503 inventory items).

**Central Pacific Readjustment Managers, 1899–1907** [404]

The small amount of archives consists of accounting papers 1899–1907.

Size: *C.* 0.30 running metre (numbered PA 618).

**Firma H.G.Th. Crone (Firm of H.G.Th. Crone), 1790–1959** [405]

In 1790 H.G.Th. Crone established a trading concern at Amsterdam, which specialized in the import of indigo from Germany, from 1830 import of indigo from the Dutch East Indies, and also import of coffee, cocoa, kapok, sugar (from the West Indies) and grains (from the Baltic area), and exported luxury and medical goods to the Dutch East Indies; from 1885 the firm conducted the management of several estate companies, especially of tobacco in Java and Sumatra; from 1911 they held agencies in Surabaja, Batavia, Semarang, after 1945 also in Makasar, Sambas, Singkawang and Singapore; from 1957 the firm no longer had any concern with Indonesia.

The archives consist of contracts, correspondence, accounting papers, papers of the agencies and estate companies, reports of tours of inspection, archives of the firm of C. Pfingsten (1850–1888), photographs. For the names and subjects mentioned in the inventory see the index.

Entry: Inventory.

Size: *C.* 10 running metres (numbered PA 609, 549 inventory items).

**Exploratie Syndicaat "Pagoeat" (Exploration Syndicate "Pagoeat"), 1897–1921** [406]

The syndicate was established in 1897. The object was to obtain or take over permits for mining prospecting and concessions for development of mines. The syndicate possessed the concessions Pagoeat and Ilota at Paguat in Celebes, where copper and gold was found. In 1915 the syndicate was liquidated.

The archives consist of minutes, articles of association, annual reports, papers received and despatched, papers about concessions, accounting papers, papers about the liquidation up to 1921, documentation, photographs. For the names and subjects mentioned in the inventory see the index.

Entry: Inventory.

Size: *C.* 2.40 running metres (numbered PA 660, 106 inventory items).

**Fabriek van Stoom- en andere Werktuigen "Kalimas" (Steam and other Machine Tool Factory "Kalimas"), 1907–1922** [407]

The factory was established in 1889 at Surabaja and in 1913 an amalgamation was made with the Machine Factory "Amsterdam". From that time forward the board of directors had their seat at Amsterdam.

The archives consist of minutes, papers received and papers despatched by S.C. van Musschenbroek, reports of W.H.D. Segboer, papers about shares of J. and W.W. Campbell and J.W.D. Francken, accounting papers, papers for J. Muysken, annual reports, articles of association.

Entry: Inventory.

Size: *C.* 0.25 running metre (numbered PA 658, 29 inventory items).

**Handelsmaatschappij voorheen Kerkhoff & Co (Trading Company late Kerkhoff & Co), 1905–1925** [408]

The archives consist of papers received and despatched 1905–1925, papers of the agencies at Kota Radja, Sibolga, Medan and Sabang, accounting papers. For the names and subjects mentioned in the inventory see the index. With these archives have been desposited the archives of:

*Technisch Handelsbureau J.H. van den Wall Bake (Firm of Technical Dealers J.H. van den Wall Bake), 1905–1926*
The archives consist of papers received and despatched 1905–1926, papers of several companies including the Bataafsche Petroleum Maatschappij (Batavian Petroleum Company) 1921–1923.

Entry: Inventory.

Size: *C.* 7 running metres (Trading Company, numbered PA 665, 199 inventory items), 4.50 running metres (Firm of Technical Dealers, numbered PA 665A, 82 inventory items).

**Van Heekeren & Co, 1880–1938** [409]

The seat at Benkulen was established in 1927. The object included supervision and administration of estate companies, management of several limited companies and sale of the products of those companies, granting advances of money, obtaining undertakings and concessions, running of undertakings, effecting insurances and conducting banking business.

The archives consist of papers from 1880, papers received and despatched, papers concerning financial matters, papers of and about several estate companies and firms. For the names and subjects mentioned in the inventory see the index.

Entry: Inventory.

Size: *C.* 57 running metres (numbered PA 584, 1292 inventory items).

**Firma Hoffmann Schöffer (Firm of Hoffmann Schöffer), 1838–1939** [410]

These archives, without an inventory, consist of papers 1838–1939 concerning coffee and tea trade.

Size: *C.* 57 running metres (numbered PA 57).

**India Verzekeringsmaatschappij (India Insurance Company), 1866–1874** [411]

This small amount of archives consists of a copybook of A.H. van Bosse, note books and accounting papers 1866–1874.

Entry: List.

Size: *C.* 0.10 running metre (numbered PA 645).

**Oliefabrieken Insulinde (Oil Factories Insulinde), 1916–1936** [412]

The factories were engaged in crushing oil from copra. These archives, without an inventory, consist of papers 1916–1936.

Size: *C.* 40 running metres (numbered PA 590).

See also no. 121.

**Nederlandsch-Indische Portlandcementmaatschappij (Dutch East Indies Portland Cement Company), from 1910** [413]

The company was established in 1910 at Amsterdam. The object was to establish and run portland cement factories and to trade in the products made there.

These archives, without an inventory, contain, besides the papers of the company, also papers of the Padang Portland Cement Company.

Size: *C.* 1 running metre (numbered PA 536).

**Oostersche Hypotheekbank (Oriental Mortgage Bank), 1888–1931** [414]

The bank was established in 1888 at Amsterdam; their branch bank at Singapore was Hooglandt & Co, the name of the bank was there Oriental Mortgage Bank; the bank was liquidated in 1928.

The archives consist of minutes, annual reports, correspondence, accounting papers, registers of shares etc., papers of J.C. de Vries, E.J. Labarre, Labouchère

Oijens & Co, B.W. Blijdenstein & Co, Wurfbain & Zoon, lists of Chinese firms in Singapore, map and photographs of Singapore.

Entry: Inventory.

Size: *C.* 3.50 running metres (numbered PA 595, 106 inventory items).

**Schaap & Van Veeren** [415]

From 1862 Schaap & Van Veeren, tobacco brokers.

These archives, without an inventory, consist of papers 18th up to 20th century, including papers about the trade in Sumatra, Java and Borneo tobacco.

Size: *C.* 0.20 running metre (numbered PA 605).

**Tanahwangie Maatschappij (Tanahwangie Company), 1854–1907** [416]

In 1854 A. Salm rented waste lands in Malang for the cultivation of tobacco, in 1856 the instrument was executed, the head office was at Amsterdam; from 1866 increasingly more coffee was grown, and also rice; in 1907 the company was liquidated.

The archives consist of minutes, accounting papers, correspondence received and despatched, annual reports and letters from W. Blokhuis Izn, M.E.H.R. van de Kerkhoff, A. Salm, P. Smithuysen, F. Boggard, J. van der Made, H.J. Huffnagel, D. Nuysink, R. Lohmann, C. Abring Hesseling, Fraser Eaton & Co, Kreglinger & Co, Van Eeghen & Co.

Entry: Inventory.

Size: *C.* 4.75 running metres (numbered PA 115, 337 inventory items).

**Vereeniging voor den Kapokhandel (Federation for the Kapok Trade), 1915–1972** [417]

The Federation was established in 1915 at Amsterdam. The object was to promote the trade in kapok.

These archives, without an inventory, consist of papers about the import of Java kapok.

Size: *C.* 0.20 running metre (numbered PA 602).

**Firma W.F. Westermann (Firm of W.F. Westermann), 1856–1960** [418]

Spice brokers business and companies under the firm name of W.F. Westermann and W.F. Westermann & Co, also companies under the firm name of Wolff and Westenberger, A. Valenkamp & Co and H.A. van Overzee.

The archives consist of papers about the above mentioned companies, papers about the Kina-Bureau (Quinine Bureau), Java Kinabast (Java Cinchona Bark),

Java Coca, Crediet- en Handelsvereeniging Banda (Credit and Trading Association Banda: tours of inspection Bandan plantation Awaya in Ceram, articles by R. Broersma), papers about mace, nutmeg, cinnamon and cocoa.

Entry: Inventory.

Size: *C.* 0.50 running metre (numbered PA 666, 112 inventory items).

**Transport organizations** [419]

These archives consist of:

a. *Zeehaven en Kolenstation "Sabang" (Seaport and Coaling Station "Sabang"), 1893–1959*
The company was established in 1905. They had an arrangement with the Ministry of Colonial Affairs to perform the operation, and all that was connected with it, of the seaport and the coaling station, on behalf of the East Indies government.

The small amount of archives consists of minutes, articles of association, reports, correspondence, papers concerning financial matters and preliminary papers from 1893. For the names and subjects mentioned in the inventory see the index.

b. *Atjeh Transport Maatschappij (Atjeh Transport Company), 1904–1946*
The company was established in 1904. The object was transport of persons and goods in Atjeh and the storage of goods in the harbours of Atjeh and its dependencies.

This part of the archives consists of minutes, correspondence and papers about financial matters. For the names and subjects mentioned in the inventory see the index.

Entry: Inventory.

Size: *C.* 0.60 running metre (numbered PA 662, 78 inventory items, Sabang and 12 inventory items, Atjeh).

## Other private organizations [420–426]

**Doopsgezinde Vereeniging tot Bevordering der Evangelieverbreiding in de Nederlandsche Overzeesche Bezittingen (Mennonite Association for the Promotion of the Evangelization in the Dutch Overseas Possessions), 1847–1957** [420]

The association was established in 1847 at Amsterdam; 1851 P. Jansz. as first protestant missionary to Japara, founded a Javanese-Dutch school; 1881 projects for land reclamation in Java by P.A. Jansz.; 1908 medical mission by H. Bervoets; from 1869 also mission in Mandailing.

The archives consist of papers from 1847, both of papers of a general nature like minutes, annual reports and letters received and despatched, and of papers of a special nature like papers about the administration, personnel, mission, accounting, miscellaneous papers, the archives of the Section of Amsterdam, the archives of the Eerste en Tweede Doopsgezinde Vrouwenhulpvereenigingen (First and Second Mennonite Women's Aid Associations), documentation. For the names and subjects mentioned in the inventory see the index.

Entry: Inventory.

Size: *C.* 11.50 running metres (numbered PA 305, 412 inventory items).

**Nederlandsch Hoofdcomité der Rijnsche Zending (Dutch Central Committee of the Rhine Mission), 1869–1968** [421]

The Rheinische Missionsgesellschaft was established at Barmen in 1828, the Dutch Central Committee in 1869; from 1829 mission work was done in Cape Colony, from 1843 among the Dajaks in Borneo, from 1862 among the Bataks, from 1865 in Nias and from 1901 in Mentawai. (H.D.J. Boissevain. *De zending in Oost en West*. 's-Gravenhage, 1934, 1943. 2 vol.).

The archives include minutes, correspondence, articles of association of Barmen, accounting papers.

Entry: Inventory.

Size: *C.* 0.40 running metre (numbered PA 485).

**Nederlandsch Luthers Genootschap voor In- en Uitwendige Zending (Dutch Lutheran Society for Mission Home and Abroad), 1852 – *c.* 1940** [422]

The society was established in 1852, from 1882–1886 active in the Pasemah area (Benkulen) and from 1889 on the Batu Islands.

These archives, without an inventory, contain papers about the mission in Tello and Sigata.

Size: *C.* 1.60 running metres (numbered PA 552).

**Koninklijk Zoölogisch Genootschap Natura Artis Magistra (Royal Zoological Society Natura Artis Magistra), 1838–** [423]

Artis was established at Amsterdam in 1838 by G.F. Westerman. Its most important task was to give information about the living animal, for which end, among other things, a zoological garden is maintained. They also teach, make films and edit a periodical.

These archives, without an inventory, consist of papers *c.* 1838–1940, including papers about the purchase of animals in various Asian countries.

Size: *C.* 115 running metres (numbered PA 395).

**Comité Amsterdam van de Nationale Inspanning Welzijnsverzorging Indië (Committee Amsterdam of the National Effort for Welfare Work in the East Indies), NIWIN 1947–1951** [424]

The object of the NIWIN was welfare work for soldiers in Indonesia.

The archives consist of minutes, name lists, files about various actions and committees, letters received and despatched, photographs.

Entry: Inventory.

Size: *C.* 2.25 running metres (numbered PA 175, 86 inventory items).

See also no. 371.

**Plaatselijke Commissie voor het Huldeblijk aan Generaal K. van der Heyden (Local Committee for Tribute to General K. van der Heyden), 1880** [425]

The small amount of archives consists of subscription lists and accounts.

Size: *C.* 0.05 running metre (numbered PA 42).

**Comité Plaatsing van een Gedenkteken in het Geboortehuis van Eduard Douwes Dekker (Committee for the Erection of a Memorial in the House where Eduard Douwes Dekker was born), 1887–1888** [426]

The small amount of archives consists of subscription lists, circulars, papers received.

Size: *C.* 0.05 running metre (numbered PA 58).

## Families and private persons [427–429]

Boissevain [427]

These family archives, without an inventory, consist of papers 1864–1944, including papers concerning the trade to the Dutch East Indies.

Size: *C.* 12 running metres (numbered PA 394).

**Ooster** [428]

The papers of F. Ooster are relevant for the history of Indonesia (7 inventory items).

*F. Ooster, April 2nd 1841 Amsterdam – ?*
1866 official at the disposal of the East Indies government, departure to the Dutch East Indies; 1867 at the disposal of the Algemene Secretarie (General Secretariat), 1868 clerk; 1869 clerk of the court of the circuit judge of the second district, Semarang; 1872 public prosecutor Makasar; 1872 president landraad (district council) Kraksaan; 1875 president Council of Justice Amboina; 1877 on leave to the Netherlands, 1879 retired from service.

This part of the archives consists of papers when he was an official in the Dutch East Indies and a diary and notes of his outward and homeward voyage to the Dutch East Indies.

Entry: Inventory.

Size: *C.* 0.10 metre (numbered PA 186).

**Westerman** [429]

In these family archives the papers originating from the following persons are relevant for the history of Asia.

a. *W. Westerman, November 14th 1864 Amsterdam – 1935*
1881 clerk Vermeer & Co; 1889 deputy manager with Everts Schmidt; 1892 chief accountant with the Nederlandsche Asahan Tabak Mij (Dutch Asahan Tobacco Company) in Sumatra; 1899 returns to the Netherlands, deputy manager with the Incassobank (Collecting Bank); 1902 secretary, 1904 managing director, 1908–1924 president of the Rotterdamsche Bank; also supervisory director of the Borneo-Sumatra Handelmij (Borneo-Sumatra Trading Company).

This part (52 inventory items) consists of diplomas, letters and other personal notes, photographs, printed matter and articles about Westerman. For the names and subjects mentioned in the inventory see the index.

b. *E. Tweer, July 17th 1862 Gevelsberg – 1944*
1883 sent to Deli for a Swiss-German tobacco undertaking, from which in 1889 the Senembah Mij (Senembah Company) came into existence; 1889 manager, 1896 chief manager, 1905 supervisory director in the Netherlands, 1927–1940 managing director of the Senembah Mij; also supervisory director of the Cultuur Mij De Oostkust (Estate Company De Oostkust), the Rotterdam Tapanoeli Cultuur Mij (Rotterdam Tapanoeli Estate Company), the Nederlandsch-Indische Cultuur Mij (Dutch East Indies Estate Company), the Oliepalm Cultuur Mij Tanah Itam Oeloeh (Oil Palm Estate Company Tanah Itam Oeloeh), the Cultuur Mij Batang Sangir (Estate Company Batang Sangir).

This part (26 inventory items) consists of diplomas, papers Senembah Mij, letters, photographs, water colours. For the names and subjects mentioned in the inventory see the index.

c. Papers originating from other members of the family, viz. from G.F. Westerman (1812–1892): diplomas of Natura Artis Magistra, the Indisch Genootschap (East Indies Society), the Koninklijk Instituut voor Taal-, Land- en Volkenkunde (Royal Institute of Linguistics and Anthropology) and the Internationale Koloniale en Uitvoerhandel Tentoonstelling (International Colonial and Export Trade Exhibition), and papers of J. Westerman (1866–1943): "damesbrieven" (letters from ladies) from the *Sumatra Post* 1901–1904.

Entry: J.H. van den Hoek Ostende. *Inventaris van het familiearchief Westerman.* Amsterdam, 1966.

Size: *C.* 3.80 running metres (numbered PA 399, 554 inventory items).

**Asian papers in Dutch collections** [430]

For a definition see no. 271.

a. **A. d'Ailly**

Notes 1947–1958 about the Indonesian question (numbered PA 306). Permission is necessary for consultation of the papers.

b. **Algemeene Maatschappij van Levensverzekering en Lijfrente (General Company of Life Insurances and Life Annuity)**

Papers insurances in the Dutch East Indies, office at Surabaja, J.F.L. Blankenberg 1881–1920, P.Th. van Hemert 1909–1911, G.S. Maingay 1913, W. Strick van Linschoten 1899–1911, . . . Smissaert, agency in Persia, Koninklijke Maatschappij tot Exploitatie van Petroleumbronnen in Indië (Royal Company for the Exploitation of Petroleum Wells in the East Indies) 1904–1905, Nillmij 1905–1927, P.H. Bodemeijer 1908–1911, A. Paets tot Gansoyen, E.A. van Kappen 1911–1912, NHM 1918, . . . Voûte, Suikerpensioenfonds (Sugar Pension Fund) 1917–1918 (numbered PA 580).

c. **Bond van Nederlandsche Brouwerijen (League of Dutch Breweries)**

Export East Indies 1931–1937, Japanese competition in the Dutch East Indies 1932–1934 (numbered PA 204).

d. **Brugmans**

Regulation government sugar cultures 1859–1864, papers Australia, letters from the Nederlandsch Bijbelgenootschap (Netherlands Bible Society) 1872 (numbered PA 86).

e. **Classis Amsterdam (Classis of the Protestant Parish of Amsterdam)**

Communications from the East and West Indies 1894–1896 (numbered PA 379).

f. **De Clercq**

Notes of P. de Clercq Gz. about the cost of a voyage to the Dutch East Indies 1873–1877, copies of deeds of G. de Clercq at Tjimanggung, letters from and to G. de Clercq, A.J.W. and J. van Delden, A.A. Land, F.W. van Eeden 1864–1890 (numbered PA 255).

g. **College Zeemanshoop (Sailor's Hope)**

Papers about the bankruptcy of the Oost-Indische Maatschappij van Administratie en Lijfrente (East Indies Company of Administration and Life Annuity) 1871–1873, logbooks of various ships bound for the Dutch East Indies, China, Japan, Batavia, Surabaja, by . . . Veenstra, F. van Duivenboode, P. van Duyvenbooden 1824–1864.

h. **Van Eeghen**

Papers concerning the trade to the Dutch East Indies (without an inventory, numbered PA 498).

i. **Handelshuis Van Eeghen & Co. (Trading House Van Eeghen & Co.)**

Probably papers concerning the trade to the Dutch East Indies (without an inventory, numbered PA 447).

j. **Gerzon's Modemagazijnen (Gerzon's Dress Shops)**

Papers of the "Afdeeling Indië" (Department East Indies) 1925–1956 by J. de Vries-Werkendam, R.A. Blatt, A. Davids, J. Vecht papers about the take-over by Wanita Modes (Women's Fashions) (numbered PA 539).

k. **Holland Land Company**

Papers concerning the trade between the United States and British India, the Dutch East Indies, China 1792–1799 (numbered PA 333).

l. **Internationale Handelsvereeniging (International Trade Association)**

Papers about relations of the United States with Japan 1935–1941, Indonesia 1938–1940, Semarang 1937–1939 (numbered PA 287).

m. **Kam**

Papers J. Kam in Amboina, correspondence J.G. Kam with I.H. Enklaar about J. Kam *c.* 1950 (numbered PA 287).

n. **Ketwich & Voombergh and Weduwe W. Borski**

Probably papers concerning the trade to the Dutch East Indies (without an inventory, numbered PA 602).

o. **J.W.G. Kirberger & Co.**

Register of East Indies orders 1921–1934 (numbered PA 659).

p. **Pijnappel-Menso**

Papers originating from M.J. Pijnappel, Oost-Indische Maatschappij van Administratie en Lijfrente (East Indies Company of Administration and Life Annuity), Weeskamer (Orphan Chamber) at Batavia, 19th century (without an inventory, numbered PA 572).

q. **Teixeira de Mattos**

Papers about the construction of the Transsiberian railway 1869–1892 (numbered PA 589).

r. **Den Tex**

Statements of accounts about the possessions in the Dutch East Indies of W.F. and F.A. Vriese and C.J.A. den Tex 1860–1861, letters from G. Vriese 1831–1848 (numbered PA 199).

s. **J.F.I.M. Tielens**

Papers about the bankruptcy of the Hollandsche Indische Textiel Compagnie (Dutch East Indies Textile Company) 1928–1930 (numbered PA 674).

t. **Vereeniging "Tentoonstelling De Vrouw 1813–1913" (Association "Exhibition the Woman 1813–1913")**

Papers of the Commissie Koloniën en Indische Huis (Committee Colonies and East Indies House): minutes, cash and note books, papers received and despatched, catalogue, sundries *c.* 1913 (numbered PA 89).

u. **Wertheim & Gompertz**

Papers about syndicates of East Indies estate companies *c.* 1900 (without an inventory, numbered PA 593).

## Gemeentelijke Archiefdienst Delft (Municipal Archives of Delft)

169 Oude Delft, 2611 HB Delft.
Admission free. Lending to institutions.
Photocopying equipment available.
Research finished in 1976.

**Instelling van Onderwijs in de Taal-, Land- en Volkenkunde van Nederlandsch-Indië (Institution for the Teaching of Linguistics, Geography and Ethnology of the Dutch East Indies), Indische Instelling, 1864–1901** [431]

The Indische Instelling was established in 1864 at Delft, after the government training centre for East Indies government officials had been transferred from the Delftse Akademie (Delft University) to Leyden (see no. 476). The Instelling, aided by the Municipality of Delft and the State, prepared for the "groot ambtenaarsexamen" (great examination for East Indies officials). There were classes in history, civics, geography and ethnology, religion, law, Malay, Javanese, Sunda, Madurese, Makasar, Buginese, Batak and Balinese. Though having much better results than the Leyden institutions (see no. 442), the Indische Instelling was abolished in 1901 because of financial difficulties and a decline of interest. Now the training for East Indies government official was conducted at the University of Leyden, later also at the University of Utrecht.

The archives consist of regulations, minutes, papers received, copybooks up to 1901, personal details about the students, papers about the library and accounting papers 1869–1900.

Entry: Inventory.

Size: *C.* 5 running metres (72 inventory items).

**Indologisch Studenten Corps (Indologist Students' Corporation), ISC 1895–1901** [432]

The corporation was founded in 1895 by students of the Indische Instelling (East Indies Institution); when the Indische Instelling was abolished the ISC disappeared too.

The archives consist of rules, minutes, letters, copybooks, lists of members, accounting papers, papers of several committees and sub corporations, viz. the Commissie tot Redactie van de Almanak (Committee of Editors of the Almanac) 1882–1900, the Commissie tot Behartiging van de Studiebelangen (Committee for the Promotion of the Interests of Study) 1896–1900, Indologische Vereeniging (Indologist Association) 1882–1900, Indologische Studenten Sociëteit Insulinde (Indologist Students' Club Insulinde) 1895–1900, Indologische Studenten Gymnastiek- en Schermvereeniging Sri Rama (Indologist Students' Gymnastic and Fencing Club Sri Rama) 1896–1900, Indologische Studenten Toneel- en Muziekgezelschap (Indologist Students' Theatrical and Musical Society) 1896–1900, Indologische Studenten Rijvereeniging (Indologist Students' Riding Club) 1897–1900 and Indologisch Dispuutgezelschap Saraswati (Indologist Debating Club Saraswati) 1897–1900.

Entry: Inventory.

Size: *C.* 3 running metres (145 inventory items).

**Indisch Leesmuseum (East Indies Reading Room), 1859–1865** [433]

The Leesmuseum was established in 1859 by students of the Delftse Akademie (University of Delft) with a view of promoting the interest for and knowledge of colonial matters. Delft students, Delft inhabitants and qualified East Indies government officials could become a member. After the abolition of the Delftse Akademie the Leesmuseum was dissolved.

The small amount of archives consists of minutes, regulations, annual reports, papers received, accounting papers.

Entry: Inventory.

Size: *C.* 0.30 running metre (14 inventory items).

# Gemeentelijke Archiefdienst Deventer (Municipal Archives of Deventer)

4 Gr. Kerkhof, 7411 KT Deventer.
Admission free. Lending to institutions.
Photocopying equipment available.
Research finished in 1977.

**Vereeniging Museum voor Tropische Landbouw (Association Museum for Tropical Agriculture), 1912–1967** [434]

The association was established in 1913 as Koloniaal Landbouwmuseum (Colonial Agriculture Museum) at Deventer. The object was to keep up a museum and to promote the knowledge of agriculture in the East and West Indies.

The archives consist mainly of papers after 1940 and of photographs.

Size: *C.* 3 running metres.

**Jordens** [435]

Papers concerning the history of Indonesia are originating from J.H.G. Jordens.

*J.H.G. Jordens*
1856–1859 first naval lieutenant at the disposal of the forestry department; 1860–1864 forest inspector.

This part of the archives includes papers concerning the forestry department.

Inventory is forthcoming.

## Gemeentelijke Archiefdienst Dordrecht (Municipal Archives of Dordrecht)

13 Stek, 3311 XS Dordrecht.
Admission free. Lending to institutions.
Photocopying equipment available.
Research finished in 1977.

**Miscellaneous papers** [436]

a. Letters from S. Stoop of the Dordtse Petroleummaatschappij (Dordrecht Petroleum Company) to S.M. Hugo van Gijn about the construction of the Tarik-Surabaja railway, 1895. (no. 2265a)

b. Papers of the sub-committee for a tribute to general K. van der Heyden, military governor of Atjeh, 1880–1881. (no. 3554)

**Asian papers in Dutch archives** [437]

For a definition see no. 271.

a. **Penn and Baudin**

Registers of the accountancy and drafts of outgoing letters concerning the trade to the Dutch East Indies in the 19th century.

b. **G. Hoogstraten**

Presumably data about Dordrecht undertakings in the Dutch East Indies (see E.P. de Booy and A.J. Looijenga. *Overzicht van archieven van particuliere ondernemingen berustend in openbare archiefbewaarplaatsen;* no. 175).

# Gemeentelijke Archiefdienst 's-Gravenhage (Municipal Archives of The Hague)

17 Loosduinseweg, 2571 AA The Hague.
Admission free. Lending to institutions.
Photocopying equipment available.
Research finished in 1977.

**Maatschappij tot Exploitatie der Kendalsche Suikerfabrieken (Company for the Operation of the Kendal Sugar Factories), 1893–1954** [438]

The company was established in 1893 at The Hague. The object was to operate the sugar undertakings Gemoe and Tjepiring at Semarang; besides these undertakings the company also operated the undertakings Sendang Kertoredjo, Sendang Boelak and Gebang Anom. The company was liquidated in 1954.

The archives consist of share registers.

Size: *C.* 0.40 running metre.

**Plaatselijk Comité ter Inzameling van Bijdragen ten behoeve der Noodlijdenden door de Ramp op Krakatau (Local Committee for the Collection of Contributions in Aid of the Distressed through the Calamity in Krakatau), 1883–1884** [439]

The small amount of archives was formed by D. Polak Daniëls, treasurer of the Committee, all the papers are pertaining to the collection of money.

Size: *C.* 0.01 running metre.

**Commissie voor de Koloniale Afdeeling van de Vereeniging ter behartiging van de Nederlandsche Belangen op de Tentoonstelling te Parijs in 1889 (Committee for the Colonial Section of the Association for the Promotion of the Dutch Interests at the Exhibition at Paris in 1889), 1888–1890** [440]

The small amount of archives consists of postbooks of letters received and despatched.

Size: *C.* 0.10 running metre.

**Ingenieurs-maatschap Eigen Beheer (Engineers' Partnership Eigen Beheer), 1916–1935** [441]

The partnership was executor of works and firm of engineers, also delivery office, office for the sending out of personnel and consulting firm for i.a. estate and industrial companies in Indonesia.

The archives consist of accounting papers.

Size: *C.* 11 running metres.

## Gemeentelijke Archiefdienst Leiden (Municipal Archives of Leyden)

2a Boisotkade, 2311 PZ Leyden.
Admission free. Lending to institutions.
Photocopying equipment available.
Research finished in 1977.

**Gemeente-Inrichting voor de Opleiding van Oost-Indische Ambtenaren te Leiden (Municipal Institution for the Training of East Indies Officials at Leyden), 1877–1891** [442]

After in 1876 the Rijksinstelling voor Onderwijs in de Taal-, Land- en Volkenkunde van Nederlandsch-Indië (Government Institution for the Teaching of Linguistics, Geography and Ethnology of the Dutch East Indies) had been closed because of lack of interest (see no. 476) in 1877 the Gemeente-Inrichting was established. L.M. de Laat de Kanter was president and P.A. van der Lith secretary. There were classes in Malay, Javanese, Sunda, history, religion, law and national institutions. Like the Rijksinstelling the results of the Gemeente-Inrichting were less satisfactory than those of the Indische Instelling of Delft (see no. 431). Therefore the Gemeente-Inrichting was closed in 1891.

The small amount of archives consists of minutes, drafts of outgoing letters and an album of those whose names were entered on the register 1877–1890.

Size: *C.* 0.10 running metre.

See also M.C. De Indische Inrichting te Leiden. In *Tijdschrift voor Nederlandsch-Indië*. 1886, vol. 1, p. 81–195.

**Jan Pietersz. Coen standbeeld (Statue for Jan Pietersz. Coen), 1888** [443]

At the invitation of the central committee for the erection of a statue for Jan Pietersz. Coen a subcommittee for Leyden and district was established in 1888. President was H.J. Bool, vice president F. Beijerinck, secretary P.A. van der Lith and treasurer J.J. Schneither.

The small amount of archives consists for the greater part of subscription lists, and in addition a number of letters 1888.

Size: *C.* 0.01 running metre.

**Hubrecht** [444]

Relevant for the history of Indonesia are the papers of:

*P.F. Hubrecht, 1880–1929*
Geologist, expeditions in New Guinea 1910–1913, 1920–1921. Originating from him are notes about expeditions along the Keizerin Augusta Rivier and Wilhelmsland 1910, South New Guinea 1911–1913 and North New Guinea 1920–1921. In addition there are notes about Indonesia in general.

Size: *C.* 6 running metres.

Inventory is forthcoming.

Permission is necessary for consultation of the papers.

# Gemeentelijke Archiefdienst Rotterdam (Municipal Archives of Rotterdam)

315 B Mathenesserlaan, 3021 HL Rotterdam.
Admission free. Lending to institutions.
Photocopying equipment available.
Research finished in 1977.

## Private organizations [445–446]

### Comité Adoptie Oost-Indonesië (Adoption Committee East Indonesia), 1947–1959 [445]

The committee was established in 1947 at Rotterdam. The object was to promote a good relationship, especially in the field of culture, with East Indonesia. Secretary/treasurer was C.A. 't Hart, official at the department of art affairs of the municipality of Rotterdam. The committee was abolished in 1959.

The archives consist of minutes, reports, correspondence, reports of activities in East Indonesia, accounting papers, photograph books Menado, Makasar, Kupang, Amboina, ground plans of the Fort Rotterdam. For the names and subjects mentioned in the inventory see the index.

Entry: Inventory.

Size: *C.* 1 running metre (100 inventory items).

**Koninklijke Rotterdamsche Lloyd (Royal Rotterdam Lloyd), 1880–1970** [446]

The company was established in 1883 at Rotterdam as a continuation of the firm of Willem Ruys & Zoonen; from 1875 up to 1958 a regular shipping service between Western Europe and Indonesia.

The archives consist of two parts, the papers originating from the organization in the Netherlands 1883–1970 and ships' logbooks 1880–1970, the first part including minutes, annual reports, correspondence, share registers, operating schedules, operating papers, papers agencies in Australia, New Zealand, Malaya, Thailand, Djakarta, Medan, British Borneo, economical and political papers (conferences of Den Pasar, Inter Indonesia, Linggardjati, Malino, Pangkal Pinang, RTC, police actions, Indonesian sovereignty), copies of reports Th.A.W. Ruys, war correspondence, papers KPM, Stoomvaartmij Nederland (Steamship Company Nederland), Coaling Station and Seaport Sabang, port industries in Indonesia, Japanese occupation, publicity papers.

Entry: Inventory and alphabetical lists of ships' logbooks.

Size: *C.* 400 running metres (1395 inventory items).

Permission is necessary for consultation of the papers.

**Families and private persons** [447–448]

**Muller** [447]

Papers of the following persons are relevant for the history of Asia:

a. *H. Muller Sz., July 24th 1819 Amsterdam – August 15th 1898 Wiesbaden*
Merchant of Rotterdam; 1850–1862 in partnership with his brother in law H. van Rijckevorsel, then independent; trade to the Dutch East Indies as (co-) director of the firms of Huibert van Rijckevorsel and H. Muller & Co (liquidated in 1900); also supervisory director of the NHM and managing director of several African trading companies, i.a. trade to the coast of Guinea.

This part of the archives (294 inventory items) consists of private papers, accountancy papers and business correspondence, reports, studies about i.a. the Dutch East Indies. For the names and subjects mentioned in the inventory see the index.

b. *S. Muller Hz., 1852–1915*
Son of H. Muller; municipal councillor of Rotterdam, 1886–1915.

This part of the archives (45 inventory items) consists of private and financial papers, and in addition business papers, including papers of his tour of inspection in 1897 through Java and Sumatra of the estate companies Way Lima and Kedongdong, a letter from the Lampong Caoutchouc Mij (Lampong Caoutchouc Company) 1911, photographs about the Lampongs with a letter from P.W. Janssen, maps of Way Lima and Kedongdong.

c. *Other members of the family*
Papers up to 1894 of M.C. van Rijckevorsel (wife of H. Muller), papers of H.P.N. Muller (son of H. Muller, concerned in the recruitment of Liberians for the KNIL), papers (including letters from J.B. 't Hooft and J.W.L. van Oordt) of A. van Rijckevorsel as president of the Commissie eener op te rigten Pakketvaart tusschen Nederland en Java (Committee for a Packet Service to be started between the Netherlands and Java), 1857.

Entry: F.A.M. Schoone. *Inventaris van het archief van de familie Muller, 1814–1971*. Rotterdam, 1976.

Size: *C.* 5 running metres (499 inventory items).

Permission is necessary for consultation of a part of the papers.

**Van Rijckevorsel** [448]

Papers of the following persons are relevant for the history of Asia:

a. *A. van Rijckevorsel, 1790 Rotterdam – 1864 Rotterdam*
Supervisory director of the NHM 1831–1859; president of the Kamer van Koophandel (Chamber of Commerce) Rotterdam 1838–1864; municipal councillor of Rotterdam 1839–1849; member of the Second Chamber 1841–1850; member of the First Chamber.

Originating from him are, besides papers about Dutch affairs and the Coast of Guinea, letters 1813–1864, including letters from D.F. van Alphen, A. des Amorie van der Hoeven, J.C. Baud, J. de Bosch Kemper, P.P. van Bosse, E. Canneman, A.J. Duymaer van Twist, A.R. Falck, H.F.Ch. Forstner van Dambenoy, I.D. Fransen van de Putte, J. Heemskerk Azn, W.G. Herklots, H. s' Jacob, J.K.J. de Jonge, U.G. Lauts, S. Muller, J.Th.H. Nedermeijer van Rosenthal, Ch.F. Pahud, J.J. Rochussen, J. Tideman, J.E. de Vrij, letters to H. Muller 1863, papers when he was a member of the First Chamber about import and export duties and the budget of the Dutch East Indies, 1853–1861, portrait of W.R. van Hoëvell.

b. *E. van Rijckevorsel, 1845 Rotterdam – 1928 Rotterdam*
Physicist; 1873–1877 as meteorologist i.a. in the Dutch East Indies to do magnetic observations; 1886–1904 municipal councillor of Rotterdam.

Some papers concern the Dutch East Indies, viz. papers about the estate company Way Lima, papers about his voyage to Indonesia 1870–1874 with letters from C.H.D. Buys Ballot, maps, photographs, diplomas Internationale Koloniale en Uitvoerhandel Tentoonstelling (International Colonial and Export Trade Exhibition).

Entry: B. Woelderink. *Inventaris van het archief van de Rotterdamse tak van de familie Van Rijckevorsel*. Rotterdam, 1969.

Size: *C.* 3 running metres (232 inventory items).

## Miscellaneous papers [449–450]

### Memorabilia [449]

A collection of small archives originating from committees for festivities and public subscriptions. Relevant for the history of Indonesia are:

a. Commissie tot Aanbieding van een Geschenk aan de Commissaris des Konings in Zuid-Holland Mr. [J.] Loudon ter Gelegenheid zijner Benoeming tot Gouverneur-Generaal van Nederlandsch-Indië, 1871. (Committee for the Presentation of a Gift to the King's Commissioner for South Holland [J.] Loudon on the Occasion of his Appointment as Governor General of the Dutch East Indies, 1871). Inventory item 24.

b. Sub-commissie te Rotterdam voor de Inzameling van Gelden voor de Noodlijdenden in Nederlandsch-Indië na de Uitbarsting van de Krakatau, 1883. (Sub Committee of Rotterdam for the Collection of Money for the Distressed in the Dutch East Indies after the Eruption of the Krakatau, 1883). Inventory item 37.

Entry: Inventory.

### Manuscripts [450]

a. Papers concerning the loss of the Dutch East Indies and the attitude of the press, 1946–1955. (Catalogue number 2388).

b. Ship's logbook of the frigate "Honingbij" (captain A.M. van Marion) kept by M.D. Pot, Rotterdam-Batavia 1856–1857. (Catalogue number 3314).

c. Ship's logbook of the "Zwerver" about the voyage Rotterdam-Hong Kong, 1860–1861. (Catalogue number 3315).

d. Regulations for the installation of the ventilation arrangements in ships during the transport of cargoes of coffee and rice from Indonesia, 1880. (Catalogue number 3319).

Entry: J.G.B. Nieuwenhuis. *Catalogus van de handschriften-verzameling*. Rotterdam, 1970, 1971. 2 vol.

## Asian papers in Dutch collections [451]

For a definition see no. 271.

a. **Baelde Jansen**

Papers about Lombok, 1894.

b. **Van der Hoeven**

Papers 1865–1877 of J. des Amorie van der Hoeven, consul at Macao.

c. **Van Vollenhoven**

Letters and notes about Multatuli of J. Schüffner 1920, letters to the minister of Colonial Affairs and drafts about Multatuli collected by J. van Vollenhoven, 1901–1904; journals and reports of tours of inspection to Surabaja by H. van Vollenhoven, 1920–1925.

d. **Kamer van Koophandel en Fabrieken voor Rotterdam (Chamber of Commerce for Rotterdam)**

Data about businesses which had connections with Asia and Oceania.

# CHAPTER VI

# UNIVERSITIES

## Gemeenteuniversiteit Amsterdam (Municipal University of Amsterdam)

## Universiteitsbibliotheek (University Library)

425 Singel, 1012 WP Amsterdam.
Admission free. Lending to students whose study has a scientific character (on introduction to students out of the University of Amsterdam) and to institutions. Lending of manuscripts and maps is only permitted to institutions with a repository specially adapted for them.
Photocopying equipment available.
Research finished in 1979.

### Library [452]

1,800,000 books (including the institutions of the departments) and 42,275 periodicals.

Directed to all subjects whieh are taught at the University.

### Manuscripts and collections [453–462]

#### Introduction to the collection of manuscripts [453]

The collection consists of 500,000 manuscripts in total and *c.* 150 objects bearing on the history of the art of writing. The acquisitions which were deposited at the University Library before 1917 are accessible by means of the *Catalogus der handschriften;* 7 vol. The papers which have been acquired after 1917 are accessible by means of an alphabetic and systematic catalogue on index cards. Among the private collections of manuscripts of the University Library are relevant for this guide: the Multatuli archives, the Frederik van Eeden collection and the J.A. Dortmond collection.

The papers relevant for the history of Asia and Oceania are mentioned below with each catalogue.

#### Catalogue of the manuscripts; vol. I, Amsterdam, 1899 [454]

This catalogue includes an inventory of a collection of autographs, given by P.A. Diederichs in 1875.

The papers are included in the catalogue in alphabetic order to name of author with an alphabetic index to name of addressee. For the history of Asia and Oceania may be relevant: a letter from W.A. Alting to F.H. Engelberts 1785, letters from J.Chr. Baud to the brothers Diederichs, . . . Michaud, C.J. Wolterbeek and an extract from the register of the proceedings and resolutions of the governor general for the commander of the squadron in the East Indies 1819–1846, letters from C.L. Blume to the brothers Diederichs and . . . van Emden 1829–1841, G.A.G.Ph. van der Capellen to the brothers Diederichs, M.G. Bieber, M.C. van Hall, J.H. van Kinsbergen, C. Kramer, G. Brolik, C.J. Wolterbeek 1810–1842, E. Douwes Dekker, C.Th. Elout to J.A.K. van Hasselt and others 1813–1827, N. Engelhard to Th. van Swinderen 1824, A.R. Falck to J.H. van Kinsbergen and others 1810–1843, C.M. Hafner 1835, W.R. van Hoëvell, D. van Hogendorp, J.W. Janssens, H.M. de Kock, J. Olivier, Ch.F. Pahud, C.G.C. Reinwardt, A. van Riebeeck, J.J. Rochussen, Raden Saleh, H. Schlegel, J.A. Schneither, Ph.F. von Siebold, P.J. Veth and C.J. Wolterbeek.

**Catalogue of the manuscripts; vol. II, Amsterdam, 1902** [455]

This catalogue contains a description of the manuscripts of the Stedelijke Bibliotheek (Municipal Library) of Amsterdam, with the subsequent acquisitions.

The catalogue is systematically classified, with an alphabetic index to name of author. With the enumeration of the papers relevant for Asia and Oceania the catalogue-numbers are mentioned in brackets. Relevant may be: "Oosterse Handschriften" (Oriental Manuscripts) and lontars in Persian, Arabic, Kawi concerning i.a. the koran, Gorontalo, Bali, Hindustan, Borneo (no. 38–65, 1564, 1565), notes concerning the Dutch East Indies originating from C.A. Duker (no. 1292), J.N. Vosmaer. Brief description of the south eastern peninsula of Celebes, 1835 (no. 1293), S. Fuller. Journals of his voyages Boston-East Indies and Cape of Good Hope-Port Jackson-Amboina 1799–1800 (no. 1309, 1310).

**Catalogue of the manuscripts; vol. IV – Letters. Amsterdam, 1919** [456]

This catalogue contains a description of a collection of letters. The descriptions are alphabetically arranged to name of author, with an alphabetic index to name of addressee. For the history of Asia and Oceania may be relevant the letters written by: J.Chr. Baud, C. Busken Huet, Ch.M. van Deventer, C.Th. van Deventer, M.L. van Deventer, E. Douwes Dekker, A.R. Falck, C.J. Hasselman, J.K. Hasskarl, F.J. Haver Droeze, H.M. de Kock, U.G. Lauts, J. Loudon, J. van Maurik, J. Olivier, J. Pijnappel Gz, J.K.W. Quarles van Ufford, O. van Rees, C.G.C. Reinwardt, J. Schneither, C. Snouck Hurgronje, *Tijdschrift voor Nederlandsch-Indië,* P.J. Veth, W.H. de Vriese.

**Catalogue of the manuscripts; vol. VI, Amsterdam, 1917** [457]

This catalogue contains a description of papers originating from P.A. Diederichs, viz. the "kleinere afdelingen" (smaller sections). After the enumeration below, the number of the place in the store where the papers can be found is mentioned.

For the history of Asia and Oceania may be relevant: an indictment (in Chinese) of Jao Hao Shu 1711 (136 Bu) and a letter from Said Bey to Bessan Bey 1836 (136 Bp).

**Catalogue of the manuscripts; vol. VII, Amsterdam, 1923** [458]

This catalogue contains a description of the manuscripts which, by virtue of loan-contract, are deposited at the University Library of Amsterdam. After the enumeration below, the number in the catalogue is stated.

For the history of Asia and Oceania may be relevant: texts from the koran (no. 376–378) and Mohammed Al-Djazari, Arabian memorial book (no. 379).

**Catalogue on index cards** [459]

Included are those papers which are not mentioned in the catalogues above. After the description the number of the place where the papers can be found is mentioned.

For the history of Asia and Oceania may be relevant:

a. Letters, notes, anthropological notes, with drawings of . . . Sasse. (VIII B 8–14)

b. Communications about culture in the East Indies. (Cw 53)

c. Translation of an inscription on a monument in Malaya. (Cw 63–64)

d. Sketches and scenes in and about Java, by J. Hageman JCz, with a biography of him by P.J.A. Meersmans, *c.* 1837. (VIII E 12)

e. Drafts of outgoing letters of J. Hageman JCz. (X F 28)

f. East Indies letters, beginning 19th century. (Gi 7)

g. Medical topography of the Dutch East Indies, originating from . . . van Leent. (Brievendoos Cw 71)

h. Resolutions and instructions concerning the Dutch East Indies, originating from W.E. Bom. (VIII A 37)

i. Papers about geographical names of the Dutch East Indies, originating from A.G. Bon. (XII B 1 a–g)

j. Malay, Arabic and Javanese manuscripts. (XI G 12–18)

k. Tobabatak and Mandailing manuscripts. (XI H 10–12)

m. Javanese manuscripts. (XII B 8)

n. Notes of Ph. van Falkenburg of the lecture on "colonial constitutional law" by Professor Van der Lith, *c.* 1888. (X A 10)

o. Drafts of outgoing letters of C. Busken Huet, 1876–1879. (XV F 24)

p. Letter from E. du Perron to M.J. Premsela, 1925. (Eu 1–61)

q. "Mohammed der Prophet und seine Frauen", by B. Schidlof. (I C 57)

r. "Shanghai kennt keine Gerechtigheit", by J.G. Rieger. (I C 56)

s. "Oostersche geschiedenis", by P.A. Pijnappel. (V A 10a)

t. "Ontdekkingsreizen", by P.A. Pijnappel. (V A 13a)

u. Letter from E.B. Kielstra to P.A. Pijnappel. (Gf 32)

v. Arabic manuscript. (III H 54)

w. Arabic manuscript. (Gm 7)·

x. Letters, notes and regulations concerning the Koninklijk Instituut voor Taal-, Land- en Volkenkunde (Royal Institute of Linguistics and Anthropology), originating from P.A. Pijnappel. (Gm 54)

y. Letter from E. Littré to S.E.W. Roorda van Eysinga. 1878. (Gn 22)

z. "Een zaaier; studiën over Multatuli", by C. Vosmaer, 1874. (XX A 22)

aa Arabic manuscript originating from Java. (XIV C 43)

ab Persian manuscript, with notes by A.L. Wensinck. (XIV C 43)

ac "De oden van Hafiz Shamzeddin", by Diven Hafiz. (XV F 12)

ad Arabic manuscript. (XV F 12)

ae Eulogy of Mohamed, Arabic manuscript, with descriptions of the manuscripts mentioned sub aa up to ae by C. Snouck Hurgronje. (XV F 11)

af Manuscript on tree bark. (XX O 12)

ag Personal notes of P.A. Pijnappel. (XXIII D 1 b)

ah Arabic manuscript. (XXIII D 4)

ai Manuscript from India. (Gx 5)

aj Arabic manuscripts. (XXV C 2, 3)

ak "Bijdragen tot de geschiedenis en de beoefening des koophandels", by A. Horstman, mainly concerning Japan, *c.* 1814–1831. (XXV C 32)

am Notes about the constitutional law of the Dutch East Indies, by Ph. Kleintjes. (Gd 30)

an Lontar in Kawi. (XII C 14 a)

ao Serat pawukon, by Adjar Artandriya. (XI G 18)

ap Hikayat Radja Djumdjuma. (XI G 13)

aq Hikayat mi'radj naik kalangit. (XI G 14)

ar Koran exegesis in Persian and Arabic. (XII B 12)

as Paper about the history of religion, Malay text in Arabic characters. (II H 58)

at Letters (in Malay) from Hadji Mohammed Tengku Udjung Pulan and Ma'Fatima to Putjut Abd'urrahman at Mụmala Batu, i.a. 1910 (1289 hidjrah). (Bb 30 a, b)

au Malay religious text. (II H 39)

av Notes about the history of the literature of British India, by A.G.C. de Vries, *c.* 1845. (XIII G 13–17)

aw Rahden Sapri, Dutch translation with transcription. (Dk 127)

ax Arabic religious texts. (XI H 13)

ay Lontar in Kawi. (XII B 8)

az Arabic manuscript of Ibn Wajjara and Ahmed bin Jusuf. (VIII E 29)

ba Arabic primbon. (Cn 11)

bb Translation of the Koran into Dutch, with a preface by S. Swigger. (XIV F 10)

bc "Alt iransche Wörterbuch", by C. Bartholomae. (XV B 33)

bd Notes about the four books of Confucius, by H.J. Klaproth. (Dp 101)

be Chinese characters, with transcription by K.F. Neumann. (Ed 138 a–d)

bf Turkish, Arabic and Persian texts. (Cn 12)

bg Description of Ceylon. (Be 74 b)

bh Notes about the geography and anthropology of the Dutch East Indies, by . . . van der Hell. (VIII F 39)

bi Communications about culture, manners and customs in Padang. (C 1053 a–c)

bj Travel story of . . . Tromp, travelling on commission to Singkel and back to Padang. (Ag 109)

bk "Beschrijving van den throon van den Grooten Mogol te Haan Daarchach". (Bf 74 c)

bm Regulations of the private landed property west of the Tjimanuk, by . . . van der Hell, copied from the *Staatsblad van Nederlandsch-Indië*. 1836, no. 19. (VIII F 38)

**Multatuli archives** [460]

These archives are partly deposited at the University Library, partly at the Multatuli-Museum (20 Korstjespoortsteeg, Amsterdam).

The archives consist of papers by and about Multatuli (pseudonym of E. Douwes Dekker).

*E. Douwes Dekker, March 2nd 1820 Amsterdam – February 19th 1887 Nieder-Ingelheim*
1838 to the Dutch East Indies; 1840 second clerk Algemene Rekenkamer (General Audit Department); 1842 comptroller West Coast of Sumatra; 1846 at the disposal of the resident of Krawang; 1846 clerk Bagelen; 1848 secretary Menado; 1851 assistant resident Amboina; 1852 on leave to Europe; 1856 assistant resident Lebak; after difficulties with his official superiors 1856 honourable discharge; 1857

to the Netherlands, publishes the *Max Havelaar* in 1860, start of his literary career.

The archives are chronologically arranged and range from 1838 up to 1887.

Size: *C.* 1.50 running metres.

**F.W. van Eeden Collection** [461]

For the history of Asia may be relevant the papers to and/or from: Amsterdam Langkat-Company, H. van Booven, H. Borel, G. Brom, B. Canter, A. Cohen, A. Coomaraswamy, L. Couperus, Ch.M. van Deventer, A. Dharma Pala, E. Douwes Dekker, N.J.G. Engelkamp-Caesar, J. Greshoff, O. Knaap, Koloniaal Instituut (Colonial Institute), M.Chr. Kooy-van Zeggelen, W.J. Leyds, A.P. Prins, J. Sawanudra, Sosrokartono, S. Sundar Singh, A. Surjosuprapto, Rabindranath Tagore, K. Uchimura, Vaderlandsche Club (Fatherland Club), J.D. Veegens, B. Veth, M. Vierhout, H.P.L. Wiessing and A.A.H. de Wit.

**J.A. Dortmond Collection** [462]

Based on this collection the script-museum J.A. Dortmond has been built up. By means of objects and manuscripts the history of calligraphy is illustrated from *c.* 3000 before Christ till now.

The collection includes Arabic, Persian, North and South Indian, Batak, Javanese and Balinese manuscripts.

**Maps** [463]

The collection of maps comprises *c.* 110,000 map sheets, partly owned by the University Library, partly on loan from the Koninklijk Nederlands Aardrijkskundig Genootschap (Royal Dutch Geographic Society). Besides the modern maps and series of maps a large portion of the collection consists of old maps. Of the Dutch East Indies all official series of maps are present, and they are practically complete. The collection of atlases comprises *c.* 2500 old and modern (national, subject, aerial-photograph etc.) atlases.

## Artis-Bibliotheek (Artis Library)

45A Plantage Middenlaan, 1018 DC Amsterdam.
Admission by appointment.
Research finished in 1979.

**Archival documents** [464]

In the library is a packet of letters originating from W. de Haan and written by the members of the Natuurkundige Commissie van Nederlandsch-Indië (Physical

Committee of the Dutch East Indies): H. Boie, J.C. van Hasselt, P.W. Korthals and P. van Oort 1819–1837. Also deposited at the library are drafts of articles on poisonous snakes and articulate animals by E. Jacobson.

**Drawings** [465]

The collection consists of drawings of i.a. J.J. van den Bosch, P. Camper, J. van Haastert, M.S. Merian (Surinam), A. Schouman, G. Schouten, K.N. Swierstra, C.B. Voet, for the greater part representing animals and plants in Indonesia.

## Instituut voor Taxonomische Zoölogie (Institute of Taxonomic Zoology)

53 Plantage Middenlaan, 1018 DC Amsterdam.
Admission by appointment.
Research finished in 1979.

**Siboga expedition** [466]

The archives are originating from the leader of the expedition (1899–1900) M.W.C. Weber. They include correspondence 1877–1910, alphabetically arranged, about zoological and oceanographic research in the Indonesian archipelago. However, the archives also contain papers originating from other persons and organizations and about other expeditions, for example the Snellius expedition.

Total size: *C.* 5 running metres.

## Vrije Universiteit Amsterdam (Protestant University Amsterdam)

Historisch Documentatiecentrum voor het Nederlands Protestantisme (Centre of Historical Documentation on Dutch Protestantism) (1800– )
1105 De Boelelaan, 1081 HV Amsterdam.
Admission by appointment.
Consultation of the archives only for scientific research.
Research finished in 1979.

**Introduction** [467]

The Documentatiecentrum (Centre of Documentation) is in charge of the archival documents from 1800 forward, which are deposited at the Vrije Universiteit. The documentation is directed towards the history of Protestantism in the Netherlands.

## Private persons [468–470]

### H. Colijn, June 12th 1869 Haarlemmermeer – September 18th 1944 Ilmenau [468]

1892 as second lieutenant to the East Indies, 1894 Lombok expedition; 1895–1905 with some interruptions in Atjeh, 1896 first lieutenant, 1901 captain, also 1900–1901 civil commander Tapatuan, 1901 adjutant to J.B. van Heutsz, 1904 on leave, 1904 civil commander Padang Tidji, 1905 Gajo- and Alas area; 1905 at the Departement van Binnenlands Bestuur (Department of East Indian Civil Service) for the organization of the administration of the possessions outside Java and Madura; 1907 leaves the service as major and becomes assistant government commissioner for the decentralization; 1908 adviser outside districts; 1909 to the Netherlands and member of the Second Chamber; 1911 minister of War; 1913 voyage to China and the Dutch East Indies; 1914 director BPM; 1922 chief editor of *De Standaard;* 1923 minister of Finance; 1925 prime minister; 1926 member of the First Chamber; 1933–1939 prime minister; in 1933 and 1935 i.a. also minister of Colonial Affairs (Supplements *RAvNI*).

The collection consists of papers chronologically arranged from 1892. Superficial tests showed that the collection includes letters to Colijn's wife, written by Colijn from Atjeh, also papers J.W.G. Gunning about protestant and catholic mission (1904, 1905, 1907), Native and European education i.a. in Tondano, Sangihe and Talaud (1905, 1906, 1908, 1909), Native government (1905), government in Atjeh, Palembang, Benkulen, Kutai (1900–1909), coal-mines in Pulu Laut (1912), colonization in north west Borneo (1913), Sarekat Islam (1913), papers about E.F.E. Douwes Dekker, Tjipto Mangunkusumo and J.G. van Hoven (1913), papers of governor general A.W.F. Idenburg, Tapanuli (1916), [W.V.] Smeets (1917), [C.] Lulofs, coolie-labour ordinance by A.F. van Blommestein (1918), . . . Creutzberg about administrative reform (1920), Djambi (1921), rubber (1933), R.M. Notosutarso (1933), [B.C.] de Jonge (1934–1936), also papers about Surinam and Curaçao (1934).

The collection is chronologically arranged.

Size: *C.* 5 running metres.

### A.W.F. Idenburg, July 22nd 1861 Rotterdam – February 28th 1935 The Hague [469]

1882 as second lieutenant at Batavia, 1883 first lieutenant; 1885 Surabaja; 1888 Atjeh; 1892 Batavia; 1894 on the half-pay list; 1896 chief of the cabinet of the general staff, also 1898–1901 vice president of the Vereeniging tot Oprichting en Instandhouding van Christelijke Bewaar- en Zondagsscholen in de Residentie Batavia (Association for the Establishment and Maintenance of Christian Infant- and Sunday Schools in the Residency of Batavia); 1901 to the Netherlands and member of the Second Chamber; 1902 minister of Colonial Affairs; 1905 governor of Surinam; 1908 minister of Colonial Affairs; 1909–1916 governor general of the Dutch East Indies; 1918–1919 minister of Colonial Affairs; 1920 member of the First Chamber; 1924 till his death in 1935 member of the Raad van State (Council of State). (Supplements *RAvNI*).

The collection consists of correspondence 1902–1935, with a few copies from 1893; files 1901–1935; autobiography 1882–1913; miscellaneous papers (mainly in print) 1910–1935. The chief points of this collection are his three ministerial terms and his office of governor general. For the subjects and names of places and persons mentioned in the inventory see the index.

Entry: Inventory.

Size: *C.* 3 running metres, *c.* 150 inventory items.

**L.W.C. Keuchenius, October 21st 1822 Batavia – December 17th 1893 The Hague** [470]

1844 lawyer and official to the procureur-generaal (public prosecutor) in the High Court of Justice at Surabaja; 1848 member of the Raad van Justitie (Council of Justice) at Batavia; 1850 advocaat-generaal (solicitor general); 1851 counsel of the Hooggerechtshof (High Court of Justice); 1854 to the Netherlands and secretary general of the Ministry of Colonial Affairs; 1859 member of the Raad van Indië (East Indies Council); 1865 member of the Second Chamber; 1867 editor of the *Nieuw Bataviasch Handelsblad;* 1869 lawyer and solicitor with the Hooggerechtshof (High Court of Justice) at Batavia; 1879 member of the Second Chamber; 1888 minister of Colonial Affairs; 1890 till his death in 1893 member of the Second Chamber.

Concerning the Dutch East Indies this very small collection contains only papers about vaccination, viz. a few letters by . . . Fabius *c.* 1880.

Entry: Inventory.

Size: 1 folder (19 inventory items).

See also no. 592.

**Asian papers in Dutch collections** [471]

For a definition see no. 271.

**A. Kuyper**

Papers about colonial policy, mission, correspondence with A.W.F. Idenburg.

**Photographs** [472]

Collection of photographs concerning A. Kuyper and H. Colijn.

# Rijksuniversiteit Leiden (State University Leyden), 1575

## Universiteitsbibliotheek (University Library)

70–74 Rapenburg, 2311 EZ Leyden.
Admission free. Lending on introduction (books published before 1800, manuscripts and maps only to scientific libraries and archives repositories).
Photocopying equipment available.
Research finished in 1971.

**Library** [473]

1,700,000 books; 6500 periodicals.

Directed to all subjects which are taught at the university. The library of the former Ministry of Colonial Affairs is being included in the library (catalogue in course of preparation, see also no. 25 C).

For literature concerning the UBL see i.a. P.C. Molhuysen. *Geschiedenis der Universiteits-Bibliotheek te Leiden*. Leiden, 1905.

### Western manuscripts [474–496]

**Introduction to the "Westerse Handschriften" (Western Manuscripts)** [474]

The "Westerse Handschriften" of the UBL are divided into three parts:

Manuscripts owned by the university, with the centre of gravity on the classical antiquity and Middle Dutch literature. They are subdivided into:

a. Collections still bearing the name of the original owner.

b. Manuscripts and collections included in the open series Bibliotheca Publica Latina (BPL).

Only the open series BPL contains collections and papers concerning Asia 1796–1949.

Entry to the Asian papers: *Catalogus compendiarius;* continens codices omnes manuscriptos qui in Bibliotheca Academia Lugduno-Batavae asservantur. Leiden, 1932–1935. 3 vol. (Catalogue [of the] Bibliotheca Academiae Lugduno-Batavae, vol. XIV, XXII, XXVI).

Vol. 1(XIV). *Codices manuscripti Bibliothecae Academiae Lugduno-Batavae.* Including (p. 97–161) BPL 1–2267 so far as not included in vol. 2. The remaining numbers (BPL 2268–2772), which have been consulted for this research, are mentioned in manuscript in the journal.

Vol. 2(XXII, XXVI). *Epistolaria et alba amicorum.* 2 vol. (Alphabetic index to important authors of letters and album inscriptions).

Manuscripts given on loan to the university. Of this part only the property of the Maatschappij der Nederlandse Letterkunde (Society of Netherlands Literature) is relevant (see no. 493–496). The collection of the Maatschappij is for the greater part directed to Dutch literature.

Entry: *Catalogus der bibliotheek van de Maatschappij der Nederlandsche Letterkunde.* Leiden, 1887. Including the manuscript numbers Ltk 1–1085 so far as not included in vol. 2. The remaining numbers (Ltk 1086–2048) which have been consulted for this research are not published in a printed catalogue.

The second part of vol. 2(XXVI) of the *Catalogus compendiarius* contains the alphabetic index for the most important authors of letters and album inscriptions of Ltk 1–1085 under the title: "Naamlijst van de brievencollectie in het bezit van de Maatschappij der Nederlandsche Letterkunde".

The archives of the university deposited with the UBL. These archives may be relevant for biographical data of former students of the Leyden university.

Entry: H. Hardenberg. *De archieven van Senaat en faculteiten, benevens het archief van de Academische Vierschaar der Leidsche universiteit.* Zaltbommel, 1935.

To all manuscripts present in the UBL the following entries have been made:

Alphabetic catalogue to authors and addressees of letters and authors of album inscriptions, consisting of 2 parts:

a. Numbers listed in the printed catalogues (see vol. 2 of the *Catalogus compendiarius* and "Naamlijst van de brievencollectie").

b. Numbers not listed in the printed catalogues.

Alphabetic author catalogue of text manuscripts, for example travel reports.

Systematic catalogue of anonymous manuscripts, including also archives of associations etc.

The research at the UBL was based on the written journals made to the collection of manuscripts of the UBL and the Maatschappij der Nederlandsche Letterkunde. This is done because these journals are more comprehensive than the printed catalogues mentioned above. However, since the alphabetic author catalogue contains names of authors and addressees of letters, which were not mentioned in the journals, one may, by means of this catalogue, find letters concerning Asia 1796–1949, which are not included in this guide. In the description of the collections the manuscript numbers will be mentioned in brackets. Bundles smaller than 0.10 running metre are regarded as miscellaneous papers.

## Organizations (BPL) [475–476]

### Internationales Archiv für Ethnographie, 1887– [475]

In 1887 J.D.E. Schmeltz, conservator of the Rijksmuseum voor Ethnografie (State Museum for Ethnography) at Leyden (the future Rijksmuseum voor Volkenkunde

(State Museum for Anthropology)), took the first step for the establishment of the two monthly periodical *Internationales Archiv für Ethnographie.* The first issue was published in January 1888. The object was to make the ethnographical material in European museums better known by means of scientific articles. From 1967 the periodical was published under the title *Tropical Man.*

The small amount of archives consists of copybooks (copies of letters despatched) 1887–1898 by the editor J.D.E. Schmeltz with some letters received (gaps in the years 1896 and 1897).

Entry: Chronologically arranged.

Size: *C.* 0.25 running metre (numbered BPL 2404, 9 vol.). Originating from A.W. Nieuwenhuis (see no. 485) but not included in his collection.

**Rijksinstelling voor Onderwijs in de Taal-, Land- en Volkenkunde van Nederlandsch-Indië te Leiden (State Institution for the Teaching of Linguistics, Geography and Ethnography of the Dutch East Indies at Leyden), 1864–1877** [476]

When, in 1863, the Koninklijke Akademie te Delft (Royal Academy of Delft) was transformed in the Polytechnische School (Polytechnic School), it was provided that the training for East Indies officials should be given at the Rijksinstelling (State Institution) of Leyden, which was established in 1864. President was T. Roorda, after his death in 1874 succeeded by H. Kern, secretary was P.J. Veth. There were classes in Javanese, Malay, geography and ethnography of the Dutch East Indies, Javanese law and institutions of the Islam by Roorda, Kern, Veth and J. Pijnappel Gzn. The course took two years. The Rijksinstelling was abolished in 1877, because it was apparent that the Indische Instelling (East Indies Institution) of Delft, also established in 1864, attracted far more students. With the closing of the archives, library and manuscripts were transferred to the university.

The archives consist of minutes of the Raad van Bestuur (Court of Governors) August 15th 1864 – September 18th 1877; letters received 1864–1877 mainly from the Ministry of Home Affairs; copies of letters despatched 1864–1877 mainly to the Ministries of Home and Colonial Affairs, with draft rules of the examination of 1871; accounts 1871–1877; album of the students 1864–1877.

Size: *C.* 0.15 running metre (numbered BPL 1057).

See also no. 431 and 442.

**Private persons (BPL)** [477–489]

**C. Busken Huet, December 28th 1826 The Hague – May 1st 1886 Paris** [477]

1844–1849 student of divinity at Leyden; 1851 minister of the Walloon church at Haarlem; 1862 on the staff of the *Opregte Haarlemsche Courant,* from 1863 also of *De Gids;* 1868 sent on a government mission to the East Indies to inquire into the freedom of the press, editor in chief of the *Java-Bode;* 1872 manager-editor in chief of the *Algemeen Dagblad van Nederlandsch-Indië;* 1876 to Europe, where he, besides his literary activities, also wrote articles for the *Algemeen Dagblad van Nederlandsch-Indië* as European correspondent.

The collection consists of two parts: letters 1859–1874 mainly from E.J. Potgieter (BPL 2197) and papers of various persons (BPL 2206) viz. letters 1861–1902 from J. Bosboom, L.G. Bosboom-Toussaint, J.P. Heye, J. Israëls and H.P.G. Quack to Busken Huet, G. Busken Huet and E.J. Potgieter; letters 1861–1876 from C. Busken Huet; sermons, literary manuscripts, draft articles and Malay notes.

Size: *C.* 0.10 and 0.30 running metre (numbered BPL 2197, 2206)

See also *Brieven van E.J. Potgieter aan C. Busken Huet;* uitgegeven door G. Busken Huet. Haarlem, 1901–1902. 3 vol.
*Brieven van Cd Busken Huet;* uitgegeven door zijne vrouw en zijn zoon [G. Busken Huet]. Haarlem, 1890. 2 vol.

**M.J. de Goeje, August 13th 1836 Dronrijp – May 17th 1909 Leyden** [478]

1854 student of divinity, from 1856 of classical literature at Leyden, graduated in 1860 on an Arabic text concerning the geography of North Africa; 1859 second adjutor Legati Warneriani UB Leiden, continued on the staff up to 1909; 1862–1879 district school inspector in South Holland; 1866 extraordinary, 1869–1906 ordinary professor of Arabic, Syric and Chaldean languages, from 1877 only teaching Arabic; in 1906 succeeded by C. Snouck Hurgronje. (*Levensberichten der afgestorven medeleden van de Maatschappij der Nederlandsche Letterkunde.* 1909/1910, p. 34–59).

The collection consists of letters 1860–1909 directed to him. For the correspondents mentioned in the alphabetic catalogue see the index.

Entry: Alphabetic arrangement.

Size: *C.* 1.50 running metres (numbered BPL 2389).

**C.F.M. de Grijs** [479]

1856 from Batavia to China to study the dialect of Emoi, he presumably stayed there till 1865; 1865–1885 interpreter for the Chinese language at Semarang, also from 1866 extraordinary member of the Wees- en Boedelkamer (Orphan Chamber) and from 1868 secretary of the second inspection department for education. (J.J.C. Francken and C.F.M. de Grijs. *Chineesch-Hollandsch woordenboek van het Emoi dialect.* Batavia, 1882/*RAvNI*).

The collection consists of translations of papers concerning Formosa, jurisdiction, opium; penal code and law of inheritance of the Chinese; papers and translations concerning dicing games, opium, Chinese oath, calendar, annual festivals, doctrine of causality, peace treaty between China and Great Britain and France 1860, treaty between China and the Netherlands 1863; the sacred books of the Chinese; the book of cause and working in Dutch and French translations.

Size: *C.* 0.25 running metre (numbered BPL 1780–1784).

**H.A. Hamaker, February 25th 1789 Amsterdam – October 7th 1835 Neerlangbroek** [480]

Student of Arabic with J. Willmet at Amsterdam; 1815 professor at Franeker; 1817 extraordinary, 1822 ordinary professor of Semitic languages at Leyden.

The collection includes papers concerning Arabic, Sanskrit, Persian languages; letters 1815–1835 from L.Ph.C. van den Berg, A.R. Falck, J.W.J. Juynboll, Ph.P. Roorda van Eysinga; Arabic and Jewish texts, texts from the coast of Malabar.

Size: *C.* 0.10 running metre (numbered BPL 2013).

**Hoffmann/Geerts** [481]

This collection consists of the papers originating from J. [J.] Hoffmann and A.J.C. Geerts.

*J. [J.] Hoffmann, February 16th 1805 Würzburg – January 19th 1878 Leyden*
1830 to the Netherlands, after having met Ph.F. von Siebold study of Japanese and Chinese languages at Leyden; 1846 translator Japanese with the Dutch government; 1855 professor Japanese. (H. Kern. Levensbericht van J.J. Hoffmann. In *Jaarboek van de Koninklijke Akademie van Wetenschappen*. 1878, p. 1–17).

*A.J.C. Geerts, c. 1843 – August 30th 1883 Yokohama*
1868 teacher of chemistry at the military medical school at Utrecht; 1869 teacher of chemistry at Nagasaki; 1874 adviser of the Ministry of Health at Tokio, was concerned in the establishment of medical laboratories at Kioto, Tokio and Yokohama, and was from 1874 on the staff of the laboratory at Yokohama. (Yamada Akiyoshi. Le docteur A.J.C. Geerts. In *Annales de l'Extrême Orient et de l'Afrique*. 1883/84 (vol. 6), p. 134–136).

The collection consists of "handschriften van het Nipponarchief" (manuscripts of the Nippon archives) including Japanese and German texts, texts for a dictionary, maps; "rapporten van Hoffmann over de Chinezen opstand" (reports of Hoffmann about the revolt of the Chinese) in Borneo 1850–1853, including a Japanese translation of the Bible and treaty the Netherlands-Liu Kiu 1859; "Japanse en Chinese bibliografie" (Japanese and Chinese bibliography) with name lists of Japanese and Chinese; "aantekeningen, dagregisters e.d. betrekkelijk de kwekelingen van de Yama-mayu" (notes, daily records etc. concerning the pupils of the Yama-mayu); "correspondentie over Yama-mayu" (correspondence about Yama-mayu) 1862–1875 with Japanese texts, Chinese grammar; "Hoffmanniana" offprints *c.* 1855; "nagelaten papieren van Geerts" (papers of Geerts left after his death) including translations of Japanese natural historical works, translation of Honrō Kō moka; "album" including Japanese manuscripts and drawings with notes and newspaper cuttings of Hoffmann and Geerts.

Size: *C.* 0.40 running metre (7 bundles and 1 exercise-book numbered BPL 2186).

**F.W.T. Hunger, April 3rd 1874 Amsterdam – June 3rd 1952 Leyden** [482]

Botanist; 1899–1905 employed with 's-Lands Plantentuin (National Botanical Garden) at Buitenzorg; then in the Netherlands lecturer of botany at the Utrecht University; 1906–1910 manager of the Vereenigd Proefstation voor Bergcultures (Associated Research Station for Upland Cultures) in Java; from 1917 employed in the Netherlands, i.a. lecturer of the history of botany at the State University of Leyden and manager of the Instituut voor de Geschiedenis van de Medicijnen en de Biologie (Institute for the History of Medicine and Biology) at Leyden.

The collection consists of papers concerning biologists and biological subjects, with correspondence 1925–1944.

Size: *C.* 0.90 running metre (numbered BPL 2596, no. 1–33).

See also no. 210.

**U.G. Lauts, May 19th 1787 Amsterdam – July 25th 1865 Utrecht** [483]

1821 teacher Dutch literature at Brussels; 1830–1840 professor of the Koninklijk Instituut voor het Zeewezen (Royal Institute for Maritime Matters) at Medemblik; publications from 1830 about the Dutch colonies, mainly in Asia.

The collection consists of diplomas; papers concerning his period in Brussels and Medemblik; notes and manuscript concerning linguistics; notes and manuscript publications concerning i.a. the Dutch colonies in Asia, viz. *Handleiding voor de aardrijkskunde van Neêrlands Oost- en West-Indische bezittingen*. Utrecht, 1834. *Onderzoek naar geest en strekking van het [Londens] traktaat van den 17den Maart 1824*. Amsterdam, 1837. *Geschiedenis van de vestiging, uitbreiding, bloei en verval van de magt der Nederlanders in Indië*. Groningen etc., 1852–1860. With correspondence. For the names and subjects mentioned in the list of component parts see the index.

Entry: List of component parts of the collection, drawn up during this research.

Size: *C.* 2.50 running metres (numbered BPL 2245).

**H.G. Nahuys van Burgst, March 28th 1782 Amsterdam – January 12th 1858 Breda** [484]

1798 law student at Harderwijk, after taking his degree in 1803, lawyer at Amsterdam up to 1805; 1805 to the East Indies with the committee C.Th. Elout, [Ch.H. van] Grasveld as councillor of finance and public lands, returns in 1807; 1809 to the East Indies to hand over despatches to H.W. Daendels, becomes inspector of the roads and postal services and in 1810 administrator of the timber woods; 1811 captured by the British, in 1813 to the Netherlands by way of Bengal and England; 1814 instruction to prepare the sending of troops to Java, returns in the same year; 1816 as lieutenant colonel to Java to prepare the arrival of the commissioners general, becomes resident of Djokjakarta; 1818 commissioner of the administration of Borneo; 1819 resident of Djokjakarta and Surakarta, leaves for the Netherlands in 1823 after a disagreement with G.A.G.Ph. van der Capellen about letting of land; 1826 as resident of Surakarta and commissioner of Djokjakarta to the East Indies; 1831 returns to the Netherlands; 1836–1839 for the last time in the East Indies as councillor of the Indies and general major; 1841 discharge.

The collection consists of papers 1741–1852, besides papers concerning the Principalities and Borneo also papers concerning i.a. Bali, Batavia, Cheribon, Kediri, Kedu, Palembang, Patjitan and outside the East Indies concerning Cape of Good Hope, Siam and the West Indies, mainly pertaining to administration, finance, letting of land and military affairs, besides papers in Dutch also papers in Javanese, Malay and Chinese. For the subjects and names of persons and places mentioned in the inventory see the index.

Entry: [J.J.F.] Wap. Bronnen voor de taal-, land- en volkenkunde van Neêrlandsch Indië. In *BKI.* 1864 (vol. 11), p. 179–191.

Size: *C.* 2.50 running metres (numbered BPL 616, 22 inventory items).

**A.W. Nieuwenhuis, May 22nd 1864 Papendrecht – September 22nd 1953 Leyden** [485]

Medical studies at Leyden and Freiburg; 1889 physician at Freiburg; 1891 army surgeon at Sambas; 1894 till his return to the Netherlands in 1901 travels through Borneo; 1904–1934 professor of ethnology of the Dutch East Indies at Leyden; he was president of the Nederlandsche Anthropologische Vereeniging (Dutch Anthropologist Association) and the Instituut voor Tropische Geneeskunde (Institute for Tropical Medicine), honorary member of the Bataviaasch Genootschap van Kunsten en Wetenschappen (Batavian Society of Arts and Sciences) and the Koninklijke Natuurkundige Vereeniging in Nederlandsch-Indië (Royal Physical Association in the Dutch East Indies) and (chief) editor of the *Internationales Archiv für Ethnographie* (succeeded J.D.E. Schmeltz in 1910 up to 1947), *Janus*, *Indië* (1917–1927) and *Tropisch Nederland* (1928–1940). (Supplements *RAvNI/Keesings historisch archief.* 1953/54).

The collection consists of six parts:

printed matter;

manuscripts of published work;

notes taken during his travels through Borneo 1894–1900, university lecture notes, notes and manuscript articles concerning geography and ethnology in the Dutch East Indies and outside the East Indies;

archives of the *Internationales Archiv für Ethnographie,* including minute books 1894–1944, financial supporting vouchers, correspondence, reviews and articles received (see also no. 475);

archives of *Janus;* Archives Internationales pour l'Histoire de la Médicine et la Géographie Médicale, consisting of minutes 1896–1903, correspondence and articles received;

miscellanies, consisting of letters and other papers 1911–1940, concerning several committees, associations and periodicals, including *Indië* and *Tropisch Nederland* and personal correspondence.

For the subjects and names of persons, committees, associations and magazines mentioned in the journal see the index.

Entry: List of component parts of the collection included in the journal.

Size: *C.* 1.75 running metres (numbered BPL 2591, 42 items).

**C.G.C. Reinwardt, June 5th 1773 Lütringhausen – March 6th 1854 Leyden** [486]

1787 to Amsterdam, study of chemistry, botany and medicine; 1801 professor of biology; 1808 manager of a zoological garden to be established successively at Soestdijk, Haarlem and Amsterdam; 1810 professor of chemistry, pharmacy, botany at Amsterdam; 1816 to the East Indies with the commissioners general G.A.G.Ph. van der Capellen, C.Th. Elout and A.A. Buyskes, becomes director of Agriculture, Arts and Sciences and later of Education, and also commissioner for the civil medical service, engaged in the establishment 1818 of 's-Lands Plantentuin (National Botanical Garden) at Buitenzorg, to which he is succeeded by C.L. Blume, travels, i.a. with Van der Capellen, through Java and in 1821–1822 to the Moluccas, Timor and North Celebes; 1823 professor of chemistry, botany, geology and mineralogy at Leyden, 1844 only botany, 1845 retired.

The collection consists of two parts:

university lecture notes, discourses, notes etc. concerning botany, geology, mineralogy, medicine, chemistry and pharmacy (BPL 593–615);

service records, correspondence, travel reports, notes concerning his stay in the East Indies etc. (BPL 2425).

For the subjects and names of places and persons mentioned in the journal see the index.

Entry: List of the parts of the collection included in the journal.

Size: *C.* 2.60 running metres and 0.20 running metre (numbered BPL 593–615, BPL 2425, no. 1–23).

See also W.H. de Vriese. *Reis naar het Oostelijk gedeelte van den Indischen archipel, in het jaar 1821;* door C.G.C. Reinwardt; uit zijne nagelaten aanteekeningen opgesteld; met een levensbericht en bijlagen vermeerderd. Amsterdam, 1858.

**M. Schaalje** [487]

1864–1872, 1875–1888 interpreter of the Chinese language in Riouw, also member of the school committee at Tandjungpenang, the orphan chamber and the committee for the management of the cemeteries for Europeans and persons possessing the same status; 1890–1896 interpreter of the Chinese language on the East Coast of Sumatra (Medan), also secretary of the Immigranten-Asyl (Immigrants Home) at Medan; Schaalje was a member of the Bataviaasch Genootschap van Kunsten en Wetenschappen (Batavian Society of Arts and Sciences) *(RAvNI)*.

The collection consists of an Emois-Dutch Dictionary; reports and laws concerning secret Chinese societies 1877–1890; papers of and/or concerning E.H. Röttger, J.W. Yong and W.P. Groeneveldt; papers concerning the Chinese language and the Emoi dialect; family names of Chinese; translations of Chinese stories; papers in Chinese.

Size: *C.* 0.45 running metre (numbered BPL 2104–2106).

**C. Snouck Hurgronje, February 8th 1857 Oosterhout – June 26th 1936 Leyden** [488]

Study of divinity and literature at Leyden, 1880 doctor of Semitic languages; 1881–1887 teacher at the Indische Instelling (East Indies Institution) at Delft, 1884/85 journey to Djeddah and Mecca; 1889 to the East Indies to study the Islam; 1891 adviser oriental languages and Mohammedan law, travels to Atjeh, i.a. on instructions of C. Pijnacker Hordijk; 1898–1906 adviser for Native and Arab affairs, up to 1903 alternately in Sumatra and Batavia, i.a. engaged in the expeditions of J.B. van Heutsz in Atjeh and the Djambi expedition; 1907–1927 professor Arabic at Leyden, succeeding M.J. de Goeje, also advising in the Netherlands on the hadji policy of the East Indies government, supporters C. van Vollenhoven and C.Th. van Deventer; 1909 stay in Constantinople; 1920–1922 secretary of the senate and than rector magnificus (principal) of the Leyden University; 1927 established the Oostersch Instituut (Oriental Institute) with the money that had been presented to him when he retired, by a committee with i.a. A.W.F. Idenburg, after he retired he remained teacher of Achinese and Arabic; was i.a. president of the Bataviaasch Genootschap van Kunsten en Wetenschappen (Batavian Society of Arts and Sciences), president (1918–1920) of the Koninklijke Akademie van Wetenschappen (Royal Academy of Sciences), honorary member of the KITLV, president (from 1927) of the Oriental Institute. (Supplements E. Gobée and C. Adriaanse. *Ambtelijke adviezen van C. Snouck Hurgronje, 1889–1936*. 's-Gravenhage, 1957–1965. 3 vol. (RGP; kleine serie no. 33–35)/*RAvNI*).

The collection consists of letters received and drawn up by Snouck in Dutch, Malay and Arabic from *c.* 1880, besides papers received and drawn up by Snouck and originating from the Oosters Instituut mainly concerning Atjeh and the Islam and the drafts of his official advices. For the names and subjects mentioned in the papers originating from the Oosters Instituut see the index.

Entry: Inventory of the letters is forthcoming, provisional entry to the papers originating from the Oosters Instituut: shelflist.

Size: Letters *c.* 1 running metre, "stukken Oosters Instituut" (papers Oriental Institute) *c.* 1 running metre. The collection of letters may be consulted no sooner than 1986, the collection is deposited at the Legatum Warnerianum of the UBL.

**P.J. Veth, December 2nd 1814 Dordrecht – April 14th 1895 Arnhem** [489]

1832 study of literature and divinity at Leyden; 1838 lecturer English and Malay at the KMA at Breda; 1840 doctor honoris causa because of the publication of an Arabic text; 1841 professor of oriental languages, Hebrew and Old Testament at Franeker, 1843 Amsterdam; 1864 professor at the Rijksinstelling voor Onderwijs in de Taal-, Land- en Volkenkunde van N.I. (State Institution for the Teaching of Linguistics, Geography and Ethnography of the Dutch East Indies) at Leyden; 1877–1885 professor of geography and ethnology at the university of Leyden; from 1843 he was a member of the board of the NBG, editor of *De Gids* (1844–1876), *De Indiër* (1850–1852 with D.C. Steyn Parvé) and the *Tijdschrift voor Nederlandsch-Indië* (1867–1871), member of the committee for regulation of the training of East Indies officials, engaged in the establishment (1873) of the KNAG, of which he was president up to 1884, concerned in the Sumatra-expedition (1877–1879), concerned in the colonial exhibition at Amsterdam (1883), publications about i.a. Java, Western district of Borneo, Atjeh, J.F.G. Brumund, W.R. van Hoëvell, J.K.J. de Jonge, C.G.C. Reinwardt and T. Roorda.

The collection consists completely of letters directed to him.

Entry: Alphabetically arranged.

Size: *C.* 0.25 running metre (numbered BPL 1756).

See also no. 502.

**Asian papers in Dutch collections (BPL)** [490]

For a definition see no. 271.

a. **H. van Roijen, (BPL 318–333)**

Besides papers concerning the Navy at Flushing a "Historieel verslag van het geschilstuk over de Oost-Indische zaken, met een voorafgaand extract uit de geschiedenis van D. van Hogendorp" 1801 and papers concerning the trade to the colonies 1801–1802.

b. **J. Valckenaer, (BPL 1030–1043)**

Papers drawn up when ambassador to Madrid (1799–1801) concerning the expedition to Java.

**Collections of letters (BPL)** [491]

These collections consist both of miscellaneous letters and of small collections of letters, which have been subdivided according to author. They have been alphabetically arranged to author and through author to addressee. The UBL contains two of those collections, a closed (BPL 885) and an open collection (BPL 1886). The collections include the following "Asian letters":

a. *BPL 885*, letters *c.* 1840–1870 from H.N. van der Tuuk to W. Fockens, . . . Meyer and J. Roos, letter 1883 from P. Jansz. to H.N. van der Tuuk.

b. *BPL 1886*, letters *from* H. Kern, C. Leemans (1885), P. Lieftink (1952), H.J. Oosting (1880), C.G.C. Reinwardt (1823–1851), T. Roorda (1843), W. Roosegaarde Bisschop, C. Snouck Hurgronje (1917), F.L. Splitgerber (1838), E.C.E. Steinmetz (1848 about the Bali expedition), W.H. de Vriese (1854–1861), A.R. Wallace (1909).

*To* J.T. Bodel Nijenhuis, J.C.E. Buschmann, J. van der Hoeven, A.P.H. Hotz, W. von Humboldt, S.H. Koorders, Ministry of Home Affairs, Ministry of Colonial Affairs, N.W. Posthumus, V. de Stuers, H.N. van der Tuuk, W.H. de Vriese, H. van der Wijck.

NB. Through the alphabetic author catalogues to letters (see no. 474) presumably more "Asian letters" may be found.

**Miscellaneous papers (BPL)** [492]

The Asian papers 1796–1949 mentioned below are arranged according to number of manuscript.

a. *BPL 619*, "Baten en lasten der residentie Bezoeki", 1829.

b. *BPL 770*, "Observatiën genomen aan boord van den Corvet The William, gecommandeerd door den Kapt. Luit. Pieter Fereris in Oost-Indiën", 1802, 1806.

c. *BPL 928*, "Generaal rapport der Commissie tot Inspectie der Gezamelijke Jaccatrasche Bovenlanden en Preanger Regentschappen", composed by C.W. Thalmen, P.F. Smith and J.G. Bauer, by order of H.W. Daendels, 1808.

d. *BPL 929*, discourse on free labour in Java by H. de Bruyn at Surabaja, 1859.

e. *BPL 935*, papers concerning the latter years of the VOC and the establishment of the Comité tot de Zaken van de OI Handel en Bezittingen (Committee for the Affairs of the East Indies Trade and Possessions), 1795–1796.

f. *BPL 972*, "Hortus Indicus", drawings of Indonesian plants with Portuguese and Indonesian names, 1871.

g. *BPL 1053*, statistics of Pasuruan, 1831–1836.

h. *BPL 1054*, F.C.F. Deeleman. "Beschrijving van de dorpshuishouding der Javanen"; (copy 1847) with 2 letters from the Minister of Colonial Affairs to T. Roorda, 1847.

i. *BPL 1055*, papers originating from H.L. Wichers: "De Javaansche wetten; of, De Hangger-Pradotto, Hangger-Sepoeloeh en Hangger Hageng", 1745–1818;

memorandum and resolution concerning "Voorschriften ten dienste van de regters in de Landraden en Regtbanken, en voor de administratie der Crimineele Justitie", 1824, 1825;

"Compendium der voornaamste Javasche wetten, naukeurig getrocken uit het Mahomedaansche wetboek", s.a.;

"Compendium der Indiesche, Mahomedaansche en Chineesche wetten", including extract general resolutions of Batavia, 1717–1766.

j. *BPL 1056,* M.W. Scheltema. Lectures on public law and the administrative system in the Dutch East Indies, delivered in the Rijks-Instelling voor Onderwijs in de Taal-, Land- en Volkenkunde van Nederlandsch-Indië (State Institution for the Teaching of Linguistics, Geography and Ethnology of the Dutch East Indies), with notes about monetary circulation, land tax, trade tax, opium and farmed out means, *c.* 1870.

k. *BPL 1164,* A.B. Cohen Stuart. Stamlijst van Javaansche vorstenhuizen naar het Javaansche handschrift van Raden Ngabehi Ronggo Warsito (Genealogical register of Javanese princely·families based upon the Javanese manuscript of Raden Ngabehi Ronggo Warsito). Surakarta, 1852.

l. *BPL 1218,* J.G. de Waldkirch Ziepprecht. Register der predikanten en gemeenten in Nederland (Register of the ministers and parishes in the Netherlands). Numbered a–ah:

o. East Indies ministers 1629–1794;
p. East Indies ministers 1816–1881;
q. East Indies parishes with their ministers;
af Namelist of Roman Catholic ecclesiastics in the East Indies.

m. *BPL 1249,* C.A.L. van Troostenburg de Bruyn. Alfabetische naamlijst van Oost-Indische predikanten, opgemaakt in 1876 en verwerkt in *Biographisch woordenboek van Oost-Indische predikanten.* 1893.

n. *BPL 1346,* letters 1824–1832 to J.B. de Salis (assistant resident of Djokjakarta, resident of Rembang) from J.A. Batiest, C.M. Baumhauer, L. [P.J.] du Bus de Gisignies, H.M. de Kock, J.B. de Linge, A.W. de Man, J.J. Melvill van Carnbee, H.G. Nahuys van Burgst, J.F.W. van Nes, R. Scott, J.I. van Sevenhoven, F.G. Valck, W.J. Waterloo, with some papers of himself.

o. *BPL 1382,* notes of the lectures of G.A. Wilken about ethnography of the East Indies archipelago, taken by A. Paets tot Gansoyen, 1888–1890.

p. *BPL 1383,* notes of the lectures of G.A. Wilken about physical geography of the East Indies archipelago, taken down by A. Paets tot Gansoyen, [1888–1890].

q. *BPL 1497,* C.G.C. Reinwardt. Dictata in rem herbarium, s.a.

r. *BPL 2102,* papers of I. Titsingh concerning Japan, viz.: "Lijkpligten, Feesten ter eere van de goden ingesteld, Bericht wegens het land Doria, Jeso-Ki, beschrijving van Jeso, Beschrijving van het naalde steeken en Moxa branden, Inleiding tot de huwelijksplechtigheden, Huwelijksplechtigheden bij landlieden, ambachts- en kooplieden", s.a.

s. *BPL 2120,* notes of the lectures of P.A. van der Lith about colonial constitutional law, *c.* 1875.

t. *BPL 2126,* Sincernus. Sieben Jahre auf Java und Sumatra, *c.* 1850.

u. *BPL 2176,* "Lexicon sinico-russicum", *c.* 1850.

v. *BPL 2180,* J. [J.] Hoffmann and L. Serrurier. Japanese-Dutch-English dictionary, *c.* 1875. 30 vol.

w. *BPL 2190,* W.H. de Vriese. Dictata in organographiam phytotoniam, 1841–1842.

x. *BPL 2205,* collection of autographs, including "Eenige aanteekeningen over Ficus Indica" by C.G.C. Reinwardt.

y. *BPL 2216,* notes of the lectures of J.M.M. de Groot about geography and ethnology of the Dutch East Indies and the inhabitants of Java, end 19th century.

z. *BPL 2252,* letters 1870–1881 to W.N. du Rieu from . . . Robidé van der Aa, A.E. Hooykaas-Schroot, P.N. van Kampen, [P.A.] Leupe, L.D. Petit, M. de Pinto, J.W. Segaar, T.C.L. Wijnmalen and others with notes of Du Rieu concerning the *Repertorium op de literatuur betreffende de Nederlandsche koloniën,* by C. Hooykaas.

aa *BPL 2281,* H.W. Tydeman. "Koloniën of bezittingen in andere werelddelen", s.a.

ab *BPL 2302,* manuscript index of "Ancient Buddhism" by M.W. de Visser, *c.* 1930.

ac *BPL 2317,* notes of the lectures of J.T. Buys about colonial constitutional law, *c.* 1900.

ad *BPL 2321,* letters 1867–1868 from E. Douwes Dekker to C. Busken Huet.

ae *BPL 2386,* "Verslag van de Commissie van Advies inzake Aanpassend Middelbaar (voorbereidend hooger) Onderwijs in Nederlandsch-Indië", 1916, with notes of C. Snouck Hurgronje and letters from W.B. Kristensen, J.H. Kern and others.

af *BPL 2396,* notes of the lectures of G.K. Nieuman about geography and ethnology of the Dutch East Indies, taken down by D.J. Cramer, 1891–1892.

ag *BPL 2397,* notes of the lectures of L.W.C. van den Berg about Muslim law, taken down by D.J. Cramer, *c.* 1890.

ah *BPL 2398,* notes of the lectures of J.A. van den Broek about the history of the Dutch East Indies, taken down by D.J. Cramer, *c.* 1890.

ai *BPL 2399,* notes of the lectures of [H.] Kern about East Indies archaeology, copied from the notes of N. Adriani by [P.S. van Ronkel], s.a.

aj *BPL 2400,* notes of the lectures of [H. Kern] about general linguistics of the Dutch East Indies, taken down by [P.S. van Ronkel], *c.* 1890.

ak *BPL 2472*, "Relaas van den oorlog met de oproerige Chinezen ter West-Kust van Borneo, beginnende in 1850", [by . . . van Prik, 1853].

al *BPL 2476*, drawings by I.J. Nanlohy of Amboinese plants, animals, villages etc. *c*. 1850.

am *BPL 2487*, letters 1843–1883 to R.P.A. Dozy (for the names of the correspondents mentioned in the alphabetic author catalogue, see the index).

an *BPL 2488*, list of governors, residents etc. who [from 1638–1876] were attached to the government of Makasar.

ao *BPL 2701*, bundle of "Multatuliana", including letters 1863, 1864 from E. Douwes Dekker, M.F.C. Douwes Dekker-Hamminck Schepel, . . . van Stockum, W.I. van Dijck, C.Th. van Deventer to E.L. Beyers, A. Hotz and J.C.P. Hotz with "Een herinnering aan Multatuli" by H.W. Cornelis.

ap *BPL 2728*, "Verslag van een reis naar Japan, Formosa, Hongkong, de Philippijnen, Indo-China en Singapore, in opdracht van de regeering van Nederlandsch-Indië aan W.Th. de Vogel", *c*. 1920, with photographs.

aq *BPL 2731, 2732*, decrees of appointment 1834–1880 concerning J.J. Hasselman with papers of and about him and letters 1854–1875 to him from C. Busken Huet, J.P. Cornets de Groot, A. van Delden, E. Douwes Dekker, W.C.E. de Geen, C.J. Heemskerk Azn., L.W.C. Keuchenius, W. van Limburg Stirum, A. de Ruyter de Wildt, R.J. Schimmelpenninck van Nijenhuis with some Indonesian letters.

**Organizations (Ltk)**

**De Gids, 1837–** [493]

The first number of this periodical was published in 1837, with E.J. Potgieter and C.P.E. Robidé van der Aa as editors, under the title *De Gids*; nieuwe vaderlandsche letteroefeningen, from 1839 published under the title *De Gids*. Although the contents were with time liable to change, the following subjects appeared to get the most attention: linguistics, literature, art, theology, political science, history, geography and ethnology (see the systematic registers 1837–1936).

In the systematic registers the articles concerning Asia are inserted principally under the following headings: "Letterkundige geschiedenis en kritiek; Oosterse letteren" (Literary history and reviews; Oriental literature). "Taalkunde; Oosterse talen" (Linguistics; Oriental languages). "Godsdienstwetenschap; Oosterse godsdiensten" (Theology; Oriental religions). "Godsdienstwetenschap; zendingswezen" (Theology; the mission). "Staatswetenschap; algemeen en buitenland" (Political science; general and abroad). "Staatswetenschap; Nederland en zijn koloniën" (Political science; the Netherlands and its colonies). "Staatswetenschap; slavernij en emigratie" (Political science; slavery and emigration). "Krijgswezen" (Military matters). "Zeewezen" (Maritime matters). "Handel en scheepvaart" (Trade and shipping). "Geschiedenis; nieuwere tijd; verschillende volken en staten" (History;

more recent times; several nations and states). "Geschiedenis; nieuwere tijd; Nederlanders in andere werelddelen" (History; more recent times; Dutchmen in other continents). "Land- en volkenkunde; Azië en Australië" (Geography and ethnology; Asia and Australia). "Land- en volkenkunde; Nederland en zijn koloniën" (Geography and ethnology; the Netherlands and its colonies) (headings from the register 1837–1876). The following editors have written many contributions concerning Asia: P.J. Veth (1844–1876 editor), H.C. Millies (1848–1850 editor), C.Th. van Deventer (1901–1915 editor) and Ch.M. van Deventer (1914 editor).

The archives consist of papers 1840–1939, viz. letters received 1840–1939, copybook of letters despatched 1843–1851, postbook of articles received 1883–1938, cash books 1909–1932.

Entry: Chronological arrangement of letters received, with author index.

Size: *C.* 4 running metres (numbered Ltk 1888).

**Asian papers in Dutch collections (Ltk)** [494]

For a definiton see no. 271.

a. **R. Fruin, (Ltk 1555–1560)**

Notes concerning Formosa with letters 1887, 1890 from J.A. Hingman and F. Louw (Ltk 1556, no. 38), memorandum . . . Jonckbloet concerning Atjeh 1874 (Ltk 1558, no. 30).

b. **A.C. Kruseman, (Ltk 1389–1510)**

Letters from J.H.C. Kern (Ltk 1444).

## Collections of letters (Ltk) [495]

For a definition see no. 491.

This collection includes the following letters concerning Asia: Letter 1843 from W.L. de Sturler to the Maatschappij der Nederlandsche Letterkunde (Society of Netherlands Literature); postcards 1935 from A. Verwey to J.J.L. Duyvendak (Ltk 1004).

**Miscellaneous papers (Ltk)** [496]

The Asian papers 1796–1949 mentioned below have been arranged according to number of manuscript.

a. *Ltk 592*, "Alphabeth der Tamoulsche taal", s.a.

b. *Ltk 593*, "Proeve eener Japansche spraakkunst", by J.H. Donker Curtius; belonging to a despatch from the Dutch commissioner in Japan to the governor general, with an explanation by J. [J.] Hoffmann, 1855.

c. *Ltk 732,* "Ceilonsche secreete resolutiën van 25 Juli 1795 tot 15 Februari 1796"; with abstracts of letters from G. van Angelbeek to W.J. van de Graaff, 1796.

d. *Ltk 1056,* "Japansche spraakkunst door een Japanneesch in het Nederlandsch geschreven", with letter to J. [J.] Hoffmann from J.S. Bodel Nijenhuis and L.S.F. Janssen, 1856.

e. *Ltk 1119,* letters 1797–1842 from A.R. Falck to C. Vollenhoven.

f. *Ltk 1263,* R. [P.A.] Dozy. "Het Islamisme", *c.* 1875.

g. *Ltk 1537,* letters 1869–1872 from E. Douwes Dekker to H.H. Huisman.

h. *Ltk 1685,* letters 1876–1877 from E. Douwes Dekker to A.C. Loffelt.

i. *Ltk 1691,* letters 1881–1883 from E. Douwes Dekker to T.H. de Beer.

j. *Ltk 1721,* letters to T.H. de Beer from P. Brooshooft, G. Busken Huet, E. Douwes Dekker, M.F.C. Douwes Dekker-Hamminck Schepel, G.J. Dozy, J. van Maurik, F. Netscher, C. van Nievelt, P.J. Veth and others.

k. *Ltk 1722,* biographical notes i.a. concerning E.H. Baumhauer, R. [P.A.] Dozy, M.J. de Goeje, J. Heemskerk Azn., P.J. Veth, for the greater part written by themselves.

l. *Ltk 1733,* poems of Soera Rana [ps. of J. Esser jr].

m. *Ltk 1776,* letters 1855–1866 from H. Kern to M. de Vries.

n. *Ltk 1873,* documents concerning i.a. C. and G. Busken Huet.

o. *Ltk 1973,* poems of Jan Prins [ps. of C.L. Schepp].

## Oriental manuscripts (Legatum Warnerianum) [497]

More than half of this collection consists of manuscripts originating from Indonesia. They have been given or lent by i.a. the Rijksinstelling voor Onderwijs in de Taal-, Land- en Volkenkunde van Nederlandsch-Indië (State Institution for the Teaching of Linguistics, Geography and Ethnology of the Dutch East Indies), the NBG, the Koninklijke Akademie van Wetenschappen (Royal Academy of Sciences), J.G.H. Gunning, G.A.J. Hazeu, C. Leipoldt, P.A. van der Lith, Ch.A. van Ophuijsen, C. Snouck Hurgronje, H.N. van der Tuuk and L. Warner (who gives his name to the section). The catalogue on index cards (subdivided according to languages) mentions the following Asian languages: Arabic (by far the larger portion of the collection), Javanese, Malay, Achinese, Sundanese, Balinese, Buginese, Makasar, Sasak, Batak, Madurese, Persian, Sanskrit and other languages (including i.a. Avestic, Burmese, Cambodian, Chinese, Gujarati, Hindustani, Khmer, Malayalam, Pali, Sanskrit, Siamese, Singhalese, Tamil, Telugu, Tibetan, Urdu), various languages (including i.a. Amarasian, Amboinese, Bandjarese, Banggaian, Batak, Batjan, Bawean, Bimanese, Buru, Ceram, Dajak (Ladak-, Ngadju- and Sea-), Endenese, Gajo, Kisar, Kupang, Lampong, Lauranese, Letinese, Manggarai, Mentawei, Minahasan, Minangkabau, Moa, Padang (Uplands), Penang, Redjang, Rotinese, Sangirese, Sawu, Serawak, Simalur, Susu, Sumba, Sumbawa, Ternatan, Tettum, Timorese).

Entries:

Catalogue on index cards (systematic for languages).

R.P.A. Dozy. *Catalogus codicum Orientalium Bibliothecae Academiae Lugduno-Batavae.* Leiden, 1851–1877. 6 vol. (Vol. 3–6 by P. de Jong, M.J. de Goeje and M.Th. Houtsma).

C. Landberg. *Catalogue de manuscrits arabes provenant d'une bibliothèque privée à el-Medine et appartenant à la maison E.J. Brill.* Leiden, 1883.

M.J. de Goeje and M.Th. Houtsma. *Catalogus codicum Arabicorum Bibliothecae Academiae Lugduno-Batavae.* Leiden, 1888, 1907. 2 vol. (Vol. 2 by De Goeje and Th.W. Juynboll).

A.C. Vreede. *Catalogus van de Javaansche en Madoereesche handschriften der Leidsche Universiteits-Bibliotheek.* Leiden, 1892.

H.H. Juynboll. *Catalogus van de Maleische en Sundaneesche handschriften der Leidsche Universiteits-Bibliotheek.* Leiden, 1899.

J. Brandes. *Beschrijving der Javaansche, Balineesche en Sasaksche handschriften, aangetroffen in de nalatenschap van H.N. van der Tuuk en door hem vermaakt aan de Leidsche Universiteits-Bibliotheek.* Batavia, 1901–1926. 4 vol. (Vol. 4 revised by Th.G.Th. Pigeaud).

H.H. Juynboll. *Supplement op den catalogus van de Javaansche en Madoereesche handschriften der Leidsche Universiteits-Bibliotheek.* Leiden, 1907, 1911. 2 vol.

H.H. Juynboll. *Supplement op den catalogus van de Sundaneesche handschriften en catalogus van de Balineesche en Sasaksche handschriften der Leidsche Universiteits-Bibliotheek.* Leiden, 1912.

Ph.S. van Ronkel. *Supplement-catalogus der Maleische en Minangkabausche handschriften.* Leiden, 1921.

R.A. Kern. *I La Galigo;* catalogus der Boegineesche tot den I La Galigo-cyclus behoorende handschriften bewaard in het Legatum Warnerianum te Leiden, alsmede in ander Europeesche bibliotheken. Leiden, 1939. (Supplements 1954, 1961).

P. Voorhoeve. *Handlist of Arabic manuscripts in the library of the University of Leiden and other collections in the Netherlands.* The Hague, 1957.

Th.G.Th. Pigeaud. *Literature of Java;* catalogue raisonné of Javanese manuscripts in the library of Leiden and other public collections in the Netherlands. The Hague, 1967–1970. 3 vol.

Size: *C.* 16,000 manuscript numbers, of which *c.* 10,000 numbers are concerning Asia.

See also P. Voorhoeve. Indonesische handschriften in de Universiteitsbibliotheek te Leiden. In *BKI;* vol. 108, p. 209–219.

**Maps and marine charts** [498]

Approximately 2700 catalogue numbers of this collection are pertaining to Asian maps. Of these *c.* 1500 numbers have relation to Indonesia and *c.* 1200 to the rest of Asia. The most important Indonesian headings are: the East Indies (general) *c.* 200 numbers, Java *c.* 500 numbers, Sumatra *c.* 350 numbers, Celebes/Moluccas/ New Guinea/Lesser Sunda Islands *c.* 300 numbers, Borneo *c.* 150 numbers. The most important Asian headings are: Asia (general) *c.* 100 numbers, Siberia *c.* 120 numbers, Central Asia *c.* 80 numbers, China *c.* 170 numbers, Japan *c.* 70 numbers, South Asia/Nearer Asia *c.* 250 numbers, Persia *c.* 50 numbers, India/Pakistan *c.* 200 numbers, Ceylon *c.* 50 numbers, Further India (Burma, Siam, Cochin China, Malaya, Singapore) *c.* 50 numbers, Philippines/North Borneo/Sulu *c.* 50 numbers. The larger portion of these maps and charts has been made after 1800.

**Portraits, pictures** [499]

The collection of portraits consists of pictures of professors and governors of the University of Leyden, alphabetically arranged according to name. The collection of pictures consists for the greater part of engravings, lithographies etc. Included is a bundle of the Dutch East Indies and a bundle of America/Africa/Asia.

Entry collection of pictures: Alphabetic catalogue to place, alphabetic catalogue to name of artist, systematic catalogue to subject.

For the collection of pictures see: H. van de Waal. De rangschikking en catalogisering van een topografische atlas. In *Oudheidkundig jaarboek.* 1940 (vol. 9), no. 1.

## Academisch Historisch Museum (Academic Historical Museum)

73 Rapenburg, 2311 GJ Leyden.
Admission by appointment.
Research finished in 1979.

**Historisch Indisch Dispuut "Dwipantara" (Historical East Indies Debating Society "Dwipantara")** [500]

The small amount of archives consists of minutes, annual reports, lists of members and texts of lectures 1939–1952.

Size: *C.* 0.05 running metre.

**Indologisch Werkgezelschap (Indologist Working Company)** [501]

The small amount of archives consists of correspondence and reports 1937–1950.

Size: *C.* 0.05 running metre.

**P.J. Veth, December 2nd 1814 Dordrecht – April 14th 1895 Arnhem** [502]

1832 study of literature and divinity at Leyden; 1838 lecturer English and Malay at the KMA at Breda; 1840 doctor honoris causa because of the publication of an Arabic text; 1841 professor of oriental languages, Hebrew and Old Testament at Franeker, 1843 at Amsterdam; 1864 professor at the Rijksinstelling voor Onderwijs in de Taal-, Land- en Volkenkunde van NI (State Institution for the Teaching of Linguistics, Geography and Ethnography of the Dutch East Indies) at Leyden; 1877–1885 professor of geography and ethnology at the Leyden university; from 1843 he was a member of the board of the NBG, editor of *De Gids* (1844–1876), *De Indiër* (1850–1852 with D.C. Steyn Parvé) and the *Tijdschrift voor Nederlandsch-Indië* (1867–1871), member of the committee for regulation of the training of East Indies officials, engaged in the establishment (1873) of the KNAG, of which he was president up to 1884, concerned in the colonial exhibition at Amsterdam (1883), publications about i.a. Java, western district of Borneo, Atjeh, J.F.G. Brumund, W.R. van Hoëvell, J.K.J. de Jonge, C.G.C. Reinwardt and T. Roorda.

The bundle of papers consists of correspondence with his son D.D. Veth in the Dutch East Indies 1877–1885 and with a few others, articles, appointments.

Size: *C.* 0.05 running metre.

See also no. 489.

## Katholieke Universiteit Nijmegen (Roman Catholic University Nijmegen)

Katholiek Documentatie Centrum (Catholic Centre of Documentation), KDC 1969.
36 Erasmuslaan, 6525 GG Nijmegen.
Admission free, permission is necessary for consultation of the archives.
Photocopying equipment available.
Research finished in 1979.

**Introduction** [503]

The KDC is housed in the University Library of the Roman Catholic University of Nijmegen and uses all attendant facilities. The KDC makes collections and documentations pertaining to Christianity and Catholicism in general and Dutch Catholicism from the 19th century in especial. A guide of the archives present at the KDC is forthcoming.

**Publications** [504]

*Jaarboek van het Katholiek Documentatie Centrum.* Nijmegen, 1971– .

Also published in *Archief voor de geschiedenis van de Katholieke Kerk in Nederland;* vol. 13– .

*Archieven van het Katholiek Documentatie Centrum;* vol. 1– . Nijmegen, 1973– .

## Organizations [505–508]

### Roomsch Katholieke Staatspartij (Roman Catholic State Party), RKSP 1891–1941 [505]

*C.* 1870 at first a combination of Catholic electoral associations; 1891 the Catholic members of the Second Chamber consolidated, through intermediary of H.J.A.M. Schaepman, into the Roomsch Katholieke Kamerclub (Roman Catholic Chamber Club), the future Tweede Kamerfractie (group of Catholic members of the Second Chamber); 1904 Algemene Bond van Roomsch Katholieke Kiesvereenigingen (General Federation of Roman Catholic Electoral Associations); 1918 Algemene Bond van Roomsch Katholieke Rijkskiesorganisaties (General Federation of Roman Catholic National Electoral Organizations); 1926 Roomsch Katholieke Staatspartij; 1937 Stichting Indisch Documentatie Bureau (Foundation East Indies Documentation Bureau) of the RKSP (its task was to provide information about and documentation of colonial affairs); 1941 liquidation of the Party owing to the German occupation; 1945 re-established with the name of Katholieke Volkspartij (Catholic People's Party). (See no. 506).

The archives consist of papers originating from the actual Party from 1918 forward, such as annual reports, postbooks, minutes, papers of the party council, party leaders, party executive and several committees. In addition the deposited archives of the Kamerclub 1891–1939, papers originating from C.M.J.F. Goseling about the Indisch Documentatie Bureau, formations of new governments and E.J.C. Goseling 1937–1939, and papers originating from H.A.M.T. Kolfschoten i.a. about the liquidation of the RKSP in 1941.

Entry: W.A.A. Mes. Inventaris van [het archief van] de Roomsch Katholieke Staatspartij, 1918–1945. In *Jaarboek van het Katholiek Documentatie Centrum;* vol. 1. 1971, p. 10–52.

Size: *C.* 2 running metres (184 inventory items).

**Katholieke Volkspartij (Catholic People's Party), KVP 1945–** [506]

The Party was established in 1945 as a continuation of the RKSP (see no. 505).

The archives consist of papers 1945–1975, including papers of the Kamerfracties (groups of Catholic members of the Chambers), the party bureau, papers about the Indonesian question, Linggardjati, mission, New Guinea. Moreover there are archives of KVP documentation, including papers about the [Ch.J.I.M.] Welter question, relations with Indonesia, New Guinea and papers about Amboinese people.

Entry: Provisional lists.

Size: *C.* 46 running metres.

**Sint Melania-Werk (Saint Melania Work), SMW 1921–1969** [507]

The object of the association, which was established in 1921, was 'To promote the Catholic education of the native woman for her family and society in our colonies'; the inspiration came from R.A. Kartini; the tasks were education, medical care and emancipation of women; at first these tasks were especially performed by W. Creutz Lechleitner, later by J. Schmutzer (deputy in the Dutch East Indies 1926–1928) and E.A. van Arcken; 1969 embodied into the Landelijk Katholieke Vrouwengilde (National Catholic Women's Guild). (See also J.H. Roes. Het Sint Melania-Werk; leerschool voor ontwikkelingshulp en missie. In *Jaarboek van het Katholiek Documentatie Centrum;* vol. 1. 1971, p. 111–137.)

The archives consist of minutes of the general meetings, meetings of the central committee and the executive committee, rules, annual reports, correspondence i.a. with Java, South Sumatra and Palembang, accounting papers, documentation including photographs of projects in Java.

Entry: T.T.A.B.M. van der Aalst. Inventaris van het archief van het Sint Melania-Werk, 1921–1969. In *Jaarboek van het Katholiek Documentatie Centrum;* vol. 1. 1971, p. 66–85.

Size: *C.* 3 running metres (139 inventory numbers).

**Apostolisch Vicariaat Batavia (Apostolic Vicariate Batavia)** [508]

Established as Apostolische Prefectuur (Apostolic Prefecture) in 1807; 1842 Apostolisch Vicariaat; territory: Bantam, Buitenzorg, Preanger, Semarang, Kudus, Pati (Rembang), Grobogan (Blora), Magelang and Temanggung (Kedu), Surakarta, Klaten and Djokjakarta; the care for the Vicariaat was entrusted to the Jesuits Fathers. *(Jaarboek [van] de Katholieke missie in Nederlandsch Oost-Indië).*

The archives which are in Jakarta in the archbishopric there, have been microfilmed for the period 1807–1949. The micro fiches may be consulted at the KDC. They include papers about article 117 of the government regulations, trade-union movement, mission organizations.

Entry: Inventory is forthcoming.

Size: *C.* 3000 micro fiches.

## Asian papers in Dutch archives and collections [509]

For a definition see no. 271.

a. **P.J.M. Aalberse**

Correspondence with and papers about the Ministry of Colonial Affairs, H. Colijn, Linggardjati, M. [J.M.] van Poll, East Indies pensions (1933–1934), A.W.F. Idenburg, League of Nations, government regulations, Ch.J.I.M. Welter.

b. **G. Brom**

Papers about his studies on Dutch East Indies art (i.a. *Java in onze kunst* and Multatuli) with letters from J. Schmutzer, J. Wils, [Ch.J.I.M.] Welter, T. Witlox, J. Hoogveld, the Bond van Nederlandsch-Indische Kunstkringen (Federation of Dutch East Indies Art Societies), R. van Geelkerken, D. Mataleupan, W. Asselbergs, . . . Baner, H. Endepols, H. Knippenberg, Th. Meltz, J.W. Meyer Ranneft, W.O.J. Nieuwenkamp, J.Ph. Vogel, A. Zimmerman, J. Nieuwhuis, [C.C.] Berg, C.W. Wormser, [W.] Creutz Lechleitner, L. van Rijckevorsel, E.J. Bergh van Eijzinga-Elias, C.C. ter Haar, K. ter Laan, B. Musterd, J. Pollmann, G. Stuiveling, A. te Winkel, M.C. Conley, B. Momme Nissen, Multatuligenootschap (Multatuli Society), E.J. Hoek, W. Pompe, Th.H. Sengers, Chr. Teeuwen.

c. **Katholieke Jeugdbeweging (Catholic Youth Movement), KJB**

Correspondence with Indonesia 1948–1972, Asian countries 1955–1972, Australia and New Zealand 1955-1972.

d. **Katholieke Radio Omroep (Catholic Broadcasting Service), KRO**

The collection of glass negatives contains the text of a lecture by A. Steger "China en de Chinezen".

e. **W. Lutkie**

Letters from i.a. J. van Ginneken 1920, H.H.A. van Gybland Oosterhoff (with "dagverhaal" 1918) 1931–1936, J. Last, M.J.M. van Poll.

f. **W.H. Nolens**

Papers about article 123 of the government regulations about the admission of priests into the Dutch East Indies 1922–1923, L.J.M. Feber 1925, colonial issues dealt with at the 14th Internationale Arbeidsconferentie (International Labour Conference) 1930, letters from i.a. P.J.M. Aalberse 1901–1931, H. Colijn 1926, L.N. Deckers 1925, 1930, L.J.M. Feber 1924, D. Fock 1923–1929, Th. Heemskerk 1912–1928, A.W.F. Idenburg 1903, 1928, C. Snouck Hurgronje s.a., Ch.J.I.M. Welter.

g. **H.J.A.M. Schaepman**

Letters from J.H. Gunning 1872–1877, O. van Rees 1885.

h. **Sint Adelbert-Vereniging (Saint Adelbert Association)**

Correspondence with the Katholieke Wetenschappelijke Kring (Catholic Scientific Society) at Batavia 1949, Commissie Katholiek Hoger Onderwijs in Indonesië (Committee for Catholic Higher Education in Indonesia) 1965, Ikatan Serdjana Katolik Indonesia 1963–1965, documents about a department to be established in New Guinea.

i. **Unie van Katholieke Studenten in Nederland (Union of Catholic Students in the Netherlands), UKSN**

Papers of the Roomsch Katholieke Studenten Vereeniging St. Petrus Canisius (Roman Catholic Students Association Saint Petrus Canisius) at Bandung 1930–1931, Roomsch Katholieke Studenten Vereeniging St. Bellarminus (Roman Catholic Students Association Saint Bellarminus) at Batavia/Djakarta 1931–1953 (from 1951 Perhimpunan Mahasiswa Katolik Republik Indonesia, PMKRI), Katholieke Studenten Vereeniging Sanctus Thomas Aquinas (Catholic Students Association Sanctus Thomas Aquinas) at Bandung 1948, 1951, Federatie van Katholieke Studenten Verenigingen in Indonesië (Federation of Catholic Students Associations in Indonesia) 1949–1951, Pax Romana Asian Seminars at Madras 1954, Manilla 1959, Bangkok 1961, Saigon 1967, papers Asian Catholic Student Conference at Colombo 1966, assistance to students in various overseas territories 1937–1953, communism in China, Islam, Indische Beweging (East Indies Movement) 1920–1940, correspondence with F.H.M. Grapperhaus 1949–1950, Ikatan Mahasiswa Katolik Indonesia (IMKI) 1952–1958, association Sintetis 1966–1968.

## Sound archive

**Kommissie Missie Memoires (Committee for Mission Memoirs)** [510]

The Committee forms archives of tape recordings of interviews with *c.* 1000 former missionaries.

# Rijksuniversiteit Utrecht (State University Utrecht)

## Universiteitsbibliotheek (University Library), 1584

9–11 Wittevrouwenstraat, 3512 CS Utrecht.
Admission free. Lending to students and libraries. Lending of archival documents to institutions within the Netherlands.
Photocopying equipment available.
Research finished in 1979.

**Library** [511]

1,200,000 books and periodicals.

Directed to all subjects which are taught at the university.

### Archival material [512–514]

**Introduction to the collection of manuscripts** [512]

The collection consists of *c.* 3000 manuscripts. They are accessible by means of the *Catalogus codicum manuscriptorum bibliothecae universitas Rheno-Trajectinae.* Utrecht, 1887, 1909. 2 vol. The acquisitions of after 1909 are accessible by means of a catalogue on index cards.

Papers relevant for the history of Asia and Oceania are mentioned below with each catalogue.

**Catalogus codicum** [513]

With the enumeration of the papers the catalogue numbers are mentioned in brackets.

a. Daily record kept by the governor of the Moluccas H. Tielenius Kruythoff on the "hongitocht" (raid) which was made in 1821. (1315)

b. Arabic manuscripts. (1430–1465)

c. Persian manuscripts. (1466–1473)

d. Papers about the languages in Hindustan, Persia, Malabar, Bengal, Malaya, Formosa and China originating from A. Reland, J. Haafner, J.C. Swijghuizen Groenewoud, F. Valentijn and A.J. Spengler. (1478–1483)

e. Papers about the church in the East and West Indies, originating from H.C. Millies. (1704, 1712–1714)

f. Notes originating from D. Koorders about Islamitic and Javanese law. (1756, 1757)

g. Journal by D. Koorders of his journeys through the Sunda area; Bantam (published in *BKI*, 3rd series, vol. IV, 1869, p. 303–335, 341–367), Badui, Sukapura. (1769)

h. Vocabularies of several Sundanese dialects with travel notes, originating from D. Koorders. (1864)

i. Letters to F.A.W. Miquel (professor of botany) from i.a. the Royal Botanical Gardens in Calcutta, S. Buitenzorg at Buitenzorg, P. Bleeker at Batavia, 1885, [C.] L. Blume, J.H. Croockewit at Buitenzorg, 1854, Dr. Leichhardt Committee at Melbourne, [A.J.] Duymaer van Twist at Batavia, 1851, J.K. Hasskarl, 1845–1870, Th. Horsfield, 1854–1859, [Q.M.R.] Ver Huell, 1857, F.W. Junghuhn, 1849–1857, R. Krone in China, 1856, F. von Muelke at Melbourne, 1858–1869, J. Pijnappel, 1859, C.G.C. Reinwardt, 1834–1845, R.H.C.C. Scheffer at Buitenzorg, 1866–1871, P.F. von Siebold, J.E. Teijsmann at Buitenzorg, 1856–1869, W.H. de Vriese, 1836–1856, . . . Wassink at Batavia, 1850, 1851, H. Zollinger, 1846–1856. (1873)

**Catalogue on index cards** [514]

With the enumeration of the papers the shelf numbers are mentioned in brackets.

a. Manuscripts in Javanese, Batak, Palinese, Cambodian, Burmese and Persian. (1 D 8–21)

b. Manuscripts in Persian. (3 L 13)

c. Papers about the Buitenzorgfonds (Buitenzorg Fund) established in 1887, with letters from P. de Boer, J. Heemskerk Azn., C.A.J.A. Oudemans, N.W.P. Rauwenhoff, W.F.R. Suringar and M. Treub. (6 A 21$^{2}$)

d. Description of the coast of Ceylon, 18th century. (6 A 23)

e. Batak grammar, vocabulary and letters, presumably originating from J.J.A. Müller, *c.* 1890. (8 B 22)

f. Persian manuscripts. (8* C 20–21)

g. J.M. Roosenschoon. "De Westerse cultures op Java voor 1870". Thesis, 1945. (8 F 3)

h. Caland collection, consisting of Sanskrit texts. (11 A–B)

i. Manuscripts in Persian, Sanskrit and Malay. (11 D 5–8)

j. Papers originating from J.H. Gunning JHzn. about the mission in Nias, with papers of J. Bieger, O. Rudersdorf, A. Homeyer at Gunung Sitoli, K. Krafft, E. Müller, A. Pieper, J. Rauws, W.F. Schröder at Pulau Telo, A.J.C. Seters, F.H. Wahle-Lagemann and R. Wegner, 1925–1929. (G 25 E 59–61)

See also "Het eiland Nias". In *Pniël;* vol. 35 (1926), p. 159–282. Consultation will be possible after 1990.

k. F. Yurino. "Graphik der Artikulation in der japanischen Sprache", *c.* 1925. (28 A 144)

l. Papers about the Maatschappij tot Bevordering van het Natuurkundig Onderzoek van Oost- en West-Indië (Society for the Promotion of Physical Research of the East and West Indies), from 1946 called Maatschappij ter Bevordering van het Natuurkundig Onderzoek der Nederlandse Koloniën (Society for the Promotion of Physical Research of the Dutch Colonies), 1946–1955, originating from F.A. Vening Meinesz. (28 E 5). Consultation will be possible after 2016, or with permission in writing.

m. Papers about the Third World Petroleum Congress, which took place from May 28th up to June 6th 1951, originating from F.A. Vening Meinesz. (28 E 9). Consultation will be possible after 2016, or with permission in writing.

n. "Nieuwe statuten van Batavia", 1761–1766, presumably originating from U.G. Lauts. (29 A 10–12)

## Bureau van de Rijksuniversiteit Utrecht (Bureau of the State University Utrecht)

29 Kromme Nieuwe Gracht, 3512 HD Utrecht.
Admission by appointment.
Research finished in 1979.

**Indologische Faculteit (Faculty of Indology), 1923–1956** [515]

The Foundation "Fonds ten behoeve van Indologische Studiën aan de Rijksuniversiteit te Utrecht" (Fund in behalf of Studies of Indology at the State University of Utrecht) was established in 1924; 1950 decision upon liquidation, effected in 1956.

The archives consist of papers concerning the previous history of the Fonds 1923–1925, minutes and correspondence of the College van Curatoren (Board of Governors) 1925–1956, minutes and correspondence of the Wetenschappelijke Raad (Scientific Council) 1925–1950, correspondence with and about teachers 1925–1956, newspaper cuttings 1925–1935, folder "Indologische studiën" (Studies of Indology).

Size: *C.* 0.50 running metre.

Permission is necessary for consultation.

# CHAPTER VII

# OTHER SCIENTIFIC INSTITUTIONS

## Internationaal Instituut voor Sociale Geschiedenis (International Institute of Social History), IISG 1935

262–266 Herengracht, 1016 BV Amsterdam.
Admission free. Lending out (except periodicals, pamphlets and books published before 1945 and archival documents).
Photocopying equipment available.
Research finished in 1978.

**Library** [516]

400,000 books; 2500 currently received periodicals; 50,000 closed periodicals. Directed to the history of labour movements throughout the world.

For literature concerning the IISG see i.a.
*International Institute of Social History;* history and activities. Assen, 1968.

G.J.A. Rieshuis. Bibliotheekbeschrijving; de bibliotheek van het IISG. In *Open;* vol. 10 (1978), no. 6, p. 314–318.

**Publications** [517]

*Bulletin of the International Institute for Social History.* 1936–1940, 1950–1955.
Continued under the title: "International Review of Social History". 1955– .
Including: "Bibliographical Part of the International Review of Social History".

**Private organizations** [518–520]

**Sociaal-Democratische Arbeiderspartij (Social Democratic Labourers' Party), SDAP 1894–1946** [518]

Socialistic political party, established at Amsterdam in 1894 and in 1946 converted into the Partij van de Arbeid (Labour Party), see no. 519.

The archives include some correspondence with individual socialists in the East Indies and with the socialistic movement there and papers of the colonial committee of the SDAP. (*C.* 1930).

Inventory of the archives is forthcoming.

Size: *C.* 80 running metres.

**Partij van de Arbeid (Labour Party), PvdA 1946–** [519]

The Party is the continuation of the SDAP (see no. 518).

The archives contain *c.* 1 running metre of papers concerning Indonesia 1945–1950.

There is no inventory to the archives.

The total size is unknown.

**Chung Hwa Hui, 1911–1962** [520]

The organization of Chinese students in the Netherlands was established in 1911; the object was to provide facilities for study to Chinese students from Indonesia in the Netherlands. The organization was in the first place a socio-cultural club, but also politically active in the field of the Chinese struggle for emancipation in China and Indonesia; the ambition was to return to Indonesia after the study; after 1962 there were no more activities, though the organization was not discontinued.

The archives consist of minutes, correspondence, dossiers and accounting papers.

Size: *C.* 3 running metres.

## Private persons [521–530]

**H.A. Aa, May 22nd 1913 – December 14th 1962** [521]

The collection consists of reports about economic and political subjects in Indonesia during the first years after 1949.

Size: *C.* 8 running metres.

**Ch.G. Cramer, October 20th 1879 Sidoardjo – 1976 Overveen** [522]

Study for engineer Delft; 1903–1904 teacher of mathematics at Semarang; 1904–1923 (chief) engineer with the Burgerlijke Openbare Werken van Nederlandsch-Indië (Civil Public Works of the Dutch East Indies), 1904 Cheribon, 1906 Pasuruan-Besuki, 1915 on leave, 1916 Batavia; 1918–1923 member of the Volksraad (People's Council), also member of the Technische Onderwijscommissie (Technical Education Committee) 1918–1920, Commissie in het belang van het Irrigatiewezen op Java en Madoera (Committee in the Interest of the Irrigation in Java and Madura) 1918–1923, Commissie tot Herziening van de Grondslagen der Staatsinrichting van Nederlandsch-Indië (Committee for the Revision of the Basis of the Constitution of the State of the Dutch East Indies) 1920, Bestuursverkavelings-Commissie (Committee for Partition of the Administration) 1922–1923, Kiesrechtcommissie (Electoral Committee) 1922–1923; in the Netherlands for the SDAP member of the First Chamber 1923–1925 and of the Second Chamber 1925–1937; up to 1942 journeys i.a. through Indonesia, 1946 return to the Netherlands. (Supplements *RAvNI*).

The collection consists of a draft of an autobiography, with a documentation and some letters.

Size: *C.* 0.50 running metre.

**D.M.G. Koch, October 1st 1881 Woudsend – February 25th 1960 Utrecht** [523]

Study for engineer Delft 1901–1907; up to his departure for the East Indies active in the socialist movement; 1907 editor of the *Nieuws van den Dag van Nederlandsch-Indië* at Batavia, founder, together with E.F.E. Douwes Dekker, of *Het Vrije Woord;* 1908–1910 editor of the *Soerabaiasch Handelsblad;* 1911–1912 founder and editor in chief of *De Indische Kroniek;* 1912 secretary of the Handelsvereeniging Semarang (Trade Association Semarang); 1913 official at the Department of Agriculture, Industry and Trade; 1916 founder, together with A.B. Cohen Stuart, P.J.S. Cramer and S. Koperberg, of *Koloniale Studiën;* 1917–1919 editor in chief of *De Nieuwe Courant* at Semarang; 1919–1922 official at the Algemene Secretarie (General Secretariat) at Buitenzorg; 1922 editor in chief of the *Indische Courant* at Surabaja; 1926–1927 editor of *De Vakbeweging,* also up to *c.* 1942 secretary of the Verbond Vereeniging van Landsdienaren (Union Association of Public Servants) at Bandung; 1927 official with the Post-, Telegraaf en Telefoondienst (Posts, Telegraphs and Telephones); 1929–1938 librarian of the central library of the Department of Government Industries; 1938–1950 (with an interruption from 1942–1947) editor of the periodical *Kritiek en Opbouw*; up to 1954 Koch stayed in Indonesia. (*D.M.G. Koch;* levensschets en bibliografie; ter gelegenheid van zijn 70ste verjaardag; samengesteld door D. de Vries. Bandung, 1951).

The collection consists mainly of documentation concerning Indonesia and letters.

Size: *C.* 1.50 running metres.

**H.H. van Kol, May 23rd 1852 Eindhoven – August 22nd 1925 Remouchamps** [524]

Study for engineer Delft; 1876–1892 engineer with Waterstaat (Water Management) in the East Indies, 1876 Batavia, 1877 Besuki, 1885 on leave, 1887 Preanger, 1888 Tegal; in the Netherlands active in the socialist movement; 1897–1909 member of the Second Chamber for the SDAP, in 1902 journey, i.a. through the Dutch East Indies; 1913–1924 member of the First Chamber. (Supplements *RAvNI*).

The collection consists of documentation, of which a small part concerns Indonesia, Japan and Persia, and in addition some correspondence from 1897 of N. van Kol, J.T. Cremer, Comité Nieuw Indië (Committee the New Indies), J.E. Stokvis, Noto Soeroto.

Size: *C.* 2 running metres.

**S. Koperberg** [525]

1918 official at the disposal of the Department of the Binnenlands Bestuur (Indies Civil Service), employed at the Encyclopaedisch Bureau (Encyclopaedic Bureau), also assistant secretary of the municipal council of Semarang; 1925–1942 secretary/treasurer of the Java-Instituut (Java Institute) and the Museum Sono Budojo in Djokja. *(RAvNI).*

The collection consists mainly of documentation about Indonesian political affairs after 1945.

Entry: List drawn up when given on loan.

Size: *C.* 1.25 running metres.

Permission is necessary for consultation of the papers.

**W. Middendorp, February 15th 1886 Deventer – ?** [526]

1911 aspirant comptroller Tarutung; 1912 comptroller Tarutung (Samosir); 1914 Telukbetung (Kota Agung); 1915 Karo area (Kota Djahe); 1920 on leave; 1923 assistant resident Tapanuli (Sibolga); 1924 Department of Binnenlands Bestuur (Indies Civil Service); 1924 Sumba (Waingapu); 1925–1928 Boni (Watampone); 1927–1930 member of the Volksraad (People's Council); 1930 assistant resident Tapanuli (Balige); 1932 official Volkscredietbank (People's Credit Bank); 1933 to the Netherlands, where he was a member of the municipal council of Renkum 1935–1953 and a member of the Provincial States of Gelderland 1939–1954. (Supplements *RAvNI*).

The collection consists of a draft of a study by Middendorp of the district round Garut titled "Welvaart en schuld".

Size: 1 bundle.

**H.J.F.M. Sneevliet, May 13th 1883 Rotterdam – April 12th 1942 Amersfoort** [527]

1900–1910 railway official, 1902 member of the SDAP; 1910 president of the Nederlandsche Vereeniging van Spoor- en Tramwegpersoneel (Dutch Association of Railway and Tramway Employees); 1913–1917 secretary of the Handelsvereeniging Semarang (Trading Association Semarang); 1914 founder of the ISDV (in 1920 PKI); 1917 up to his banishment in 1918 secretary of the Nederlandsch-Indische Vereeniging van Spoor- en Tramwegpersoneel (Dutch East Indies Association of Railway and Tramway Employees); in the Netherlands from 1919–1920 active in the Nederlandsche Federatie van Transportarbeiders (Dutch Federation of Transport Workers); 1921–1924 in the service of Komintern (China); 1924 president of the Nationale Arbeiderssecretariaat (National Labourers' Secretariat); 1929 founder Revolutionaire Socialistische Partij (Revolutionary Socialist Party); 1931 member of the Provincial States of North Holland, up to 1933 member of the Hooge Raad van Arbeid (High Council of Labour); 1933–1935 member of the Second Chamber for the RSP, 1935–1937 RSAP; 1940 till he was shot in 1942 member of the Dutch resistance against the Germans.

The collection includes one half running metre of papers concerning Indonesia. Publication of these papers is forthcoming.

Size: *C.* 4 running metres.

Permission is necessary for consultation of the papers.

**J.E. Stokvis, February 23rd 1875 The Hague – December 30th 1951 The Hague** [528]

Journalist with the *Haagsche Courant*, *Dagblad voor Zuid-Holland*, *De Telegraaf*, *Het Vaderland*, *Het Volk;* 1911–1917 editor in chief of *De Locomotief;* 1923–1931 member of the Volksraad (People's Council), also editor of the *Indische Volk* 1922–1928 and member of the College van Gedelegeerden (Board of Deputies) 1928–1929; 1923–1928 president ISDV; 1935 member of the Provincial States of South Holland; 1937–1946 member of the Second Chamber for the SDAP, after 1949 for the PvdA.

The collection consists of documentation in the form of cuttings. For the names and subjects mentioned in the inventory see the index.

Entry: Inventory.

Size: *C.* 4.50 running metres.

**J. Toot, November 2nd 1887 Uitgeest – 1963** [529]

1915 teacher in Gorontalo; 1916 School tot Opleiding van Inlandsche Artsen (School for the Training of Native Physicians); 1919 Kweekschool tot Opleiding van Chineesche Onderwijzers en Onderwijzeressen (Training College for Chinese Teachers, both male and female) at Meester Cornelis; 1921 till his discharge in 1932 secretary and member of the council of education at Weltevreden, also member of the Commissie voor de Vaststelling van de Catalogus van Leermiddelen voor het Lager Onderwijs (Committee for the Settlement of the Catalogue of School Materials for Elementary Education) at Batavia.

The collection consists of data about education collected when he was inspector, including a collection of photographs.

Size: *C.* 0.50 running metre.

**J.L. Vleming, ? – 1965 Hattem** [530]

1922–1926 head of the government accountancy of the Inland Revenue at Batavia, one of the East Indies specialists of the SDAP.

The collection consists of documentation and correspondence about his work as Revenue officer in the outer islands.

Size: *C.* 3 running metres.

## Asian papers in Dutch archives and collections [531]

For a definition see no. 271.

a. **De Bosch Kemper**

Originating from J.M. Kemper letters 1811–1814 from D. van Hogendorp; originating from J. de Bosch Kemper letters from the Bataviaasch Genootschap van Kunsten en Wetenschappen (Batavian Society of Arts and Sciences) 1840,

1874–1875, Indisch Genootschap (Indies Society) 1869, J.H. and J.W. Gunning, J. Heemskerk Azn. 1863–1874, C. Busken Huet 1862, 1874, Institut d'Afrique, KITLV, D.J. Mackay 1863, 1868, H.G. Nahuys van Burgst 1842–1845, J.F.W. van Nes 1869–1872, D.J. Steyn Parvé 1865, A.A. de Pinto 1842–1847, O. van Rees 1857–1865, C.G.C. Reinwardt 1851, [J.J.] Rochussen 1873–1874, D.J. Veegens 1850, P.J. Veth 1861, E. de Waal 1868, photographs.

b. **"De Dageraad"**

Papers 1856–1961, including minutes, postbooks, accounting papers, lists of members, letters from R.C. d'Ablaing van Giessenburg 1880, P. Appatioray Bhetty at Madras 1883, C.Th. van Deventer 1910, E. Douwes Dekker jr. 1910, H. [H.] van Kol 1876, 1916, G.P. Rouffaer 1883, Vereniging Nederland-Indonesië (Association the Netherlands-Indonesia) 1945–1947, papers concerning Multatuli. Permission is necessary for consultation of the papers.

c. **F. Domela Nieuwenhuis**

Papers of and/or concerning H. [H.] van Kol, T. Ottolander, J.D. Veegens, A. Cohen, E. Douwes Dekker, M. Douwes Dekker–Hamminck Schepel, L.M. Keuchenius, N. van Kol. S.E.W. Roorda van Eysinga, photographs from *c.* 1860.

d. **F. van der Goes**

Papers *c.* 1880–*c.* 1935 of and/or concerning J.G. Gleichman, J. Heemskerk Azn., J. de Kadt, S. Koperberg, imperialism, "neo-ethiek" (neo-ethics) in Indonesia, C. Busken Huet, H.M. van Andel, war danger in South East Asia, Ch. [M.] van Deventer, E. Douwes Dekker, M. Douwes Dekker–Hamminck Schepel, Multatuli Museum, M.T.H. Perelaer, the mutiny on "De Zeven Provinciën", the SDAP and Indonesia, colonial policy, H. [J.F.M.] Sneevliet, A.C. Wertheim, mission.

e. **International Union of Socialist Youth Organizations**

Papers 1907–1947 from Australia of M. Lazarus at Melbourne of the Labour Guild of Youth, Socialist Party of Victoria at Melbourne; from China of the Parti Social-Démocrate Chinois, Shanghai College, Tsan Wan, correspondence Nangking-Berlin, Woe Kek Kong, Pei-Chuan, Kwei-shien; from India of the socialist youth congress at Calcutta; from Japan of Kokusai Shobo at Tokio and Sangyo Rodo Chosasho at Tokio; from Korea of Shidai Ilbo and the *Korea Times* at Seoul; from Manchuria of the Ministry of Foreign Affairs at Hsingking.

f. **International Union of Socialist Youth**

Papers 1946–1969 of which the majority after 1949, including reports from and/or concerning Peking, Hongkong, colonialism and imperialism, conflict China-Russia, Vietnam, Djakarta, West Irian, Japan, Korea, Mongolia, Singapore.

g. **B. Kautsky**

Letters from E.F.E. Douwes Dekker 1913–1915, H.H. van Kol 1895–1921, concerning Japan and about colonialism 1907.

h. **R. Kuyper**

Papers *c.* 1900–*c.* 1935 concerning imperialism, India, Japan, Indonesia, mutiny on "De Zeven Provinciën".

i. **Labour and Socialist International**

Papers 1923–1940 i.a. concerning Australia, China, India, Japan, Mauritius and New Zealand.

j. **League Against Imperialism**

Dossiers and resolutions concerning imperialism in Asia 1927, 1929, 1931.

k. **K. Marx and F. Engels**

Originating from Marx documentation and abstracts from various publications concerning China, the English trade in East Asia, India and Ceylon by J. Phear 1880, products (including cotton), administration finance, population of India, South India and Bombay by M. Wilks, J.F. Royle, H. Murray, G. Campbell, J.P. Fallmerayer, J. Chapman, J. Sullivan, J. Dickinson, Prince Saltykow, J.R. Mac Culloch, A.H.L. Heeren 1810–1853, the Grand Mogul by F. Bernier 1830, Asian monarchies by R. Patton 1801, slavery by W. Blair 1833, Java by Th.S. Raffles and J.W.B. Money 1817 and 1861, the war against Persia 1856; originating from F. Engels the Commonwealth by W. Godwin 1824.

l. **J. Motteler**

Letters 1890–1891 from Adelaïde from H. Schmid member of the Südaustralischer Allgemeiner Deutscher Verein, A. Timmerman, C. Wiese and A. Bröse.

m. **W. van Ravesteyn**

Correspondence with J.W. Meyer Ranneft and H.J.F.M. Sneevliet.

n. **J.W. Riemens**

Papers about the Vereniging Nederland-Indonesië (Association the Netherlands-Indonesia), *c.* 1945–1950.

o. **J. Romein**

Correspondence and dossiers from *c.* 1910 i.a. from and about R. Abdulgani, R.M. Ali, Antara, Bandoengsche Kunstkring (Bandung Art Society), Budoyo, Centraal Comité voor Hulpverlening aan de Burgerbevolking van China (Central Committee for Aid to the Civil Population of China), Hulpcomité voor

Noodlijdende Chinezen in Amsterdam (Emergency Committee for Distressed Chinese in Amsterdam), M.A. Djuhana, Djojopuspito, R.W. Dwidjosugondo, Go Dien Tjoa, G. Gonggrijp, Hadi, Hardjono, embassy India, Indisch Instituut (East Indies Institute), embassy Indonesia, J. de Kadt, D.M.G. Koch, J. Last, M.A.P. Meilink-Roelofsz, J.W. Meyer Ranneft, R. Nieuwenhuys, Nur Tugiman, Ong Hien Hoey, Ong Sioe Tjiang, L.N. Palar, J. Panikkar, Kedaulatan Rakjat, *Oriëntatie,* G.J. Resink, Rotary Club Semarang, C. Snouck Hurgronje, Sudjadmoko, Suleiman, Sunarjo Djokokusumo, Sundoro, Supomo, Sribudoyo, A. Teeuw, Tjen A Kwoei, Tjioe Tiong An, Tan Malakka, Publishing House Ganaco, Publishing House Pembangunan, Universitas Indonesia, Vereniging Nederland in de Vreemde (Association the Netherlands in Foreign Countries), Vereniging Nederland-Indonesië (Association the Netherlands-Indonesia), B.H.M. Vlekke, C. van Vollenhoven, E. [de Willigen-]Vuyk, B. IJzerdraad. Permission is necessary for consultation of the papers.

p. **Socialistische Arbeiders Internationale (Socialist Workers International)**

Including correspondence from and about Australia, China, India, Japan, New Zealand, *c.* 1925–1940.

q. **Socialistische Gids**

Minutes, articles in manuscript, letters from *Chung Hwa Hui Tsa Chih;* orgaan der Chineezen-Vereeniging 1917, Cultuurbond (Cultural Union) in Bandung 1922, J. de Waard of the Handelsvereeniging (Trade Association) at Medan 1919, Het Java-Instituut (The Java Institute) 1921, D.M.G. Koch 1921, H.H. van Kol 1915–1924, *Koloniale Studiën* 1917–1924, *De Naamloze Vennootschap;* Nederlandsch-Indisch maandblad 1922, Vereeniging voor Studie van Koloniaal-Maatschappelijke Vraagstukken (Association for Study of Colonial-Social Problems) 1920, 1923, *Het Indische Volk* 1921, the Volksraad (People's Council) 1918–1922, M.C. van Zeggelen.

r. **Socialistische Internationale (Socialist International)**

Reports of various congresses, papers of the administration, secretariat, committees, correspondence mainly after 1950, up to 1950 papers concerning China, India and Japan. Permission is necessary for consultation of the papers.

s. **P.J. Troelstra**

Papers *c.* 1880–*c.* 1925 of and/or concerning imperialism, H. [J.F.M.] Sneevliet, H. [H.] van Kol, Ch.G. Cramer, E. Douwes Dekker, D.M.G. Koch, J. Last, Multatuli Museum, J.E. Stokvis.

t. **R. de Waard**

Some letters i.a. from Noto Soeroto, T. Ali, *Indische Verlofganger,* J.C. van der Kaaden, [J.] H. [A.] Logemann, M. Ovink-Soer, *c.* 1925–1930.

u. **H.P.L. Wiessing**

Papers *c.* 1900–*c.* 1950 of J. Last, Lim Boan Tai, Notosuroto, H.J.F.M. Sneevliet, Surianingrat, T.K. Tan, H. Colijn.

**Photographs, posters, pictures** [532]

The visual material has been arranged to country, persons alphabetically to name.

The collection contains *c.* 20,000 posters, 50,000 photographs and 10,000 cartoons.

## Kamer van Koophandel en Fabrieken voor Amsterdam (Chamber of Commerce for Amsterdam), 1811

13 Koningin Wilhelminaplein, 1062 HH Amsterdam.
Admission by appointment.
Research finished in 1979.

**Archives** [533]

The archives consist of correspondence and other papers including data concerning Dutch businesses which were in contact with Asia and Oceania.

Size old archives (1811–1922): *C.* 30 running metres.

## Koninklijk Instituut voor de Tropen (Royal Institute for the Tropics), KIT 1910

63 Mauritskade, 1092 AD Amsterdam.
Admission after identification. Lending out of books or other material only to institutions.
Photocopying equipment available.
Research finished in 1977.

**Introductory data** [534–536]

**Library** [534]

Centrale Bibliotheek (Central Library), containing 90,000 vol. and 11,000 periodicals, of which 7500 currently received; in addition department libraries of Sociaal-Wetenschappelijk Onderzoek (Socio-Scientific Research), Tropische Hygiëne (Tropical Hygiene) and Agrarisch Onderzoek (Agricultural Research). Directed to the tropics and subtropics, developing countries, before 1950 mainly to the applied sciences in the Dutch colonies.

Into the Central Library are embodied the libraries of the Nederlandsch-Pacific Instituut (Dutch Pacific Institute), the Nederlandse Vereeniging van de Rubberhandel (Dutch Association of the Rubber Trade), the Koloniaal Museum (Colonial Museum) of Haarlem, Atjeh Institute, Bali Institute, Batak Institute, Minangkabau Institute, Moluccas Institute, Dutch Java Institute, East Coast of Sumatra Institute and South Sumatra Institute.

For literature about the KIT see i.a.
C.J. Hasselman. *De Koninklijke Vereeniging Koloniaal Instituut te Amsterdam;* wording, werking en toekomst. Amsterdam, 1924. 3rd ed. 1935.

A.D.A. de Kat Angelino. *Het Koninklijk Instituut voor de Tropen en zijn nieuwe taak en doelstelling.* Amsterdam, 1954.

*Twee richting verkeer;* Koninklijk Instituut voor de Tropen, 1910–1960. Wormerveer, 1960.

H. Offerhaus. *De realisering van de taakuitbreiding van het Koninklijk Instituut voor de Tropen.* Amsterdam, 1961.

**Publications** [535]

*Koninklijk Instituut voor de Tropen;* jaarverslag; vol. 1– . Amsterdam, 1910– .

*Tropical abstracts;* vol. 1–22. Amsterdam, 1953–1974. Continued as "Abstracts on tropical agriculture"; vol. 1– . Amsterdam, 1975– .

*NEDO-abstracts;* vol. 1– . Amsterdam, 1974– .

**Introduction to the archival material** [536]

The Vereeniging Koloniaal Instituut (Association Colonial Institute) was established at Amsterdam in 1910, from 1923 called Koninklijke Vereeniging Koloniaal Instituut (Royal Association Colonial Institute), from 1945 Koninklijke Vereeniging Indisch Instituut (Royal Association East Indies Institute) and from 1950 Koninklijk Instituut voor de Tropen. The object was to collect and spread knowledge about the Dutch overseas provinces and particularly to promote commercial agricultural, industrial and other interests of the mother country and colonies by advancing the economical development of the peoples in the Indonesian archipelago. The initiative for the establishment came from J.T. Cremer and H.F.R. Hubrecht. Motive was the project for expansion and improvement of the Koloniaal Museum (Colonial Museum) at Haarlem, by means of a central colonial institution, which could take over the task of the Koloniaal Museum. The possessions of this Museum were transferred to the Koloniaal Instituut January 1st 1913.

Virtually from the beginning were, besides the Algemene Secretariaat (General Secretariat), to which belonged also the Centrale Bibliotheek (Central Library) and the photographic department, the following departments united with it: Handelsmuseum (Commercial Museum) 1912, at present Agrarisch Onderzoek (Agricultural Research), Volkenkunde (Ethnology) 1912, at present Sociaal

Wetenschappelijk Onderzoek (Social Scientific Research) and Tropen Museum (Museum for the Tropics) and Tropische Hygiëne (Tropical Hygiene) 1913. From 1928 the following Hulp-Instituten (Branch Institutes) were connected with the Afdeling Volkenkunde (Department for Ethnology): Bataksch Instituut (Batak Institute) 1908, Atjeh-Instituut (Atjeh Institute) 1914, Bali-Instituut (Bali Institute) 1915, Oostkust van Sumatra Instituut (East Coast of Sumatra Institute) 1916, Zuid-Sumatra-Instituut (South Sumatra Institute) 1916, Minangkabau-Instituut (Minangkabau Institute) 1917, Molukken-Instituut (Moluccas Institute) 1918 and the Nederlandsch Java-Instituut (Dutch Java Institute) 1927. In addition there were connected with the Instituut: Comité voor Indische Lezingen en Leergangen (Committee for East Indies Lecture Meetings and Courses of Instruction) 1917, West-Indische Kamer (West Indies Chamber) 1928, Encyclopaedisch Bureau (Encyclopaedic Bureau) 1931, Vereeniging Pacific Instituut (Association Pacific Institute) 1934, F.W. van Eeden Fonds (F.W. van Eeden Fund) 1912, M. Greshoff's Rumphius Fonds (M. Greshoff's Rumphius Fund) 1912, Eykman-Medaille-Fonds (Eykman Medal Fund) 1923, Dr. H.F.R. Hubrecht-Fonds (Dr. H.F.R. Hubrecht Fund) 1928, later Hubrecht-Janssen Fund, S.P. van Eeghen-Stichting (S.P. van Eeghen Foundation) 1930 and the Schüffner-Laboratorium (Schüffner Laboratory) 1937.

Papers about the institutes mentioned above are to be found in the Centraal Archief (Central Archives) and in some cases in the department archives. These archives are occasionally, by permission, open for scientific research. Of a number of Institutes archives kept separately have been preserved. These are described below (see no. 537–543, 552).

## Centrale Bibliotheek (Central Library) [537–551]

### Private organizations [537–545]

#### Hulp-Instituten (Branch Institutes), 1928–1951 [537]

*C.* 1925 7 regional institutes were, in conjunction with the Koloniaal Instituut (Colonial Institute), at work collecting data and making them of service for the populations of the Batak area, Atjeh, Bali, East Coast of Sumatra, South Sumatra, Minangkabau and the Moluccas.

At the initiative of the Zuid-Sumatra-Instituut (South Sumatra Institute) attempts were made to expand the field of action to all of Indonesia in 1926.

This resulted in the establishment of the Nederlandsch Java-Instituut (Dutch Java Institute) in 1927.

From 1928 they co-operated, besides with the department of Volkenkunde (Ethnology), also with the departments of Handelsmuseum (Commercial Museum) and Tropische Hygiëne (Tropical Hygiene).

From 1945 only the Nieuw-Guinea Studiekring (New Guinea Study Circle) of the Moluccas Institute was still active and on a small scale also the Atjeh Institute and the East Coast of Sumatra Institute.

The small amount of archives consists of lists of members, minutes and rules 1939 of the Atjeh Institute, Groote-Oost-Instituut (Great East Institute) as successor of the Moluccas Institute and the Batak Institute, correspondence of the East Coast of Sumatra Institute 1934–1954.

Entry: Shelflist.

Size: *C.* 0.05 running metre.

See also no. 538–543.

The archives of the Atjeh Institute and the South Sumatra Institute have not been found.

**Bataksch Instituut (Batak Institute), 1908–1952** [538]

The Institute was established in 1908. The object was to collect as many data as possible about the Batak area and to make these data of service for the inhabitants of the Batak area. From 1924 there existed a close co-operation with the Koloniaal Instituut (Colonial Institute).

The archives consist of correspondence of the governing board 1908–1925 with post-book; correspondence of the secretariat 1913–1952; correspondence between C.W. Janssen and H.M. Botje 1910–1917; documentation: copies of notes, reports, travel accounts *c.* 1860–1908 numbered C 1–200, papers about education, medical care, religion, mission, Handel Mij Güntzel & Schumacher (Trading Company Güntzel & Schumacher), papers Tapanuli, plans, maps, cuttings.

Entry: Shelflist Hulp-Instituten (Branch Institutes).

Size: *C.* 2 running metres.

**Bali-Instituut (Bali Institute), 1915–1938** [539]

The Institute was established in 1915. The object was to collect data about Bali in co-operation with the Koloniaal Instituut (Colonial Institute) and to make them of service for the population of Bali.

The small amount of archives consists of correspondence 1915–1938 with postbook.

Entry: Shelflist Hulp-Instituten (Branch Institutes).

Size: *C.* 0.05 running metre.

**Oostkust van Sumatra Instituut (East Coast of Sumatra Institute), 1916–1960** [540]

The Institute was established in 1916. The object was to collect and adapt data about the East Coast of Sumatra and to promote the welfare of the population there.

The archives consist of rules and regulations, correspondence 1916–1960, minutes 1916–1965, annual reports 1916–1964; documentation: Deli, tobacco, rubber, tea, palm, oil, sisal, opium, Belawan, harbour Batavia, Deli Plantersvereeniging (Deli

Planters' Association), statute labour, labour issues, AVROS, Internationale Vereeniging voor Rubber- en Andere Cultures (International Association for Rubber and Other Cultures), ONRANI, emigration, coolie ordinance, penal sanction, J. Cramer, Chinese, Javanese, British Indians, health care, education; papers L.C. Westenenk, G.L. Tichelman, . . . Kleingrothe; photographs.

Entry: Shelflist.

Size: *C.* 8 running metres, of which *c.* 5 running metres documentation.

**Minangkabau-Instituut (Minangkabau Institute), 1917–1938** [541]

The Institute was established in 1917. The object was to collect data about the Minangkabau in co-operation with the Koloniaal Instituut (Colonial Institute) and to make them of service for the population of Minangkabau.

The small amount of archives consists of minutes 1917–1938 and a memorandum Binnenlands Bestuur (Indies Civil Service) *c.* 1920.

Entry: Shelflist Hulp-Instituten (Branch Institutes).

Size: 1 folder.

**Molukken-Instituut (Moluccas Institute), 1918–1952** [542]

The Institute was established in 1918. The object was to collect data about the Moluccas in co-operation with the Koloniaal Instituut (Colonial Institute) and to make them of service for the population of the Moluccas. Projects to convert this Institute into the Groote-Oost-Instituut (Great East Institute) in 1939, were not realized. After 1945 only the Nieuw-Guinea Studiekring (New Guinea Study Circle) was active. In 1951 the Institute was closed.

The small amount of archives consists of minutes 1918–1939, correspondence 1918–1950, accounting papers 1918–1952, administrative memorandum 1781 concerning Amboina, memorandum Buru, memorandums coconut growing *c.* 1920, papers about circumcision at Amahai, colonial reports about Ternate.

Entry: Shelflist Hulp-Instituten (Branch Institutes).

Size: *C.* 0.15 running metre.

**Nederlandsch Java-Instituut (Dutch Java Institute), 1927–1943** [543]

The Institute was established in 1927. The object was to collect data about Java in co-operation with the Koloniaal Instituut (Colonial Institute) and to make them of service for the population of Java.

The small amount of archives consists of minutes and rules 1927–1943 and a postbook 1927–1938.

Entry: Shelflist Hulp-Instituten (Branch Institutes).

Size: *C.* 0.05 running metre.

**Jan Pieterszoon Coen Stichting (Jan Pieterszoon Coen Foundation), 1919–1938** [544]

The object of the Stichting was the establishment and maintenance of a home at Batavia for school children of all races and ranks.

The archives consist of a memorial book of the first JPC voyage 1925 and about eighty travel reports 1925–1938 of teachers and pupils, viz.: L.J.C.G.M. de Bosch van Drakesteyn, G. Duykinx Sander, J. van Haselen, H.F. Heerkens Thyssen, Y.A. Schuller tot Peursum, N. Maas, L. de Block, A.D. Bouma, P.F.J.A. Julien, J. Oosterhoff, C.J. Vos, P.C. van Lier, Ch.G.L. Behrens, J.H. Faber, H.J. de Kanter, J. van der Linden, C.F.C. Menser, P.A.M.G. Mignot, J. Nienhuys, A.K. Kuipers, W.L. Beekman, L. Cohen, P. Eibergen, H.J. Slyper, H.E. Noordman, E. Boer, A.M. Vakton, M. Zeldenrust, J.E. Blankenberg, L.N.J. Holleman, W. Kalff, A.W. Romkes, W.A. Semeijns de Vries van Doesburgh, A. van Deursen, L. Eskes, S.J. de Groot, J.J. Hendriks, E. Rijpman, F.A.C. Bloem, A.A.J. Guepin, F. Hoogewerf, J.H.H. Hülsmann, J. Melchior, J.A. Mulder, W.G. Schrijver, J.B. Wilmink, K.J. van der Giessen, L.W. Houttuyn Pieper, C.W. de Visser, H. Marwitz, A. Schierbeek, J. van Vroonhoven, P.J. van Winter, J.J. Kloppert, W.J.A. Schouten, G.W. Bolkestein, J.J. Swart, E. Frenkel, J. Jensma, J. Hoek, A.F.M. Reynders, S.R. van Asperen de Boer, L. Bouma, C. Kleywegt, C.J.J. van der Maas, J. Verseveldt, F. Brouwer, K. Bouma, W.M.D. Heijs, J.H. Sebus and P.C. van Steenwijk.

Size: *C.* 1 running metre (numbered R 821).

**Vereeniging Indië-Nederland (Association the Indies-the Netherlands), 1931–1953** [545]

The association was established in 1931 at Amsterdam with the object to promote a harmonious development in the East and West Indies in the field of culture, economics and social and political affairs within the relationship of the "rijkseenheid", without representing a political party. The chairman was M.B. van der Jagt. The association was abolished in 1953.

The archives consist of minutes, correspondence, accounts, papers of the drafting committee, papers Numfoor, rules of the Commissie van Advies tot Herziening van de Reglementen op het Beleid der Regeringen in de Kolonies Suriname en Curaçao (Advisory Committee on the Revision of the Regulations for the Policy of the Governments in the Colonies Surinam and Curaçao), papers of the Commissie voor de Bestudering van het Advies der Agrarische Commissie (Committee on the Study of the Advice of the Agricultural Committee), correspondence L.A. Vellenga.

Entry: Shelflist.

Size: *C.* 2 running metres.

**Private persons** [546–547]

**M.B. van der Jagt** [546]

1896 in the BB (Indies Civil Service); 1900 Banjumas; 1902 comptroller Kediri; 1904 on leave; 1907 comptroller Pekalongan; 1910 secretary Preanger; 1915 Semarang; 1916 assistant resident Kebumen; 1920 Modjokerto; 1922 resident Kedu; 1927 Surakarta; 1928 governor Surakarta; 1930 retired; from 1931 he was chairman of the Vereeniging Indië-Nederland (Association the Indies-the Netherlands).

The collection consists of diaries 1897–1948, correspondence, drafts of publications with documentation i.a. for his memoirs "Inlandsch vorstenbestuur" and about Surakarta, correspondence with several publishers, papers drawn up and received when he was chairman of the Vereeniging Indië-Nederland, photographs.

Entry: Shelflist.

Size: *C.* 1 running metre.

**J.E. de Vrij, January 31st 1813 Rotterdam – July 31st 1898 The Hague** [547]

1832 apothecary at Rotterdam; 1837 doctor of chemistry; 1857 in Java assistant to F.W. Junghuhn as chemist for the cultivation of cinchona; 1863 return to the Netherlands, took up his residence in The Hague, where he did chemical research into the cinchona bark till his death.

The collection consists of scientific correspondence, journals and notes about his chemical research, documentation.

There is no inventory to the collection.

Size: *C.* 0.80 running metre.

**Miscellaneous papers** [548–549]

**Manuscripts and papers deposited in the "ijzeren kast" (iron cabinet)** [548]

a. Alor. "Silir het heidenjongetje", manuscript entry for the Garoeda competition 1938/39.

b. Papers originating from the Uitvoerend Comité voor de Aanbieding van een Huwelijksgeschenk aan H.K.H. Prinses Juliana (Working Committee on the Presentation of a Wedding Gift to HRH Princess Juliana), 1936–1937.

c. Ballot, J. "Reglement op de landschapskassen in de residentie Oostkust van Sumatra", s.a.

d. Beers, N. "In memoriam Professor Dr. L.P. le Cosquino de Bussy (1879–1943)".

e. Papers originating from the government official G.W. Caudri: essay "Grondslagen voor de beoordeeling van de belastingdruk op de inheemsche bevolking van Nederlandsch-Indië", 1931, minutes 1936 of the administrative conference at Benkulen, diary when he was aspirant comptroller of Kehapiang 1936, diary when he was comptroller of Pontianak 1946, "Boek voor brieven en andere stukken" including decrees of appointment.

f. Decree concerning revision of self rule of Bila, by W.H. Keuchenius 1916.

g. Decree concerning revision of self rule of Kualu, by W.H. Keuchenius and S. van der Plas 1916.

h. Decree concerning regulations of the administration in Siak, s.a.

i. Bosboom, H.D.H. "Aantekeningen betreffende Batavia; met plattegronden van historische huizen te Batavia". *c.* 1900.

j. Cosquino de Bussy, L.Ph. le. "Reis door Atjeh en de Gajolanden", 1927.

k. "Chineezen werving voor Banka", 1905, letters from J. Stecher.

l. Political contract with the self rule of Deli, *c.* 1907.

m. Haver Droeze, [F.J.] "Reis naar Deli als consul", 1899.

n. Elmberg, J.E. "Nota naar aanleiding van ethnologisch onderzoek in de Nimboran, Gressi en Japsi", 1949.

o. Greshoff, M. Notes about his study of the frequency of names of plants in popular expressions and literature.

p. Greve, L.L.F. de. "Belawan, Oceaan-haven", 1923.

q. Heinoldt, C. "Emigratie van Chineezen naar Banka", 1897.

r. Hegge Spies, J.D.I. van der. "Levensschets van Habib Abdoel Rachman", *c.* 1875.

s. Hoetink, B. "Reis naar China in verband met de Deli emigratie", 1898.

t. Hoogeveen, H.J.C. Report about his voyage to Siam, 1873.

u. Marginal notes on the article of H.H. van Kol about Deli, *c.* 1900.

v. Memorandums of the chief accountant of the Kina-Onderneming Pondok Gedehlanden (Cinchona Estate Pondok Gedeh Lands), 1940–1942.

w. "Kloetuitbarstingen, De", biological and volcanological notes by Th. Wurth concerning the Kelut, Ardjuno, Raung, Proefstation Malang (Research Station Malang), with correspondence and photographs, 1910–1920.

x. Koorders, S.H. Letters, notes, articles about the flora of Indonesia, *c.* 1915.

y. Koordeman, J.P.M. "Beschouwingen over de Atjehpolitiek op de buitenbezittingen", 1908.

z. Kraal, J.W. "Van verschillende stam", manuscript entry for the Garoeda competition, 1948.

aa Kreel, . . . van. Lecture on the railways on the East Coast of Sumatra, 1929.

ab Kroesen, J.H. "Historische nota betreffende de plaatsing van bestuursambtenaren op Nederlandsch Nieuw-Guinea", 1940.

ac Lezer, L.A. "Avonturier in Tropisch Nederland", draft of a novel about the KNIL, 1959.

ad Loos. H. "Rapport betreffende een tocht van Kota Pinang naar Pasar Pangarajan en terug", 1925.

ae Lusink, M.J. "Berijmde kroniek over 1924 gehouden voor het Oostkust van Sumatra-Instituut", 1925.

af correspondence 1923–1926 of the Maatschappij tot Exploitatie der Ondernemingen (Company for the Working of the Undertakings) originating from W.A.
Baud.

ag "Maleische pantoens".

ah Mohnike, O.G.J. "Bemerkungen über einige gegenwärtige politische Verhältnisse und Zustände von Japan", *c.* 1865.

ai Papers originating from . . . Reyers: A. Couvreur "Celebes problemen", 1929, with letter from H.A.F. de Vogel and article by H.C. Zentgraaff; "Moordende kris; geschiedenis van Ken Arok voor de jeugd bewerkt", manuscript entry of Reyers for the Garoeda competition 1952.

aj Niemann, G.K. "Dictaat geografie van Nederlandsch-Indië", 1890.

ak "Nota van het historiële der buitengewone zending naar Japan in 1844", drawn up in 1851.

al "Nota inzake de overbrenging van de zetel der regering naar Bandoeng", by C.P. Mom, 1938.

am "The outlines of the political development of the Indonesians in Sumatra and the Netherlands policy towards the Indonesian political movement", *c.* 1940.

an Prinsen Geerligs, H.C. "Geschiedenis van de Java Suiker Proefstations, 1885–1910", 1927.

ao Rees, D.F.W. van. "Beschrijvingen van Nederlandsch Oost-Indië", *c.* 1915, with "Aanteekeningen van mijn reis in Azië".

ap "Reis naar en door de Kangean-archipel; reisaanteekeningen van het hoofd van de dienst der wildhoutbosschen op Java en Madoera", 1929.

aq "Rekest van den Plantersbond Oostkust van Sumatra inzake dollarkwestie", 1915.

ar "Request betreffende de spoedige aanleg van een Oceaanhaven te Belawan Deli", 1917.

as "Resumé der krijsgsgebeurtenissen te Sumatra's Westkust in de jaren 1842–1845", by F.D. Cochius.

at Reports of the Sarekat Islam congresses at Bandung, Batavia and Surabaja, 1916–1918.

au Schlager, E., and Th. Meier. "The Barong dance in Bali", *c.* 1948.

av Notes about the number of pupils 1923–1929 at the Senembah Schools.

aw Biography in Malay of Singa Mangaradja, with a report 1896 of the comptroller of Baros about the descent of the princes of Kuria Ilir.

ax Accounts, reports, 1911–1912 of the Algemeen Soerabajaasch Gezondheids Comité (General Surabaja Health Committee) ASGC.

ay "Sedjarah Melaju", history of the Malay and Palembang princes.

az Soerjonegoro. "Kawi-Javaansch woordenboek", 1864.

ba "Stamboom van de vorsten van Deli en Serdang", 1879.

bb Stecher, J. "Bemerkingen bij het rapport van J.S.C. Kasteleyn aan de directeur van de Deli Maatschappij", 1911, with "Rapport mijner reis naar de Sabang-baai in verband met de overscheping van tabak aldaar", 1904.

bc Stecher, J. "Emigratie van China naar Deli", 1905.

bd Stecher, J. "Rapport betreffende Chineezen-werving en arbeidstoestanden op Billiton", 1905.

be Sturler, W.L. de. "Verzameling van eenige in de Molukko's voorkomende houtsoorten", *c.* 1865.

bf Report about the "terugzending van zieke koelies van Banka naar China", 1894.

bg Tichelman, G.L. "De Islam in Simeloengoen", 1937.

bh Tichelman, G.L. Report about the "overdracht van het borstbeeld van C.W. Janssen in het Koloniaal Instituut van Amsterdam", 1938.

bi "Verslag betreffende de bijeenkomst ter bespreking van de Middelbare Koloniale Landbouwschool te Deventer", 1912.

bj Verspijck, [G.] M. "Politiek verslag van de residentie Zuid- en Oosterafdeeling van Borneo over het jaar 1859".

bk "Verzameling van brieven, attesten etc. over de toepassing van de 60 jarigen arbeid van C. de Mooij", 1864–1924.

bl "Verzameling Maleische handschriften", 1890.

bm Veth, D.D. "Dagboek Sumatra expeditie", 1877–1878.

bn Vogel, H.A.F. de. Papers concerning his career as comptroller public revenue, letters from and to W. de Vogel, H. Muller Sz., M. Mees, H.W. van Eeden and R. Everwijn, diary 1854–1859, papers about his pamphlet "Hinderpalen", papers about cattle farming, tobacco, coffee, sugar, indigo, cotton, Rembang, Bodjonegoro, Tuban and Surabaja, Javanese and Arabic texts, 1850–1881.

bo Vries, G. de. "Pang Paneu, de Toenonger", manuscript entry for the Garoeda competition 1941.

bp Wal, J.H.E. van der. "Het bivak van het 6de bataljon infanterie te Tjakra Nagara in Lombok", 1894, with photographs.

bq Wal, J.H.E. van der. "De expeditie naar Lombok in 1894".

br Westenenk, L.C. "Doortrekking van den asweg ter Sumatra's Oostkust", 1923.

Entry: List of papers, numbered 1–169.

**Archival material desposited in the library** [549]

The library number is mentioned behind the description.

a. Aalders, A.A. "De zogenaamde gouvernementslandschappen", *c.* 1950. (A 313.1 Gr. formaat)

b. Bousquet, . . . "Nota betreffende de rechtstoestand van de grond ter hoofdplaats Djokjakarta", 1918. (Bou 324=541 Gr. formaat)

c. "Brieven en rapporten over het voorkomen van nikkel op Celebes door onder andere J.W.H. Adam, W. Dieckman te Soroako, Verbeekgebergte en Bulu Balang", 1918–1921. (I 361.1 Gr. formaat)

d. "Chemische samenstelling van vruchten in Indonesië voorkomende". (Che 634=54)

e. Deinum, H., and L. Setijoro. "Producten van de sagopalm; cultuur-economische beschrijving en gebruik". (Dei 633.3=546)

f. Franssen Herderschee, E. "Het Midden-Sumatra spoorwegvraagstuk", 1927–1928. (H 365.15)

g. "Gegevens inzake Dairilanden, voornamelijk rapporten betreffende thee proefstations"; door H. Loos, H. Christoffel, Handels Vereeniging Amsterdam, . . . Schuitemaker, . . . Diephuis, W.G. Graafhorst, . . . Vageler, . . . Maurenbrecher, . . . Bolderhey, . . . van Veen, . . . Raadsheer, . . . Hesseling, 1927–1930.

h. Haan, J.H. de. "Soilconservation in Indonesia", 1949. (Broch Haa 631.4=54)

i. "Handtekeningen van bezoekers van het Koloniaal Museum", 1885–1923. (Han 019=41 Gr. formaat)

j. "Inleiding tot de film Het schoone Langkat". 1939. (I JJ)

k. Kostelijk, M.A. "Studie over Indonesisch fruit", *c.* 1940. (Kos 634=540)

l. "Mission Dr. Cramer et Ringoet; journal de voyage [à la Malaisie, l'Indochine, Java, Sumatra et Ceylan]", 1934. (Cra 63.41 Gr. formaat)

m. "The native states in Sumatra; by the Gunseiba General Sumatra Advisory Board Medan", *c.* 1944. (Broch Nat 322=542 Gr. formaat)

n. "Nota inzake het immigrantenasyl to Medan", 1903. (Broch Not 359.81=542 Gr. formaat)

o. "Ontwerpen voor eene zeehaven beoosten het bestaande havenkanaal te Batavia", 1874. (O 365.11)

p. Perquin, P.J. "Beschrijving van Hindoe Javaanse en andere bouwwerken van oudheidkundig belang: Panataran, Prambanan, Tjandi Ngawen, Tjandi Merak, Bhatara, Guru van Singosari, Palembang, Kesepuhan te Cheribon", 1925–1927. (Per 940=540 I–V)

q. Polder, L. van de. "Rapport sur la culture de la plante appelée toshinto, juncus communis, et juncus leschenaultie et autres plantes dont on fait nattes au Japon. – Traduction d'extraits du traité sur la culture etc. des bambous du Japon", 1893. (Broch Pol 63V(bam)=52)

r. Poldermans, H.W. "Enkele economische gegevens omtrent de geschiedenis der kapokcultuur in Nederlands-Indië", 1941. (Broch Pol 63V(kap)=54.0)

s. "Rapporten over djelutung zaken op Borneo", 1912. (Dje 63R(Cb4)=543.1)

t. "Report on autonomous bodies in Sumatra", *c.* 1940. (Broch Rep 341=542 Gr. formaat)

u. Ruiter, J.H. "Een historisch overzicht van het aandeel van de Indonesische bossen in de houtvoorziening van Singapore", *c.* 1950. (Rui 63H(E)=54)

v. Ramaer, J.W. "Japansche groot-industrie", *c.* 1935. (Ram 379.33=52 Gr. formaat)

w. Sladden, G.E. "La culture indigène du café aux Indes Néerlandaises", 1932. (Sla 63 Ko(C)=54)

x. Steinmetz, K.K.J.R. "De haven van Belawan", 1931. (S 365.11)

y. Temminck, J.T.W. "De rubbercultuur en de beteekenis ervan voor de welvaartstoestand in Nederlandsch-Indië", 1924. (Tem 63 R(Ba)=54 Gr. formaat)

z. Veer, H. de. "Inventaris Cultuurtuin Tjibinong", 1933. (B 082 NI Gr. formaat)

aa "Verslag van de bevindingen omtrent een aan den opperhoutvester C. van de Koppel tijdens zijn buitenlandsch verlof verleende opdracht om in Nederland, Duitschland, Engeland en Amerika een studie te maken van den handel in en het gebruik van copal en rotan", *c.* 1935. (Kop 63 Ha(copal) Gr. formaat)

ab "Verslagen over de rosellacultuur op Java", 1938. (Ros 63 V(hibiscus)=54)

ac Visser, H.F.E. "De weef- en ikat-kunst in Nederlandsch-Indië", 1932. (V 741)

ad Vogel, W.Th. de. "Verslag van een reis naar Japan, Formosa, Hongkong, de Philippijnen, Indo-China en Singapore in opdracht van de regeering 1920–1921". (Vog 612=501 A I–II)

**Maps and marine charts** [550]

The collection consists of *c.* 18,000 sheets, of which the main component pertains to Indonesia.

Entry: Systematic geographic catalogue.

**Photographs** [551]

The collection consists of about 500,000 photographs concerning tropical countries. Approximately one half of these concerns Indonesia.

Entry: Systematic catalogue.

## Afdeling Agrarisch Onderzoek (Agricultural Research Department) [552–553]

### Koloniaal Museum te Haarlem (Colonial Museum of Haarlem), 1864–1913, 1913–1931 [552]

The Museum was established in 1864 by the Nederlandsche Maatschappij van Nijverheid (Netherlands Society of Industry). The object was to start collections of raw materials and products of home industry, and also to build up a library, from 1897 there was a laboratory connected with the Museum; 1871 opening to the public; 1884 independent, up to 1901 director F.W. van Eeden, up to 1910 M. Greshoff, then J. Dekker; 1913 handed over to the Afdeeling Handelsmuseum van het Koloniaal Instituut (Trade Museum Department of the Colonial Institute) of Amsterdam.

The archives consist of correspondence 1867–1931 with postbooks and a separate piece of writing including documentation about opium.

Size: *C.* 2 running metres.

### W. Callman [553]

Bundle with drawings made for M. Greshoff's *Nuttige Indische planten, c.* 1900.

## Afdeling Sociaal Wetenschappelijk Onderzoek (Social Scientific Research Department) [554–555]

### Memories van overgave (Reports by colonial officials on being transferred) [554]

The "memories van overgave" were collected as part of the activities of the Encyclopaedisch Bureau (Encyclopaedic Bureau), which was working at the Institute from 1931, after the Bureau at Batavia (established in 1910) had been abolished. Because of lack of government subsidy and of assistance in money from the Hulp-Instituten (Branch Institutes) the Bureau was forced to stop its activities in 1939.

According to the annual report of 1938 of the Koloniaal Instituut (Colonial Institute) 873 "memories" had been collected. For the names and places mentioned in the list see the index.

Entry: List of microfilmed "memories van overgave".

**Archival material deposited in the library** [555]

a. Nicholson, J.A. "Staten van dienst en brieven".

b. Hikajat Kumala Bahrein. "Maleise tekst in Arabische karakters", c. 1872.

c. Kolff, G.H. van der. "Opstel over het Staatsblad 1878, no. 10".

d. Lorentz, F.A.N. "Journaal van de schoener De Vlieg", 1848–1854.

e. Schermers, C. "Aanteekeningen over: Lombok, de Molukken, Buru, Ceram, Goram, Watubela, Key, Aru, Tanimbar, Babar, Leti, Mou, Kisar, Manuk, Sarua, Nila, Wetar, Uliassers, Banda, Nieuw-Guinea, Tengerezen, koperbewerking, kunst, Minangkabau, Hindu-priesters op Bali, landbouw op Bali, aanteekeningen van H. Geurtjens", *c.* 1920. 16 vol.

f. Wilken, G.A. "Handleiding voor vergelijkende volkenkunde van Nederlandsch-Indië", *c.* 1890. In 1893 published under the same title by C.M. Pleyte.

g. Sampano. "Adatrecht in verhaalvorm door Pakih Sampano, hoofd van negeri Sidjantang, Sumatra's Westkust", 1886.

h. Blokzeyl, A.H.C. "Maleisch-Balische woordenlijst", 1856–1858.

i. Lekkerkerker, C. "Een Balineesch handschrift", *c.* 1923.

j. Schermers, C. "Lombok; land en volk", 1924.

k. Montague. "Verhaal van het leven en zendingswerk van . . . Montague, opgetekend door . . . op ten Noort", 1892.

l. Schermers, C. "Nieuw-Guinea; beschrijving van land en volk", 1923. 4 vol.

Entry: Systematic catalogue. The archival material is to be found under the heading ai.

## Afdeling Tropenmuseum (Museum for the Tropics Department) [556–557]

**Manuscripts** [556]

A large portion of the collection consists of manuscripts in non-western languages: Javanese, Malay, Sundanese, Balinese, Batak, Buginese, Achinese, Urdu, Tamil, Singalese, papers from South Celebes, Lampongs, Toradja area, Boni, Lombok, Persia, Burma, Thailand, Tibet, Ceylon, Japan and China. In addition there are some archival documents present, mainly in western languages:

a. "Een boek betreffende de verkoop van opium"; open letter from L.W.G. de Rook to H.J.A.M. Schaepman. S.a. (505, no 23)

b. Letter from G.L. Tichelman, 1935. (1898, no 1)

c. Papers originating from F.W. Stammeshaus, including letters from Teuku Umar, T. Panglima Polem and Sultan Moh. Dawot Sjach Kamala, *c.* 1895. (687, no 52–79)

d. "Ambons geneesmiddelenboekje"; by M. Laturieuw. (2705, no 1)

e. Book of drafts of outgoing letters and of accounts of G.P. King at Ampenan, 1847–1867, Malay and Balinese text. (365, no 1)

f. Letters from various persons in India: Atmu J. Kamlanian to H.G. Wells about Gandhi, 1931; Argyl to . . . Hooker 1873; Maharadja of Burdwan to H.L. Erskine 1867; . . . Crickmer to E.D. Bourdillon, *c.* 1867; J. Adams, 1867; A. de Vecchy, 1866; C.E. Bernhard to H. Bergne and F. Abel, 1861; . . . Smith, 1858; W.R. Brooke to . . . Scoble, 1899; D. Steward to . . . Mayor, 1897; W. Wedderburn to Massinghani, 1897; Lal Chandra to i.a. G. Reid, 1897; Prolapa Chandra Roy to E. Arnold, 1893; Radja Pudmanund Sing Bahadoor, 1895; G.W. Forrest, 1891; . . . Lewin to . . . Rost, 1879; J.M. Spuchly to C. Barker, 1878; Salar Jung, 1876; H. Summer Maine, 1876; S.P. Sinka to A. Fitzroy, 1919; M.A. Stein to M. Colles, 1909; Raleigh to Lady Stanley, 1902; C.A. Sherring, 1907; J. Prinsep to M.I.I. Marcel, 1835; H.H. Wilson to M. Lumsden 1814, 1820; J. Balfour, 1821; J.R. Sneyd to . . . Scott, 1825; D. Calcutta, 1845; D. Wilson, 1862; L. Darell, 1801; . . . Stanhope to J.P. Josy, 1828; P.M. Taylor, 1829; J. Nash, 1843; Dwarhananth Tagore to . . . Parrot, 1845; J.H. Stocqueller, 1847 and J. Nash. (370, no 21–58)

Entry: Systematic catalogue on index cards, classified according to geographical areas. The geography is subdivided into subjects.

The manuscripts were found under subject heading XII.

**Drawings and pictures** [557]

The collection consists of *c.* 2100 drawings and pictures, of which *c.* 1500 concerning Asia and Oceania. The vast majority of these concerns Indonesia (*c.* 1000), and in addition drawings about Japan, Indo China, India, China and some drawings about Persia, New Guinea, Thailand, Pakistan, Ceylon, Maladives and Tibet. The collection contains mainly drawings from the 19th century.

Entry: Systematic catalogue.

## Nederlandsch Bijbelgenootschap (Netherlands Bible Society), NBG 1814

366 Herengracht, 1016 CH Amsterdam.
Admission by appointment. Lending out (not of archival documents).
Research finished in 1970.

**Library** [558]

11,000 books; 22 periodicals.

Directed to translation and propagation of the bible and to the activities of other bible societies.

For literature concerning the NBG see i.a.
C.F. Gronemeijer. *Gedenkboek van het Nederlandsch Bijbelgenootschap, 1814–1914*. Amsterdam, 1914.

E. René van Ouwenaller. *De eerste twintig jaren der tweede eeuw van het Nederlandsch Bijbelgenootschap*. Amsterdam, 1934. (Pamphlet in connection with the memorial vol. of Gronemeijer).

*Avontuur met het Woord;* indrukken van honderdvijftig jaar Bijbelgenootschapswerk; onder redactie van M.G. Schenk en J.B.Th. Spaan. Amsterdam, 1964.

**Archives, 1814 –** [559]

The Bible Society was established in 1814 at Amsterdam. The Society has been active in almost all the islands of the Indonesian archipelago, often in co-operation with the protestant mission. The papers present at the NBG are the residue of those activities. However, the archives between the years 1860 and 1940 show gaps.

The archives include minutes of the central committee and annual reports; reports; drafts of translations of the bible, dictionaries, grammars and reading books i.a. in the following languages: Balinese, Batak languages, Dajak languages, Javanese, Madurese, Malay, Minahasa languages, Molucca languages, Morese, Niassese, Rotese, Sangirese, Sasak, Sawunese, Sumbanese, Sundanese, Toradja languages (the library contains printed matter in an even larger number of languages); papers 1822–1852 of the Vierde Commissie Nederlandsch Oost-Indische Zaken (Fourth Committee for Dutch East Indies Affairs); annual reports 1906–1930 of the Zendingsconsulaat (Protestant Missionary Consulate); portfolioes 1852–1860 relations with organizations in the Netherlands, "Algemene Oost-Indische Zaken" (General East Indies Affairs), "Dajaks Agentschap" (Dajak Agency), portfolioes with letters and other papers to and/or from: N. Adriani, J.H. Duyverman, W.N. Engelman, J.F.C. Gericke, H.C. Klinkert, H. Kraemer, J.M. Martens, B.F. Matthes, Ministry of Colonial Affairs, L. Onvlee, J.A. Palm, Rheinische Missionsgesellschaft, J.M. Segveld, W.G. Shellabear, J.L. Swellengrebel, B. Talar, H.N. van der Tuuk, H. van der Veen, C.J. van der Vlis, L. Waller.

Entry: Alphabetic index. A systematic arrangement according to the decimal system is forthcoming.

Size: *C.* 10 running metres.

## Nederlandsch Economisch-Historisch Archief (Netherlands Economic History Archives), NEHA 1914

218–220 Herengracht, 1016 BT Amsterdam.
Admission free.
Research finished in 1979.

**Archival material** [560]

The NEHA collects documents in the field of the economic history of the Netherlands and the former overseas territories. The more extensive collections of archives are deposited at various state and municipal archives. At the NEHA are kept, besides a library, a collection of archival documents called "kleine aanwinsten" (small acquisitions).

For the history of Asia may be relevant: papers originating from the firm of Temminck & Van Twist 1789–1813 (no. 13), printed silk from China 19th century (no. 28), papers concerning the Algemeene Koloniale en Internationale Tentoonstelling (General Colonial and International Exhibition) at Semarang 1914 (no. 45), shipowners' books of various ships 1847–1899 (no. 71), circular letter about the public revenue in Java 1817 (no. 128), open letter from [H.G.] Nahuys [van Burgst] about the granting of land in Djokja 1822 (no. 129), papers about the sale of linen cloth by the VOC 1793–1800 (no. 159), papers by H. van Rijckevorsel about the East Indies trade 1842–1847 (no. 163), priced catalogues of the Handels-Vereeniging (Trade Association) at Surabaja, W.B. Ledeboer & Co. at Macassar, E. Schellhas & Co. at Hongkong, Handels-Vereeniging (Trade Association) at Semarang, Hooglandt & Co. at Singapore, In 't Veld & Co. at Padang 1877 (no. 168), payment of taxation for the bondman Vrijdag by W. Forreman 1844 (no. 172), bill of exchange of S. Mente at Batavia 1822 (no. 173), priced catalogues Batavia 1849 (no. 174), allegorical representation by M. van Weeren 1861 (no. 175), photographs and reports about the radio communication the Netherlands–the East Indies, the Internationale Radio-Conferentie (International Radio Conference) in Washington, the installation of telegraph cables in the East Indies 1920–1928 (no. 204), papers of the Commissie van Advies inzake de Tinexploitatie op Banka en de Kopermijnen op Timor (Advisory Committee for the Exploitation of Tin in Banka and the Copper Mines in Timor) 1849–1850 (no. 206), papers of the Oost-Indische Maatschappij van Administratie en Lijfrente (East Indies Company of Administration and Life Annuity) 1844–1870 (no. 233), data about batik and the transport of goods to the Dutch East Indies *c.* 1851 (no. 246), harbour advices Batavia, Semarang, Surabaja 1831–1832 (no. 259), journal of a voyage Harderwijk-Batavia 1838–1846 (no. 344), reports of the Weduwen- en Wezenfonds van de Officieren van de Landmacht in Nederlandsch-Indië (Widows' and Orphans' fund of the Officers of the Land Forces in the Dutch East Indies) 1844–1895 (no. 391), "Fox-contract" of the Indonesian-American Corporation 1947 (no. 398), liquidation accounts of the firm of Wermuth & Co. at Semarang 1868 (no. 448), prospectuses of the Internationale, Koloniale & Uitvoerhandel Tentoonstelling (International, Colonial & Export Trade Exhibition) of Amsterdam 1883 (no. 494).

Entry: I. Vellekoop. Inventaris van de verzameling "kleine aanwinsten". In *Economisch en sociaal-historisch jaarboek;* vol. 39, 1976.

Size: Unknown number of running metres (500 inventory items).

## Rijksinstituut voor Oorlogsdocumentatie; Indische Afdeling (State Institute for War Records; East Indies Section)

28 Prins Hendriklaan, 1075 BD Amsterdam.
Admission for scientific research workers, by appointment.
Lending to institutions.
Research finished in 1977.

**Documentation** [561]

The "Indische collectie" (East Indies collection) consists of *c.* 40 running metres of documentary materials, mainly concerning the European community in the former Dutch East Indies 1942–1945. To this have been added the following items or collections: *c.* 180 diaries, *c.* 100 boxes with documents of the case of the Tokyo Trial 1945–1948, *c.* 60 running metres of library of the section, periodicals, newspapers and cuttings, *c.* 14 running metres press reviews, broadcast transmissions (out of the Dutch East Indies) and reports of listeners (monitoring-reports) 1942–1945, *c.* 10,000 photographs, *c.* 6 running metres of drawings, posters etc.

Entries: Personal and catchwords card indexes and catalogues, inventories, lists of headings.

Size: *C.* 200 running metres.

See also no. 79.

Permission is necessary for consultation of the papers.

## Rijksmuseum "Bronbeek" (State Museum "Bronbeek"), 1863

147 Velperweg, 6824 MB Arnhem.
An admission fee is charged.
Consultation of the archival documents by appointment. In principle no lending.
Research finished in 1970.

**Library** [562]

6000 books, 20 currently received periodicals.

The library was founded for the benefit of the inhabitants of "Bronbeek".

A library directed to the history of the Dutch East Indies, in special the KNIL, is being built up.

For literature concerning "Bronbeek" see i.a.
J.G.J. Smits. *Gedenkboek van het Koloniaal Militair Invalidenhuis op Bronbeek.* Arnhem, 1881.

J.G. Raatgever. *Honderd jaar Bronbeek.* 's-Gravenhage, 1963.

W.R. van der Weide. Het Koninklijk Militair Invalidenhuis op Bronbeek. In *Mars et historia,* 1968.

**Archives of Bronbeek, 1862–1955** [563]

Bronbeek was established in 1863 as a home for disabled soldiers of the colonial armies.

As far as can be estimated the archives consist of the commandantsarchief (archives of the commanders) 1862–1955 (up to 1900 for the greater part destroyed); memorandum of the Koninklijk Koloniaal Militair Invalidenhuis (Royal Colonial Disabled Soldiers Home) 1862–*c.*1900 (alphabetic index for papers received and despatched); "stamboeken" (rolls) of the inhabitants of the Disabled Soldiers Home; household regulations; papers concerning the canteen, the orchestra, management of provisions; documentation (i.a. in connection with the 75th anniversary), articles, photographs.

Entry: Inventory is forthcoming.

Size: *C.* 1 running metre.

The archives of the commander may be consulted only as a great exception.

**Miscellaneous papers** [564]

As far as can be estimated papers concerning:

a. *Atjeh:* Correspondence, regulations, subscription lists etc. 1874–1876 concerning the foundation of the Atjeh monument at Batavia in consequence of the killing of general [J.H.R.] Köhler (originating from C.J. Marinkelle); declarations of subjection to the Dutch authority by the princes of Sungai Radja 1877, Perlak 1833 (with additional declarations 1897, 1898), Djolok Besar 1884, Idi Besar 1895, Simpangulim 1895 (with additional declaration), Langsar 1897, 1898; memorandum concerning the commandoes Blangkedjeren and Lukup, comprising the subdivision Gajo Luos and Serbödjadi (military survey 1905–1912).
b. *Riouw:* Memorandums import and export duties 1880, 1881 by the comptrollers N.J. Struick and J.G. de Serière at Indragiri.

c. *New Guinea:* Journal 1919–1922 of the commander [J.] van Arkel of the detachment for the preparation of the scientific expedition to the central mountain land of New Guinea with correspondence, documentation and lectures given by Van Arkel.

d. *Celebes:* Military memorandum Parepare with annexes 1927–1931; diary of the comptroller at Parepare [R.H. Mulder] with annexes 1929–1932.

**Maps** [565]

As far as can be estimated about 1000 mainly military ordnance survey maps 1900–1949 concerning all of the archipelago.

Inventory is forthcoming.

**Drawings, paintings etc.** [566]

Besides museum pieces such as arms, banners, furniture, ethnographics etc. "Bronbeek" possesses dozens of paintings and drawings, for the greater part displayed in the museum. They are mainly portraits of soldiers and scenes of war.

**Photographs** [567]

As far as can be estimated *c.* 2000 photographs, mainly concerning military expeditions, and in addition landscapes and portraits of soldiers.

Inventory is forthcoming.

## Koninklijke Militaire Academie (Royal Military Academy), KMA 1828

10 Kasteelplein, 4811 XC Breda.
Admission by appointment.
Photocopying equipment available.
Research finished in 1978.

**Library** [568]

118,000 books; 1480 periodicals, of which 600 currently received.

Directed to military matters.

For literature concerning the KMA see i.a.
G. van Steyn. *Gedenkboek van de Koninklijke Militaire Academie, 1828–1928.* Breda, 1928.

H.J. Wolf. *Het Kasteel van Breda en de Koninklijke Militaire Academie.* 's-Gravenhage, 1958.

H.J. Wolf. *Honderdvijftig jaar Koninklijke Militaire Academie, 1828–1978.* 's-Gravenhage, 1978.

**Introduction to the documents** [569]

Practically all the documents inserted below have been described in J.D.B. Ophorst. *Catalogus van de bibliotheek der Koninklijke Militaire Academie.* Breda, 1900–1926. 1 vol. and 18 supplements. The library numbers from this catalogue are mentioned in brackets directly behind the descriptions below.

Beside this catalogue, special catalogues were published for the manuscripts and the "traditieverzameling" (historical objects collection). The relevant numbers of these catalogues are mentioned behind the descriptions of the titles.

**Manuscripts** [570]

a. Survey of the main events in Djokjakarta from the foundation of the empire in 1775 till the end of the Java War in 1830, originating from A.F.C. List. *C.* 1830. 150 p. (no. 3955)

b. Letters and reports of Javanese chiefs and station commanders written during the Java War, 1825–1830. 3 vol. Arabic, Malay and Dutch text. (no. 3962)

c. Brief account of the war in Java, 1741–1759, originating from A.F.C. List. 226 p. (no. 3947)

d. Hikajat Fathimah Sami. 75 p. Javanese text. (no. 6644)

c. Prima surata Corani, adjunctas sibi habens binas versiones scilicet Mallaeorum et Javanorum. 57 p. (no. 6646)

f. Jura maritima Malaeorum conscripta sub imperio Regis-Mahmud Sjah, anno Hezras 1066, sermone Malaico. 34 p. (no. 6619, 107B6)

g. Hikajat Nabi Musa. 2 copies originating from D.M. Valentijn. (no. 6620, 107B8)

h. Collection of manuscripts in Arabic and Malay, mainly concerning religious matters. 28 papers. (no. 6557, 107B5)

i. J.H. Fuchter. "Leerboek der militaire aardrijkskunde en statistiek van Nederlandsch Oost-Indië". (no. 693/a4)

j. J.B.A. Bruining. "Samenstelling van de Indische woning". 1925. (no. 5382/4)

Entry: H.J. Wolf. *Overzicht van de handschriften-verzameling opgenomen in de bibliotheek van de Koninklijke Militaire Academie.* Breda, 1965. (Cyclostyled publication, no. 69–71, 76–81)

**Maps** [571]

Approximately 100 catalogue numbers are pertaining to the Dutch East Indies, viz. to Java, Madura, Sumatra, Borneo, Celebes, Riouw, Bali, Lombok, Timor, the Moluccas and New Guinea. (no. 6361–6468)

**Paintings, drawings etc.** [572]

Drawings of Javanese payoongs (no. 6382), "Bataviaasch album; verzameling van een tiental gezigten van de hoofdstad van Nederlandsch-Indië" (no. 6403), portraits of K. van der Heijden by G. van der Heijden, A.V.P.M. Michiels by J.A. Kruseman, the pretender sultan of Atjeh by N. van der Waaij, paintings of the storming of Baté Ilië by J. Hoynk van Papendrecht, the expedition to Banka (1819–1821) by Th. Hans and objects (mainly arms) originating from H.C.J. ter Beek and G.M. Verspyck acquired in Tjakranegara, Atjeh, Central Sumatra, Nias and Borneo.

Entry: H.J. Wolf. *Catalogus van de traditie verzameling der Koninklijke Militaire Academie.* Breda, 1960. (Cyclostyled edition, chapter collection of paintings, no. 16, 17, 37, 44, 54, 55, chapter special collections: Ter Beek and Verspyck)

## Bureau voor de Archieven der Nederlandse Hervormde Kerk (Bureau for the Archives of the Reformed Dutch Church), 1625

100 Javastraat, 2585 AV The Hague.
Admission free.
Photocopying equipment available.
Research finished in 1979.

**Library** [573]

10,000 books; 400 periodicals.

For literature concerning the Bureau see J.P. van Dooren. *Kisten en kasten;* 350 jaar kerkelijke archiefdienst. 's-Gravenhage, 1976.

**Archives** [574]

For the history of Asia may be relevant the archives of:

a. *Commissie tot de Zaken der Protestantsche Kerken in Oost- en West-Indië (Committee for the Affairs of the Protestant Churches in the East and West Indies), 1815–1946.* These archives have been temporary deposited at the Algemeen Rijksarchief and have been described in that chapter (see no. 144).

b. *Nieuw archief van de Algemeene Synode der Nederlandsche Hervormde Kerk (New archives of General Synod of the Reformed Dutch Church), 1816–1945,* after 1945 called "Archief van de Generale Synode". These archives include correspondence about the contacts with the East Indies churches and clergymen. In total *c.* 100 running metres (1085 numbers).

c. *Oecumenische Raad van Kerken in Nederland (Oecumenical Council of Churches in the Netherlands), 1948–1968.* These include, besides reports of the Wereldraad van Kerken (World Council of Churches) 1948, also papers from 1919, i.a. of the Bond van Internationale Vriendschap door de Kerken (League of International Friendship through the Churches), the Wereldbond van Kerken (World League of Churches) and the Oecumenische Vereeniging (Oecumenical Association). In total *c.* 30 running metres.

d. *Centrale Comitee Kerkelijk en Particulier Initiatief voor Sociale Zorg ten behoeve van Gerepatrieerden (Central Committee Church and Private Initiative for Welfare of Repatriates), 1950–1969.* Though the archives, mainly consisting of alphabetically arranged personal files, fall outside the period 1796–1949, they are mentioned for completeness because they are related with the colonial period. In total *c.* 75 running metres.

e. *Commissie Interkerkelijk Contact in Overheidszaken van de Oecumenische Raad (Committee for Inter Church Contact in Governmental Matters of the Oecumenical Council), 1950–1968.* These archives are mentioned also because of their relation with the colonial period. They consist of the papers originating from the sub-committee Gerepatrieerden (Repatriates) and from the sub-committee for the Amboinese. In total *c.* 2 running metres.

Entries: *De archieven van de Nederlandse Hervormde Kerk in korte overzichten.* Leiden, 1960, 1974. 2 vol. These include inventories of archives not deposited at the Bureau.

*Kerkelijke archieven;* inventarissen. 's-Gravenhage, 1970– . Vol. 1– . (In vol. 10: inventory of the archives of the classis of Middelburg).

## Centraal Bureau voor Genealogie (Central Bureau of Genealogy), CBG 1945

26 Prins Willem Alexanderhof, 2595 AB The Hague.
Admission on payment of a fee.
Photocopying equipment available.
Research finished in 1979.

### Introduction [575]

The CBG possesses a library of some thousands of books and some hundreds of periodicals directed to genealogy, heraldry, topography and history, archival sources and inventories.

In addition the Bureau is in charge of a collection of genealogical manuscripts and a genealogical documentation.

The CBG works closely together with the Iconographisch Bureau (Iconographic Bureau), where a card-index-system is kept up to date of *c.* 100,000 index-cards and where a collection of reproductions is present of some ten thousands of portraits.

For literature about the CBG see *Overzicht van de verzamelingen berustende bij het Centraal Bureau voor Genealogie en het Iconographisch Bureau.* 's-Gravenhage, 1977. 4th ed.

**Genealogical manuscripts** [576]

The collection consists of *c.* 40,000 manuscripts, alphabetically arranged to name. Besides, the collections of Van Beresteyn and Wynaendts van Resandt may contain genealogical data about families in Asia.

**Genealogical documentation** [577]

The following collections contain data about persons and families in Asia:

a. Births, marriages and deaths columns from newspapers.

b. Births, marriages and deaths columns from the *Javasche Courant,* 1810–1875.

c. Register for the *Regeringsalmanak voor Nederlandsch-Indië,* 1817–1923.

d. Register of persons killed in the Atjeh war.

e. Micro index-cards of letters from the archives of the classis of the Reformed Dutch Church of Amsterdam, received from persons from India, Ceylon and the East Indies.

f. Micro index-cards of the "burgerlijke stand" (civil registry) compiled by P.C. Bloys van Treslong Prins and deposited at the Arsip Nasional Republik Indonesia at Jakarta. Included are data about the Dutch community, the Portuguese community, the Malay community and registrations of baptisms, marriages and deaths in Batavia 17th up to 19th century, conduct sheets of officials in various places in the Dutch East Indies 19th century, adoptions in Batavia 17th up to 19th century, lists of the European population 19th century and VOC personal notes. For the names of places and departments mentioned in the list of conduct sheets see the index.

## Koninklijk Huisarchief (Archives of the Royal Family), 1813

74 Noordeinde, 2514 GL The Hague.
Admission by appointment.
Research finished in 1978.

**Library** [578]

50,000 books; 60 currently received periodicals. Directed to the dynastic history of the Dutch royal family, national Dutch history, general history, art and travel literature, and in addition deposited collections of several members of the royal family.

See E. Pelinck. *Het Koninklijk Huisarchief;* geschiedenis en overzicht. 's-Gravenhage, 1971.

**Introduction to the archives** [579]

The archives consist of papers originating from the members of the royal family from the 13th century, with in addition papers concerning the royal househould and papers not belonging to the Oranje Nassau family.

For the archives of the Kabinet des Konings (King's Cabinet) see no. 12.

Permission in writing of the Director is necessary for consultation of the papers deposited at the Koninklijk Huisarchief. Consultation of the papers dating from the period of King Willem III is only possible in exceptional cases.

**Members of the Royal Family** [580]

For the history of Asia are relevant the papers originating from:

a. *Willem I, 1772–1843*
   Correspondence with D.F. van Alphen, J.P. and W.C. van Braam, P.A. Brugmans, [L.P.J.] du Bus de Gisignies, G.A.G.Ph. van der Capellen, [H.W.] Daendels, C.Th. Elout, [J.C.] Smissaert; papers about colonial affairs, viz. a letter from [P.A.] van Rappard to sergeant-major . . . Broekmeyer about the fight near [Kadondong, Cheribon] 1818, letters from A.R. Falck to C.Th. Elout about the eventual appointment of H.M. de Kock as successor of G.A.G.Ph. van der Capellen 1821–1822, notes about political and commercial relations between Great Britain and the Netherlands 1825, "plan der positiën van het leger in de Vorstenlanden" (plan of the positions of the army in the Principalities) by [F.V.H.A.] de Stuers 1828, "nota omtrent de geldheffingen van het Hollandsch gouvernement op Java" (memorandum concerning the levies of the Dutch government in Java) s.a., papers about the publication of the Flora Javae and Rumphia by [C.L.] Blume 1841–1850.

b. *Willem II, 1792–1849*
   Correspondence with D.J. de Eerens 1837, J.J. Garney about the prisons in the Netherlands and slavery in the West Indies and Elmina 1841, letter from J. Kruseman about the colonial surplus 1841, letters from J.J. Rochussen.

c. *Anna Paulowna, 1795–1865*
   Letters from J.J. Rochussen.

d. *Prince Hendrik, 1820–1879*
Correspondence with J.J. Rochussen 1846–1848, L. Fix, C. van den Berg, tokens of regard from the Vereeniging van Nederlandsch Indische en Oud Indische Strijders (Association of Dutch East Indies and Ex Dutch East Indies Soldiers), the Oost- en West-Indische Opvoedingsmij. (East and West Indies Society of Education), the Red Cross, papers about shipping, trade and industries viz. Ph.F. von Siebold about the foundation of a trading station in Billiton 1851, P.J. van der Does about the establishment of a trading business in Japan 1876, papers about steam packet trade to the Dutch East Indies 1857, Billitoncompany 1872, Mij. "Nederland" (shipping company) 1869–1878, Nederlandsch-Indische Droogdokmij. (Dutch East Indies Dry-Dock Company) 1872–1878, Utrechtsche Vereeniging Port Said (Utrecht Association Port Said) 1873–1878, papers about the Suez Canal and the trading establishment in Port Said 1869–1881.

e. *Willem III, 1817–1890*
Letter from J.J. Rochussen, tokens of regard from the Koninklijke Natuurkundige Vereeniging in Nederlandsch-Indië (Royal Physical Association in the Dutch East Indies), Semarang, King Willem III School, Indisch Landbouwsyndicaat (East Indies Agricultural Syndicate), honours granted by Hawaii, Siam, papers about his supreme patronage over the Koninklijk Bataviaasch Genootschap van Kunsten en Wetenschappen (Royal Batavian Society of Arts and Sciences) 1871, survey of the richness in minerals of the Dutch East Indies 1851, address of thanks from Semarang and Djokjakarta for the concession for the railway Semarang-Principalities 1863, report about Atjeh, 1873.

f. *Emma, 1858–1934*
Honours granted by Persia and Japan.

g. *Alexander, 1851–1884*
Honours granted by Hawaii.

h. *Wilhelmina, 1880–1962*
Honours and tokens of regard from Persia, Siam, Japan, China, Shanghai, Nepal, Indonesian princes, Singapore, Calcutta, Australia, exercise-books with Malay texts for the purpose of teaching Wilhelmina that language.

i. *Prince Hendrik, 1876–1934*
Diary of his voyage to British India 1894–1895, letter from Hamengku Buwono VI, honours granted by Siam, Japan, Persia, oriental manuscripts, including Javanese texts.

j. *Queen Juliana*
Tokens of regard from the Dutch East Indies.

**Papers not belonging to the Oranje Nassau Family** [581]

For the history of Asia are relevant the papers of:

a. *C.Th. Elout,* viz. decrees issued by Louis Napoleon about the recall of Elout as commissioner general 1806–1810.

b. *C.P. Schimpf,* viz. correspondence 1855–1867 with Ch.F. Pahud, P. Mijer, J.J. Rochussen, A.A.J. Sloet van de Beele, I.D. Fransen van de Putte and . . . Blanken i.a. about military matters and slavery.

c. *Autographs,* series II, letters from F.W.H.A. von Humboldt to J.H. Stone, H.M. Williams, Ch.A. Coquerel, M.H.K. Lichtenstein, . . . de Masters, A.L.Ch. Coquerel 1809–1846, letter from G.Th.F. Raynal 1789, J.J. Rochussen 1841, 1852, W.H. de Vriese to L.C. Luzac 1847.

**Maps** [582]

The collection contains *c.* 200 catalogue numbers concerning the Dutch East Indies, including manuscript maps of Banda, Makasar (Fort Rotterdam), Java and Madura, defence works in Batavia, Sidaju (Bagelen), Principalities, Bay of Amboina, seat of war in Atjeh, Upper Atjeh, Pedie (Atjeh), Lampongs and Ternate.

**Photographs** [583]

The collection contains *c.* 100 East Indies albums compiled on the occasion of jubilees or presented on visits of members of the royal family.

## Koninklijk Nederlandsch Genootschap voor Geslacht- en Wapenkunde (Royal Dutch Society for Genealogy and Heraldry), 1883

26 Prins Willem Alexanderhof, 2595 AB The Hague.
Admission by appointment. Lending to members.
Research finished in 1978.

**Library** [584]

38,000 books. Directed to genealogy and heraldry.

**Publications** [585]

*De Nederlandsche Leeuw.* 's-Gravenhage, 1883– .

**Archives** [586]

The Genootschap possesses exclusively genealogical documents, including documentation based on the papers deposited at the former Landsarchief (Arsip Nasional Republik Indonesia) in Jakarta, with in addition a department East and West Indies wills.

Entry: Alphabetic catalogue on index cards to family name.

Size: *C.* 1000 running metres (the size of the complete collection of the Genootschap).

# Koninklijke Bibliotheek (Royal Library), KB 1798

5 Prins Willem Alexanderhof, 2595 BE The Hague.
Admission free. Lending on introduction (manuscripts to libraries and archives repositories only).
Photocopying equipment available.
Research finished in 1971.

**Library** [587]

900,000 books; unknown number of periodicals.

Directed to philosophy, theology, sciences of law and politics, history, art, literature, anthropology, education, social geography, bibliography and library science.

For literature concerning the KB see i.a.
L. Brummel. *Geschiedenis der Koninklijke Bibliotheek*. Leiden, 1939.

*Gedenkboek Koninklijke Bibliotheek 1798–1948*. 's-Gravenhage, 1948.

*Gids voor gebruikers [van de] Koninklijke Bibliotheek*. 's-Gravenhage, 1969.

**Publications** [588]

*Verslag der Koninklijke Bibliotheek over 1866–* . 's-Gravenhage, 1867– .

**Central Catalogues** [589]

The KB has 3 central catalogues, stating the possession of books and periodicals of a large number of Dutch libraries:

The Central Catalogue (of books), associated *c.* 80 libraries.

The Central Catalogue of Periodicals, associated *c.* 220 libraries, also published in computerised form.

The Central Catalogue of Congresses, including proceedings and reports of congresses, symposiums etc.

**Introduction to the collection of "manuscripts"** [590]

The collection of manuscripts consists of *c.* 6500 vol. and portfolioes. The East Indies collections and miscellaneous papers have been found by means of the headings "Land- en Volkenkunde; Azië" (Geography and Anthropology; Asia), "Geschiedenis; Azië" (History; Asia) and "Brievencollecties" (Collections of Letters) in the systematic catalogue. In addition to a systematic catalogue there are also an alphabetic index (to author, title (when anonymous) and for letters to addressee) and a shelf catalogue to the complete collection. With the description of the collections the numbers of the manuscripts will be mentioned in brackets.

In addition to this collection of manuscripts the KB also contains some collections given on loan. Of these the Cornets de Groot collection includes East Indies papers.

Bundles smaller than 0.10 running metre are regarded as "miscellaneous papers".

## Families and private persons [591–593]

### Cornets de Groot [591]

The collection consists of family papers from the sixteenth century. Asian papers are originating from:

a. *A.D. Cornets de Groot sr., February 29th 1768 Bergen op Zoom – November 10th 1827 Surabaja*
In 1816 after a career in the Dutch army, to the East Indies, where he became a member of the Raad van Financiën (Council of Finance); 1820–1826 resident of Grisee; 1826 member of the Bataviaasch Genootschap van Kunsten en Wetenschappen (Batavian Society of Arts and Sciences).

   This part of the collection consists of papers originating from him or concerning him: a sketch of his life by his son H. Cornets de Groot; decrees of appointment; letters 1787–1851 mainly from and/or about members of the family; report left by him for his successor in Grisee 1826. For the names and subjects mentioned in the very comprehensive manuscript inventory see the index.

b. *H. Cornets de Groot, August 17th 1798 Groningen – April 15th 1847 The Hague*
Second son of A.D. Cornets de Groot sr.; in 1816, after a career in the Dutch navy, to Batavia as lieutenant; 1818 first lieutenant; 1819 superintendent Public Revenue Tegal; 1819 secretary direction Bosschen; 1825 assistant resident Amboina; 1826 Saparua and Haruku; 1827 resident Menado; 1828 official at the Algemeene Rekenkamer (General Audit Department); 1829 assistant resident Sumenep and Pamakasan; 1830 resident Riouw, i.a. actions against piracy; 1836 resident Besuki; 1841 to the Netherlands; in 1832 he became a member of the Bataviaasch Genootschap van Kunsten en Wetenschappen (Batavian Society of Arts and Sciences).

   Originating from him are decrees of appointment; letters from and to him 1811–1847; correspondence about the foundation of a Cornets de Groot fund in aid of education in Riouw. For the names and subjects mentioned in the very comprehensive manuscript inventory see the index.

c. *A.D. Cornets de Groot jr., April 1st 1804 Groningen – July 10th 1829 Utrecht*
Third son of A.D. Cornets de Groot sr.; 1816 with his parents to the East Indies; 1818 clerk Algemeene Secretarie (General Secretariat) Batavia; 1819 "élève voor de Javaansche taal" (pupil for the Javanese language) at Surakarta under the supervision of [A.M.T.] de Salis and H.G. Nahuys van Burgst; 1823 secretary of the residency of Surakarta; 1827 government secretary for

Inlandsche Zaken (Native Affairs); 1829 to the Netherlands; in 1825 he was charged with the compilation of a Javanese grammar and dictionary, was concerned with the foundation of the first Javanese printing establishment, from 1826 a member of the Bataviaasch Genootschap van Kunsten en Wetenschappen (Batavian Society of Arts and Sciences).

Originating from him are decrees of appointment; books of travel 1816 Batavia-Amboina, 1827 Surakarta-Surabaja; letters 1815–1829; papers 1825–1844 concerning the compilation and publication (after his death) of the Javanese grammar. For the names and subjects mentioned in the very comprehensive manuscript inventory see the index.

d. *J.P. Cornets de Groot, April 6th 1808 Groningen – July 2nd 1878 The Hague*
Fourth son of A.D. Cornets de Groot sr.; 1828 second clerk Algemeene Secretarie (General Secretariat); 1830 first clerk; 1832 principal clerk; 1834 assistant secretary; 1836 general secretary; 1842 secretary-general of the Ministry of Colonial Affairs; 1848 member of the Raad van Indië (East Indies Council); 1851 to the Netherlands, where he became minister of Colonial Affairs in 1861; 1862 member of the Raad van State (Council of State).

The larger portion of the Asian papers are originating from him. However, they are not described in the comprehensive manuscript inventory. The papers consist of letters and official papers 1808–1877, letters from members of the family, notes for a book to be published about colonial politics, official papers and letters 1857–1859 concerning colonization.

e. *Other members of the family*
Apart from the many genealogical papers and family portraits the collection contains East Indies papers of the following members of the Cornets de Groot family:

J.H. Cornets de Groot (1796–1831, eldest son of A.D. Cornets de Groot sr., 1825–1831 successively secretary of the residency at Buitenzorg, assistant resident Batavia (Krawang), assistant resident Surabaja), a sketch of his life by his brother H. Cornets de Groot;

J. Cornets de Groot (1793–1818, eldest daughter of A.D. Cornets de Groot sr., married with R.H. Cateau van Rosevelt), letters to J.P. Cornets de Groot;

L.E.A. Cornets de Groot (1802–1877, second daughter of A.D. Cornets de Groot sr., married with P.W. van der Hoop, later with B.H.A. Besier), letters to J.P. Cornets de Groot and the Besier family;

A.R.C.M. Cornets de Groot (1807–1833, third daughter of A.D. Cornets de Groot sr., married with H. Beth), letters to J.P. Cornets de Groot and the Beth family, drawings;

M.W. Cornets de Groot (1809–1868, fourth daughter of A.D. Cornets de Groot sr., married with Th. Schuurman), letters to J.P. Cornets de Groot and the Schuurman family, papers concerning the steam packet service to Java of Clavières-Schuurman;

A.J.W. Cornets de Groot (1823–1893, eldest son of J.H. Cornets de Groot, jurist, 1850–1890 at Batavia, Surabaja and Semarang, retired when he was raadsheer van het Hooggerechtshof (counsel of the High Court of Justice) at Batavia), some letters and papers;

W.A.C. Cornets de Groot (1825–1898, second daughter of J.H. Cornets de Groot, married with E. Rappard), some letters and papers;

C.J.H. Cornets de Groot (1827–1909, third daughter of J.H. Cornets de Groot, married with G.I.M. van Reede van Oudtshoorn), from them letters to J.P. Cornets de Groot;

A.A. Cornets de Groot (1828–1906, fourth daughter of J.H. Cornets de Groot, married with J. Kniphorst), from them some letters;

L.H. Cornets de Groot (1831–1904, fifth daughter of J.H. Cornets de Groot, married with I.D. Fransen van de Putte), from them some letters and papers;

S.M.J. Cornets de Groot (1833–1913, daughter of J.P. Cornets de Groot, married with J.A.H. Netscher), notes about several travels.

Entries: D.J.H. ter Horst. Overzicht van het familie-archief Cornets de Groot. In *Anti-Revolutionaire Staatkunde*. 1940. Manuscript inventory, drawn up by J.P. Cornets de Groot, with a detailed description of the papers addressed to him.

Size: *C*. 6 running metres.

**L.W.C. Keuchenius, October 21st 1822 Batavia – December 17th 1893 The Hague** [592]

1844 lawyer and official to the procureur-generaal (public prosecutor) at Surabaja; 1848 member of the Council of Justice at Batavia; 1850 advocaat-generaal (solicitor general); 1851 raadsheer Hooggerechtshof (counsel of the High Court of Justice); 1854 to the Netherlands and secretary general of the Ministry of Colonial Affairs; 1859 member of the Raad van Indië (East Indies Council); 1865 member of the Second Chamber; 1867 editor of the *Nieuw Bataviasch Handelsblad;* 1869 advocaat en procureur bij het Hooggerechtshof (solicitor at the High Court of Justice) at Batavia; 1879 member of the Second Chamber; 1888 minister of Colonial Affairs; 1890 till his death in 1893 member of the Second Chamber.

This collection consists of papers 1840–1892 concerning himself; official papers and notes 1854–1893, with some preliminary papers from 1804, i.a. concerning education, mission, several legal cases, cultures, opium, Atjeh; letters 1859–1893. For the names and subjects mentioned in the inventory see the index.

Entry: D.J.H. ter Horst. Beschrijving der schriftelijke nalatenschap van Mr. L.W.C. Keuchenius. In *Anti-Revolutionaire Staatkunde*. 1940

Size: *C*. 1.50 running metres (numbered 68 E 1–21, 68 F 1–17, 38 inventory items).

See also no. 470.

**J.F. Scheltema, May 16th 1855 Makasar – June 13th 1922 London** [593]

1874–1882 clerk Post- en Telegraafdienst (Postal and Telegraph Service), 1874 assistant clerk, 1875 clerk third class, 1876 clerk second class, 1880 postagent (sub-postmaster) at Singapore; 1882 war correspondent in Tonkin; 1884 chief editor of *De Locomotief*, dismissed in 1886 in consequence of the Tjiomas affair; 1886–1887 voyages the East Indies–Europe; 1887–1893 in the United States railway employee, representative of the press association at St. Louis; 1893–1895 travels through Europe; 1895–1898 chief editor of *De Locomotief;* 1898–1902 chief editor of the *Bataviaasch Nieuwsblad* as successor of P.A. Daum, dismissed in consequence of his criticism of the government policy respecting the opium farming, 1901–1902 also a member of the board of the Nederlandsch-Indische Vereeniging tot Bescherming van Dieren (Dutch East Indies Association for the Protection of Animals); 1903 to the Netherlands; 1904–1906 study of "Arabic and oriental relations" at Yale University; 1906–1915 employed at Edinburgh; 1915–1920 study of semitic languages at Yale University; 1920 to the Netherlands; 1921 to London. (G.A.N. Scheltema de Heere. Ter herinnering aan J.F. Scheltema. In *De Indische Gids*. 1924, II, p. 1061/*RAvNI*).

The collection consists of papers regarding his journalistic and official work 1874–1902, i.a. concerning the press and press offences and private landed property.

Size: *C.* 1 running metre (numbered 122 B 9).

The collection may be consulted after 1997.

**Asian papers in Dutch collections** [594]

For a definition see no. 271.

a. **C.F. Sirtema van Grovestins**

Letters to G.A.G.Ph. van der Capellen, R.[L.J.] van der Capellen, C. Kruseman, J.W. May and [J.J.] Rochussen. (Numbered 75 D 26–28).

**Miscellaneous papers** [595]

The papers are arranged according to number of manuscript.

a. *72 A 29,* J.A. Meesen. Album of a six years' travel through the East Indies. Utrecht, 1871. With pictures and photographs.
b. *72 E 17,* letters from Th.S. Raffles to N. Wallich (botanist at Calcutta) from Penang, Singapore and Benkulen, 1819–1824.
c. *74 H 4,* letters from A.R. Falck to i.a. P.J. Elout van Soeterwoude (1837) and G.A.G.Ph. van der Capellen (1840).

d. *75 B 48*, papers concerning Java:

1. Draft cultivation system, 1831.
2. "Geschiedverhaal van de vorsten van Java tot 1823; vertaald uit het Javaansch in het Maleisch door Tjondro Koesoemo; in het Nederlandsch vertaald door D. Landoung".

e. *76 A 9*, papers about Java, Malaya and Siam, drawings of Javanese arms and payoongs, 19th century, Javanese and Dutch text.

f. *76 B 57*, correspondence 1882–1883 between F. s'Jacob and W.M. de Brauw, with letters from W. van Goltstein, P.F. Hubrecht and A. Pruys van der Hoeven at Kota Radja.

g. *76 D 21*, papers 1856–1863 of and concerning F.W. Junghuhn, viz. photographs, catalogue of his library, travel sketches, notes and letters from K.H.Th. Friederich, I. Groneman, E.L. van Heeckeren, F.W.H.A. von Humboldt, C.A.J.A. Oudemans, M.J. Polak, F. von Richthofen, J.J. Rochussen, D. Scherzer, . . . Schnell, C. Semper and L.A.J.W. Sloet [van de Beele].

h. *76 D 33–36*, papers 1893–1908 originating from P.J.F. Louw, viz. genealogies of princes in the Vorstenlanden (Principalities), memorandum concerning Sukabumi and the pawn shop service in Java and Madura, reports and other papers 1892–1900 concerning the Java War by himself, R.J. Fruin and E.B. Kielstra in reference to the book *De Java-oorlog, 1825–1830*, letters from F. de Bas, B. Croon, C. Deykerhoff, [A.R.W.] Gey van Pittius [jr.], C. van der Gon Netscher, J.C.C. Hartevelt, Koesoemowinoto, E.Chr. Mulholland, . . . Nauta, C.B. Nederburgh, E. Nittel, . . . Oosterman, G.J. Oudemans, . . . ten Raa, . . . Rost van Tonningen, J.K.E. Schütt, . . . Schultz van Haegen, P.J. Veth, W.P.D. de Wolff van Westerrode, papers concerning G.J. Holsman and A. Pompe.

i. *78 C 83*, besides East Indies papers before 1796:

1. secret notes 1803 by W.J. Cranssen concerning Ternate;
2. summary of the codes of law angger pradoto, angger sepuluh, angger ageng, 1818;
3. study on the trade on the West and North Coast of Sumatra, 1825–1826;
4. report 1830 about the organization of the detective force;
5. report by inspector [J.B.] de Linge.

j. *78 F 25*, letters 1900–1902 from and about prisoners of war in Ceylon of the Boer War in South Africa by J.G. Bendien, H. Betist, J.A. Bouwer Izoon, H.J.C. Funke, W.C.J. van Hoogstraten, G. van Houtum, J.G. Houwink, J.A. van Kretschmar, W.J. Louw, S.M. Muller, P. Postma, J.D. Rosto, P.H. Roux, P.C. Swemer, J.S. de Villiers, B.A. Wennink, C.T.M. Wilcocks to M. Berdenis van Berlekom, Middelburgsch Dames-Comité (Middelburg Ladies' Committee), S. Suringar-de Vlugt.

k. *79 D 1,* album with signatures after the earthquake of 1923 presented by "l'Association de la jeunesse Japonaise constituée pour recueillir les signatures des habitants de Tokyo en remerciements de la sympathie qu'on a manifestée les peuples étrangers au Japon", text in French and Japanese.

l. *120 B 12,* letters from i.a. G.A.G.Ph. van der Capellen and D. van Hogendorp 1810–1823.

m. *120 C 3,* letters from i.a. E. Douwes Dekker 1885–1886.

n. *129 B 15,* "Beschouwingen over den Atjeh-oorlog 1873–1892; met als aanhangsel het zoogenaamd verraad van Toekoe Djohan Pahalawan, alias Toekoe Oemar", by L.W.A. Kessler s.a.

o. *132 G 29,* "De majoor Van Coehoorn van Houwerda; herinneringen uit Oost-Indien uit het dagboek van een Belgisch officier", 1824–1826.

p. *133 B 12,* "Reise durch Sumatra", by A.M. Müller, *c.* 1840.

q. *133 M 90,* collection of letters from L., G., R. and F. Neumann from the Netherlands, Batavia and Buitenzorg to their father F.J.E. Neumann, government official in Timor and Amboina, and to M. Neumann 1867–1898, with a letter from Raden Tumenggung Prawoto at Kudus to H. Kleyn van de Poll 1858.

r. *134 B 10,* correspondence with C.Th. van Deventer 1903–1904. For the names of the correspondents mentioned in the inventory see the index.

## Nederlands Letterkundig Museum en Documentatiecentrum (Dutch Literary Museum and Centre of Documentation), 1954

11 Juffrouw Idastraat, 2513 BE The Hague.
Admission free. Lending to university libraries.
Photocopying equipment available.
Research finished in 1979.

**Introduction** [596]

The Letterkundig Museum collects papers about and by literary men and women who published after 1750 in the Dutch language. The Documentatie (Documentation) department collects articles about authors from newspapers and weekly and monthly periodicals. The Iconografie (Iconographic) department collects mainly portraits of literary men and women. The Documenten (Documents) department collects and is in charge of archival documents, mainly letters and manuscripts. Also exhibitions are organized and periodical publications brought out.

For literature about the Museum see *Nederlands Letterkundig Museum en Documentatiecentrum.* S.l. and s.a.

**Documents Department** [597]

The department is in charge of files and collections originating from *c.* 5000 authors. For the history of Asia may be relevant the papers originating from:

H.P. van den Aardweg, G.S. Adama van Scheltema, K.J.L. Alberdingk Thijm, A. Alberts, L. Aletrino, J.L. Bakker, F. [E.A.] Batten, A.H. de Balbian Verster, J.H. Been, A.P.A.A. Besnard, J.J.Th. Boon (ps. = Tjalie Robinson and Vincent Mahieu), H.J.A. Borel, M. ter Braak, W.A. Braasem, J. ten Brink, J.W. Broedelet, G. Brom, P. Brooshooft, J. Brouwer, C.L. [van Bruggen-] de Haan, C. Busken Huet, B. Canter, A. Cohen, R.A. Cornets de Groot, L. Couperus, R. Cramer, P.A. Daum (ps. = Maurits), H.A.M.E. Dermoût [-Ingerman], Ch.M. van Deventer, L. van Doorn, E. [Douwes] Dekker (ps. = Multatuli), E.F.E. Douwes Dekker, N.[J.G. Engelkamp-] Caesar, I. Esser, E. Fabius, Jan Fabricius, Johan Fabricius, M.[D.] Ferguson [-Wigerink], J.J. Fock, H.J. Friedericy (ps. = Merlijn), R.J.W. van Genderen Stort, G.L. Gonggrijp (ps. = Opheffer), J. Greshoff, W.J. de Gruyter, R. van Gulik, H.Z.A. Gutteling, A. ter Haghe, N.[J.] Haisma, J. Hanlo, W.F. Hermans, A. den Hertog, A.Th.A. Heuting, J.M. Hondius, A.G. van der Hoogte, Th.E. Hoven, Khouw Bian Tie, J. [Kloos-] Reyneke van Stuwe, W.G. Klooster, W.S.B. Klooster (ps. = Willem Brandt), M.Chr. [Kooy-] van Zeggelen, H.R. Kousbroek, W.G. Kierdorff (ps. = Pim Hofdorp), J.C.F. Last, H.S. [van Lelyveld-] Haasse, W. Mallinckrodt, J. van Maurik, L.A.J. de Neve (ps. = Louise BB), R. Nieuwenhuys (ps. = E. Breton de Nijs), C. van Nievelt, K. Norel, Notosuroto, E.F.F.J. [Overduyn-] Heyligers, G.H. Pannekoek, Ch.E. du Perron, Z. [Postema-] Zikken, A.P. Prins, H.M. van Randwijk, G.J. Resink, J.P.R. Reyneke van Stuwe, A. [van Rhijn-] Naeff, R. Roegholt, J. Romein, A. Romein-Verschoor, M.P.C. Scheltema, A. van Schendel, C.L. Schepp (ps. = Jan Prins), A.L. Schneiders, J.B. Schuil, J.G. Schuur, J.J. Slauerhoff, N.M.C. Sloot (ps. = Melati van Java), W.H.C. Smit, E.G.F. Smit Kleine (ps. = Piet Vlugtig), A.H. van der Stok, M.H. [Székely-] Lulofs, Rabindranath Tagore, A.E.B.J. Tergast, S. Vestdijk, G. Valette, D. Verspoor, J.W. Vervoort, B. Veth, A. Viruly, H. de Vreede, L. Vroman, J.H. van der Wal, J.H.F. van de Wall, W. Walraven, J.B. Wemmerslager van Sparwoude, J.W.H. van Wermeskerken, A. W[erumeus] Buning, L.C. Westenenk, H.P.L. Wiessing, E. [de Willigen-] Vuyk, A.A.H. de Wit, C.W. Wormser (ps. = Boekan Saja), Chr. Wijnveldt, W. Zaal, J.C. Zimmerman.

Entry: Alphabetical catalogue to name of author.

Permission is necessary for consultation of the letters found in the collections.

## Orde van Vrijmetselaren (Fraternity of Freemasons), 1756

22 Fluwelen Burgwal, 2511 CJ The Hague.
Admission by appointment.
Research finished in 1978.

**Library** [598]

Unknown number of books and periodicals.

For literature concerning the Fraternity see P.J. van Loo. *De geschiedenis van de Orde van Vrijmetselaren onder het Grootoosten der Nederlanden.* 's-Gravenhage, 1967.

**Publications** [599]

The oldest publication is
*Nederlandsche Maçonieke almanak.* 1751–1841.

At present are being published
*Algemeen Maçoniek tijdschrift;* vol. 1– . 1937– .

*Toth;* tijdschrift voor Vrijmetselaren; vol.1– . 1949– .

In Indonesia was published
*Indisch Maçoniek tijdschrift.* 1895–1956.

**Archives** [600]

The first lodge in the Netherlands was established in The Hague in 1734; in 1756 a ten lodges united into what is now called the Orde van Vrijmetselaren onder het Grootoosten der Nederlanden; from here lodges were established in the Antilles, Surinam, South Africa and Brazil and also in Indonesia, India, Malaya, China and Ceylon. All the lodges are mentioned in *Lijst van loges, welke onder buitenlandse grootmachten op Nederlands gebied werken of gewerkt hebben;* [door G.F.E.W. Lowensteijn. 's-Gravenhage], 1961. See for the names of the Asian lodges the index.

For the history of Asia are relevant

a. The series of the Grand Lodge itself, such as resolutions from 1756, with annexes from 1800 and alphabetic indexes, correspondence from 1756 (with an alphabetic index to lodge up to 1840), reports of the central board from 1797, accountancy.

b. Files of various lodges, clubs, unions and associations of lodges, including Arbeid Adelt, Constante et Fidèle, Deli, Fidèle Sincérité, Mata Hari, Palembang, De Ster in het Oosten, La Vertueuse, De Vriendschap, Excelsior, Hollandia, Bataviasche Leesbibliotheek (Batavian Library), Mataram.

c. Papers originating from persons, including L.A.J.W. Sloet van de Beele, C.G. Bijleveld, A.S. Carpentier Alting, M.W. Willemstijn, W. Cool, A.W.J.H. Hoitink, W. Keuchenius (i.a. concerning J. Penning Nieuwland), J. Kouwenaar, W.D.J. van Meeteren Brouwer (i.a. about M.T.H. Perelaer), . . . van Tongeren, Th.G.G. Valette, D.J. Veegens, J.J. Voûte.

d. Lists of members.

e. Lodges archives, transferred from Indonesia (*c.* 40 running metres), viz. La Vertueuse, La Fidèle Sincérité, De Vriendschap, La Constante et Fidèle, De

Witte Roos, Provincialaat Grootloge, Princes Frederik der Nederlanden, Humanitas, Deli, Sint Jan, Malang, Mataram, Vooruitgang, De Goede Hoop, Palembang, Arbeid Adelt, De Ster in het Oosten, papers of J. Schille about the history of Freemasonry in Java.

Entry: An inventory is forthcoming in addition to the entries drawn up by the administration.

Size: *C.* 250 running metres.

Permission is necessary for consultation of the papers.

**Miscellaneous papers** [601]

The catalogue number is mentioned in brackets behind the descriptions.

a. St. Jan Lodge at Bandung. 1926 (123 E 2)

b. Farewell speech of J. Eilbracht in the lodge La Fidèle Sincérité in Batavia. 1783. (123 E 59)

c. Correspondence of J.J. Voûte and W.P. Barnaart about the effected union of the lodges La Vertueuse and La Fidèle Sincérité into De Ster in het Oosten. 1838. (122 E 48)

d. "Ritualen en bouwstukken" originating from J. Penning Nieuwland about De Ster in het Oosten. 1844–1865. (123 E 13, 17, 125, 46–47)

e. W.L. Ritter. Speech about the "vernietiging en onsterfelijkheid" (destruction and immortality) given in the lodge De Ster in het Oosten. 1847. (122 E 139)

f. Letter and bill about the amendment of the rules and regulations of the lodge De Ster in het Oosten. 1905. (122 E 92)

g. J.H. Carpentier Alting. "De doodstraf" (capital punishment), speech given in the lodge De Ster in het Oosten. 1918. (123 E 15)

h. Papers about the jubilee and leaving of A. de Geus and J.H. Delgorge. 1928. (122 C 8)

i. Papers about the history of the lodge Deli. 1941–1947. (123 E 75)

j. Speeches given in the lodge De Vereeniging in Colombo. 1795, 1800. (123 C 59, 60)

k. Regulations for the lodge De Opregtheid at Point de Galle. *C.* 1805. (123 E 29)

l. Diploma of N. Engelhard. 1798. (123 E 1)

m. Speech by H. Nienhuis at the inauguration of H.M. de Kock. 1836. (125 C 80)

n. Papers about the jubilee of Th.G.G. Valette. 1929. (123 A 2)

o. Correspondence between D.J. Veegens and G. Kloss. 1847–1848. (193 B 2)

p. Rules of the lodge La Fidèle Sincérité. S.a. (123 C 64)

q. Rules of the lodge at Kuta Radja. 1884. (122 E 94)

r. Rules of the lodge Deli. S.a. (122 E 93)

s. Rules of the lodge Veritas. 1882. (122 E 14)

t. "Bijwet" (supplementary rules) for the lodge De Vriendschap. 1824. (124 B 113)

**Pictures, photographs** [602]

The Fraternity possesses a collection of pictures and photographs of persons and buildings in Europe, Australia, South Africa, Indonesia, China and Surinam, together with masonic attributes.

Entry: Provisional inventory.

Size: Some hundreds of items.

## Instituut Kern (Kern Institute), 1925

27 Rembrandtstraat, 2711 VV Leiden.
Admission free.
Photocopying equipment available.
Research finished in 1979.

**Library** [603]

40,000 books; 400 periodicals.

Directed to the cultures and languages of India, Pakistan, Bangladesh, Sri Lanka, Nepal, Bhutan, Sikkhim, Tibet, ancient Iran and Afghanistan and to the art history and archaeology of Burma, Cambodia, Thailand, Vietnam, Laos, Malaysia and Indonesia.

**Publications** [604]

*Annual bibliography of Indian archaeology* (ABIA). Leiden, 1926– .

### Archival material [605–608]

**Archives** [605]

For this guide are counted as archives: the old archives (up to 1958) of the actual Institute and deposited archives of the Commissie van Bijstand (Advisory Committee).

a. *Instituut Kern (Kern Institute), 1925–*
The Instituut was established in 1925 and named after the Sanskrit scholar H. Kern. The object was to promote study of Indian (India and Indonesia) archaeology, for which i.a. the aforesaid ABIA was published, a library was built up and several other collections were started to support research.

The archives consist of annual reports, minutes, lists of members, papers about the personnel, correspondence, several files. For the names and subjects mentioned in the shelflist see the index.

Entry: Shelflist.

Size: *C.* 1 running metre (45 items).

Permission is necessary for consultation of the papers.

b. *Commissie van Bijstand in Nederland voor het Oudheidkundig Onderzoek op Java en Madoera (Advisory Committee in the Netherlands for Archaeological Research in Java and Madura), 1901–1922*
The Commissie van Bijstand was established in 1901 after in Batavia the Commissie in Nederlandsch-Indië voor Oudheidkundig Onderzoek op Java en Madoera (Committee in the Dutch East Indies for Archaeological Research in Java and Madura) had been established; chairman in the Indies was J.L.A. Brandes, members in the Netherlands were H. Kern, W. Pleyte, V. de Stuers, J.W. IJzerman and G.P. Rouffaer; the tasks included care of publications (with assistance of the KITLV) and providing advice and information for the minister of Colonial Affairs; as a result of the Oudheidkundige Dienst in Nederlandsch-Indië (Archaeological Service in the Dutch East Indies) becoming an independent service, the Commissie van Bijstand was abolished in 1922.

The small amount of archives consists of minutes with annexes and reports about Tjandi Tumpang, Lawadjati, Tjandi Panataran, Singasari, Tjandi Kidul, Malang, Kutabedah, Bataviaasch Genootschap van Kunsten en Wetenschappen (Batavian Society of Arts and Sciences), Tjandi Bima, Djokjakarta, Tjandi Sewu, Tjandi Mendut, Tjandi Kalibening and Semarang.

Entry: Inventory.

Size: *C.* 0.20 running metre (19 inventory items)

See also *Rapporten van de Commissie in Nederlandsch-Indië voor Oudheidkundig Onderzoek op Java en Madoera.* 1901– .

*Oudheidkundig verslag van de Commissie in Nederlandsch-Indië voor Oudheidkundig Onderzoek op Java en Madoera.* 1912– .

**J.L.A. Brandes, January 13th 1857 Rotterdam – June 26th 1906 Batavia** [606]

Study of theology at Amsterdam, East Indies languages at Leyden; 1884 graduated, departs for the Dutch East Indies as language official; 1901–1905 chairman of the Commissie in Nederlandsch-Indië voor Oudheidkundig Onderzoek op Java en Madoera (Committee in the Dutch East Indies for Archaeological Research in Java and Madura).

The collection consists of copybooks of letters despatched, official speeches, university lectures' notes of J. Pijnappel, notes, correspondence, necrologies.

Entry: Shelflist.

Size: *C.* 0.10 running metre (6 items).

**J.Ph. Vogel, January 9th 1871 The Hague – c. 1958 Oegstgeest** [607]

Study of Dutch literature at Amsterdam; 1897 thesis *Het leemen wagentje;* 1898–1900 lecturer at Amsterdam; 1901–1914 superintendent Archaeological Survey of India at Lahore; 1914 professor of Sanskrit and East Indies archaeology at Leyden; member of the committee of the Instituut Kern.

The collection consists of university lectures' notes, notes and articles, scientific correspondence about archaeology and linguistic subjects. For the names and subjects mentioned in the shelflist see the index.

Entry: Shelflist.

Size: *C.* 2.75 running metres (96 items).

**Other archival documents** [608]

The library contains

a. P.A. Tiele. Documenten voor de geschiedenis der Nederlanders in het Oosten (Documents for the history of the Dutch in the East). *C.* 1890. Draft publication. (I kd 9)

b. Letters in Arabic characters i.a. from the sultan of Ternate. (I n 82)

### Oriental manuscripts [609]

The collection consists mainly of Sanskrit texts, and besides i.a. Javanese papers and Tibetan blockprintings.

### Photographs [610]

The collection consists of *c.* 26,000 photographs and 6000 lantern views, mainly concerning Hindu-Buddhist archaeology.

## Koninklijk Instituut voor Taal-, Land- en Volkenkunde (Royal Institute of Linguistics and Anthropology), KITLV 1851

10 Stationsplein, 2312 AK Leyden.
Admission for members and by introduction. Lending to members (except of archival material, maps, drawings etc., photographs).
Photocopying equipment available.
Research finished in 1979.

# Introductory data [611–614]

## Library [611]

100,000 books; 1200 periodicals, of which 450 currently received.

Directed to South East Asia (in especial Indonesia), the Caribbean (in especial Surinam and the Antilles), Oceania.

The library of the former Indisch Genootschap (Indies Society) has been incorporated in the library of the KITLV.

For literature concerning the KITLV see i.a.

*Gedenkschrift;* uitgegeven ter gelegenheid van het 75-jarig bestaan op 4 Juni 1926 [van het] Koninklijk Instituut voor de Taal-, Land- en Volkenkunde van Nederlandsch-Indië [te] 's-Gravenhage. 's-Gravenhage, 1926.

NB. See also the manuscript H 380 (papers concerning the establishment of the Institute, originating from J.C. Baud) and the manuscripts H 781 and H 782 (papers concerning the 100th anniversary of the Institute), deposited at the archives of the Institute.

## Publications [612]

*Catalogus der koloniale bibliotheek van het Koninklijk Instituut voor de Taal-, Land- en Volkenkunde van Nederlandsch-Indië en het Indisch Genootschap.* 's-Gravenhage, 1908. 1st, 2nd supplement. 1915, 1927. 3rd supplement published under the title: "Catalogus der koloniale bibliotheek van het Koninklijk Instituut voor de Taal-, Land- en Volkenkunde van Nederlandsch-Indië". 1937. From the 4th supplement published under the title: "Catalogus van de bibliotheek van het Koninklijk Instituut voor Taal-, Land- en Volkenkunde". 1966– .

*Bijdragen tot de taal-, land- en volkenkunde van Nederlandsch-Indië.* 's-Gravenhage, 1853– . From 1949 published under the title: "Bijdragen tot de taal-, land- en volkenkunde".

*Adatrechtbundels;* bezorgd door de Commissie voor het Adatrecht. 's-Gravenhage, 1910–1955. 45 vol.

*Verhandelingen van het KITLV;* vol. 1– . 's-Gravenhage, 1938– .

*Bibliographical series KITLV;* no. 1– . 's-Gravenhage, 1955– .

*Translation series KITLV;* vol. 1– . 's-Gravenhage, 1956– .

*Bibliotheca Indonesica;* no. 1– . 's-Gravenhage, 1968– .

*Excerpta Indonesica.* Leiden, 1970– . No. 1– .

Works apart from the aforesaid series, such as grammars, dictionaries, catalogues of manuscripts, maps and drawings (for the catalogues see no. 614, 661, 662, 663).

**Documentation** [613]

Documentatie Geschiedenis Indonesië (Documentation History of Indonesia) up to *c.* 1950 (since 1964), including keeping and documentation of archives, documentation from printed material, keeping and documentation of visual material, keeping of maps and charts.

Documentatie Modern Indonesië (Documentation Modern Indonesia) from *c.* 1945 (since 1968), documentation from printed material, keeping collection of micro index cards and micro films.

**Introduction to the archival material** [614]

The archival material of the Institute falls into two parts:

The archives of the actual Institute, formed in the course of and as a result of its activities (see no. 615).

The collection of manuscripts. This collection falls into two parts:

a. Western manuscripts (manuscripts in western languages). The larger portion of this collection concerns Asia, of which about three quarters pertain to the period 1796–1949. The western manuscripts consist of collections of which the components were often described under various manuscript numbers.

   Collections larger than 0.10 running metre are described below as an integral part with reference to their manuscript numbers (see no. 616–659). Moreover the collection of western manuscripts contains a large number of miscellaneous papers (see no. 660).

   Entry: H.J. de Graaf. *Catalogus van de handschriften in westerse talen [H 1–824], toebehorende aan het KITLV.* 's-Gravenhage, 1963. Supplements were published under the title: "Handschriften in westerse talen". In *BKI;* vol. 123(1967), p. 405–408: acquisitions 1963–1967, H 825–911; vol. 125(1969), p. 300–304: acquisitions 1967–1968, H 912–980; vol. 127(1971), p. 319–320: acquisitions 1968–1970, H 981–1005; vol. 129(1973), p. 409–410: acquisitions 1970–1972, H 1006–1029; vol. 131(1975), p. 404–406: acquisitions 1973–1974, H 1030–1046; vol. 133(1977), p. 497–498: acquisitions 1975–1976, H 1047–1063; vol. 135(1979): acquisitions 1977–1978, H 1048–1081.

   Size: *C.* 65 running metres (1086 numbers).

b. Oriental manuscripts (manuscripts in non western languages), see no. 661.

## Private organizations [615–622]

**Koninklijk Instituut voor Taal-, Land- en Volkenkunde (Royal Institute of Linguistics and Anthropology), KITLV 1851–** [615]

The Institute was established in 1851 by J.C. Baud, G.A. Simons and T. Roorda as a scientific centre with a library for the benefit of teachers and students of indology of the Delftse Akademie (Delft Academy), with the object of promoting the study of linguistics, geography, ethnology and history of the Dutch overseas territories.

The archives are the residue of the activities of the Institute. They consist of rules and regulations, minutes of the general and board meetings, correspondence, papers membership records, accountancy, library, editing, special departments and sundries.

Entry: H.H. Scholtens. *Inventaris van het archief van het Koninklijk Instituut voor Taal-, Land- en Volkenkunde.* Leiden, 1979.

Size: *C.* 6.50 running metres (497 inventory numbers).

**Indisch Genootschap (Indies Society), 1854 – c. 1940** [616]

The Genootschap was established in 1854 with the object of studying socio-economic and political colonial issues of current interest and of building up a library about the Dutch East and West Indies and the Coast of Guinea; from 1878 co-operation with the KITLV became closer and closer; the Genootschap ended its activities in 1940.

The archives consist of papers up to 1928, viz. rules and regulations, minutes, annual reports, correspondence, papers membership records, accountancy, library and sundries.

Entry: H.H. Scholtens. *Inventaris van het archief van het Indisch Genootschap, 1854–1928.* Leiden, 1979.

Size: *C.* 1.75 running metres (217 inventory items).

See also *Handelingen en geschriften van het Indisch Genootschap.* 's-Gravenhage, 1854–1940.

NB. The archives were originally deposited at the archives of the KITLV, later kept separately under no. H 1075.

**Oost-Indische Kiesvereeniging/Vereeniging "Moederland en Koloniën" (East Indies Electoral Association/Association "Home Country and the Colonies"), 1898–1921** [617]

The Kiesvereeniging was established in 1898 with the object of stimulating the co-operation between those interested in colonial affairs and of promoting their interests; abolished in 1900; 1901 establishment "Moederland en Koloniën" as a continuation of the Kiesvereeniging, with the object of encouraging people to take part in political life; 1921 amalgamation with the Indisch Genootschap.

The archives consist of rules and regulations, minutes, correspondence, papers accountancy, Comité tot Bevordering van Gezelschapsreizen naar Indië (Committee for the Promotion of Conducted Tours to the East Indies).

Entry: H.H. Scholtens. *Inventaris van het archief van de Oost-Indische Kiesvereeniging/Vereeniging "Moederland en Koloniën", 1898–1921.* Leiden, 1979.

Size: *C.* 0.15 running metre (38 inventory items).

See also *Orgaan van de Vereeniging "Moederland en Koloniën"*. 's-Gravenhage, 1901–1921.

NB. The archives were originally deposited at the archives of the Indisch Genootschap, later kept separately under no. H 1076.

**Adatrechtstichting (Customary Law Foundation), 1909–1970** [618]

In 1909 the Commissie voor het Adatrecht (Committee for Customary Law) was established, within the KITLV, chairman was C. Snouck Hurgronje, secretary C. van Vollenhoven; 1917 establishment Adatrechtstichting (Customary Law Foundation); the object was to promote research into uncodified law of the native population of the Dutch East Indies; between 1955 and 1970 there were hardly any activities; 1979 activities resumed by the Nederlands Onderzoekscentrum voor het Recht in Zuid-Oost Azië en in het Caraïbisch Gebied (Dutch Centre of Research into Law in South East Asia and in the Caribbean) in Leyden.

The archives consist of minutes, correspondence and accountancy 1909–1970, documentation about persons, Java, Madura, Sumatra, Riouw, Borneo, Celebes, Moluccas, New Guinea, Lesser Sunda Islands, territories outside Indonesia. For the names and subjects mentioned in the survey see the index.

Entry: F.G.P. Jaquet. *Overzicht van het archief van de Adatrechtstichting*. Leiden, 1973.

Size: *C.* 4.30 running metres (H 1051, 120 items).

See also *Adatrechtbundels;* bezorgd door de Commissie voor het Adatrecht; vol. I–XLV. 's-Gravenhage, 1910–1955.

**(Koninklijke) Vereeniging "Oost en West" ((Royal) Association "East and West"), 1899–1971** [619]

"Oost en West" was established in 1899 at the initiative of N. van Zuylen-Tromp with the object of promoting the interests of those staying in the East and West Indies and of those returned from there; to this end a periodical was published and there were also activities in the field of housing, spread of popular knowledge, education, industry and charity; local branches carried out the activities under the control of a national secretariat; 1972 abolition of the secretariat, the branches continued in being.

The very incomplete archives consist of the papers of the secretariat. Preserved were the minutes of the members' meetings and annual reports 1899–1900, 1940–1971, minutes of the central board 1899–1912, 1941, 1942, 1945–1971, minutes executive board 1933–1938, 1945, 1950–1971, correspondence 1963–1971, papers membership records, accountancy, sundries fragmentary from 1950.

Entry: F.G.P. Jaquet. *Inventaris van het Secretariaat van de (Koninklijke) Vereeniging "Oost en West", 1899–1971*. Leiden, 1979.

Size: *C.* 0.50 running metre (H 1077, 153 inventory items).

See also *Koloniaal weekblad.* 's-Gravenhage, 1900–1934. From 1934 continued under the title: "Oost en West", discontinued in 1972.

**Federatie van Vereenigingen van NI Bergcultuurondernemingen (Federation of Associations of Dutch East Indies Upland Estate Companies), 1935–1968** [620]

The archives consist of memorandums chronologically arranged 1947–1966, together with reports originating from W.J. de Jonge about Moral Rearmament, tin fusion, growth of population, post office, pawn shop service, "volkscredietwezen" (people's credit system), traffic, costs of living, crisis measures *c.* 1940.

Size: *C.* 1 running metre (H 1031, 89 bound volumes).

NB. For the history and organization of the Vereeniging see no. 93–99.

**Ronde Tafelconferentie (Round Table Conference), RTC 1948–1949** [621]

The archives consist of minutes with annexes of the RTC, Bijeenkomst voor Federaal Overleg (Assembly for Federal Consultations) BFO 1948–1949, "Haags Overleg" (Consultation in The Hague) 1948, Federale Conferentie Bandung (Federal Conference Bandung) 1948, the papers are originating from the secretary of the RTC H.G. Schulte Nordholt.

Size: *C.* 3 running metres (H 1074).

**Indische Kulturele Kring (East Indies Cultural Circle), 1959** [622]

The Kulturele Kring was established with the object of promoting interest and understanding for cultural expressions, particularly cultural expressions originated in tropical territories. In 1964 the Kunstkring (Arts Circle) offered a prize for writing down East Indies recollections.

The collection contains the majority of the entries for the competition. They are originating from: Th. Bennink, C.M. Bouwens-Hoek, A.Th. Brandenburg-de Rooy, A.E.J. Breuning-Lambert, E.J. Bünger-Stadelman (for her novel-manuscript see H 861), M. Elesser (ps. of M. Sitsen-Russer), A. van den Ende, B. van den Ende-van Huffel, R. de Geus, A.T. Haarman-Strijland, F.D. Hansen-Raae, P. Leembruggen, H.M. Marsman-van Deventer, V.N. Moens, H.A. Nuse, B. Schmidt, M.J.E. van der Sloot-Schreiber, C.A. van Sluis, A. Timmermans, L.Th. Vossenaar-Thal Larsen, E.M. Weijgers, A.W. van der Woude-Leenart, C.M. Wijmans-Feltzer, V. van Zadelhoff-van Heemskerck.

Size: *C.* 0.40 running metre (H 837–860).

## Families and private persons [623–659]

**W.J. Beck, August 4th 1892 The Hague – ?** [623]

1914–1916 (aspirant) comptroller Tapanuli (1914 Padang Sidimpuan, 1915 Toba, 1916 Sibolga); 1916–1920 comptroller Riouw (1916–1918 Karimun, 1918–1920 Tandjung Pinang); 1920–1923 Menado (Tondano); 1926–1929 secretary of the district administration East Coast of Sumatra (Medan); 1929–1930 assistant resident Deli (Medan); 1930–1932 official Kantoor voor Bestuurshervorming (Office for Administrative Reform) Batavia; 1934–1935 assistant resident Palembang (Lahat); 1935–1938 member of the Commissie van Bijstand en Advies inzake het Boschwezen (Advisory Committee for the Care and Management of Forests); 1938–1942 resident with the provincial administration of the government of Sumatra.

The collection includes decrees of appointment, conduct sheets etc. 1911–1946; papers concerning the administration at Tapanuli, Riouw, Menado 1914–1926;colonization and the situation of the labourers in the East Coast of Sumatra and Atjeh 1925–1930; ceremonial for visits of high foreign civil and military persons 1929–1939; Commissie Bestuurshervorming (Committee for Administrative Reform) 1930–1932; self-rule and decentralization, general, Sumatra and Celebes (Gowa) 1931–1939; selection of East Indies civil servants 1936–1937; minutes and other papers of the Commissie Boschwezen (Committee for the Management of Forests) 1935–1938; report on the Japanese period in the East Coast of Sumatra 1945; memoirs as a government official.

Size: *C.* 0.45 running metre (H 900).

**P. Brooshooft, October 18th 1845 Giessendam – August 14th 1921 The Hague** [624]

1874–1876 editor of several newspapers; 1877–1880 editor in chief of the *Semarangsche Courant;* 1880–1883 editor of the *Sourabaja Courant;* 1885–1887 correspondent in the Netherlands of the *Sourabaja Courant;* 1887–1895, 1898–1904 editor in chief of *De Locomotief;* 1895–1898 correspondent and editor of several Dutch newspapers; 1904–1917 correspondent in the Netherlands of *De Locomotief.*

The collection mainly consists of articles, novels, plays, poems (both in print and in draft) written by Brooshooft and correspondence connected with his activities as a journalist. For the names and subjects mentioned in the inventory see the index.

Entry: *Inventaris van de collectie Brooshooft;* [door F.G.P. Jaquet]. 's-Gravenhage, 1966.

Size: *C.* 0.50 running metre (H 896, 87 inventory items).

**H.T. Damsté, June 28th 1874 Den Helder – January 6th 1955 Leyden** [625]

Government official 1894–1923; 1894–1896 (aspirant) comptroller at the Departement van Binnenlands Bestuur (Department of the Indies Civil Service) Batavia; 1896–1898 West Coast of Sumatra (Sidjundjung); 1898–1901 Padang (Talu); 1901–1903 Atjeh (Sigli, Lamnjong, Meulaboh); 1903–1905 South Celebes; 1906–1908 Palembang (till 1908 Musi Ilir, 1908 city of Palembang, also president of the municipal council); 1908–1913 Atjeh (1908 Kota Radja, till 1912 Idi, till 1913 Sinabang); 1913 Bali and Lombok; 1913–1914 assistant resident Celebes (Makasar); 1919–1923 resident Bali and Lombok; after 1923 in the Netherlands i.a. secretary of the Instituut Kern (Kern Institute) and the Maatschappij der Nederlandsche Letterkunde (Society of Dutch Literature), treasurer of "Oost en

West" (East and West) Leyden branch, vice president KITLV, member of the board of the Commissie voor het Adatrecht (Committee for Customary Law). (*BKI;* vol. 111 (1955), p. 113).

The collection consists mainly of papers about Atjeh, and also papers about Palembang, Makasar, Padangse Bovenlanden, Indonesian question, decentralization, land policy, population, photographs, drawings. For the names and subjects mentioned in the shelflist see the index.

Entry: Shelflist.

Size: *C.* 1 running metre (H 1084, 60 numbers).

**Douwes Dekker** [626]

This collection consists of the papers originating from A.H.E. Douwes Dekker and his son G.M.G. Douwes Dekker.

*A.H.E. Douwes Dekker, September 9th 1850 Amsterdam – March 24th 1924 Batavia*
1866–1873 clerk in the service of the government at Surabaja, after that in private service; 1873–1878 at Surabaja; 1878–1880 at Pasuruan; 1880–1891 at Batavia; from 1891 till he retired agent of the Nederlandsch-Indische Escompto Mij. (Dutch East Indies Discount Company) first at Surabaja, next at Batavia; from 1898 member of the Freemasons' Lodge at Batavia, later resigned; when retired stockbroker at Batavia.

*G.M.G. Douwes Dekker, October 20th 1883 Meester Cornelis – March 9th 1959 Schoorl*
1899 volunteer in the Boer War South Africa; then successively journalist, commercial employee, teacher in a European primary school, referendary –, head accountancy department of the Departement Gouvernements-Bedrijven (Department of Government Businesses), head financial department of the Staats-Spoorwegen in Nederlandsch-Indië (State Railways in the Dutch East Indies) till he retired in 1938. After the second World War back to the Netherlands.

The collection consists of family papers and "Multatuliana". Among the family papers were included: genealogical registers concerning the families of Douwes Dekker, Vogel and De Wit; diary 1859–1892, and cashbooks 1869–1873, 1877–1878, 1885–1888 drawn up by A.H.E. Douwes Dekker; financial papers and notes about the country houses "Fancy" and "Waspada", originating from G.M.G. Douwes Dekker; documentation concerning the Douwes Dekker family. Among the "Multatuliana" were included: articles and notes by A.H.E. and G.M.G. Douwes Dekker; articles by H.M. van Andel, M. Anderson, M. ter Braak, P.M.L. de Bruyn Prince, H.C. Douwes Dekker, J.F. Otten, J. Pée, E. du Perron, A.W. Stellwagen, J. van Vollenhoven and others; letters and photographs of E. Douwes Dekker (Multatuli); correspondence about Multatuli of J. van den Berg, M. Douwes Dekker-Hamminck Schepel, . . . Herklots, H.J. Lummel, H.C. Muller, J. Pée, E. du Perron and others.

Size: *C.* 1.50 running metres (H 923).

**E. Gobée, December 3rd 1881 Den Helder – December 7th 1954 Voorschoten** [627]

1903–1906 as naval officer active in the East Indies archipelago in connection with hydrographic observations; 1906–1908 in the Netherlands study of indology; 1908–1910 aspirant comptroller Central Celebes (Poso); 1910–1914 comptroller Batak area (Barus); 1914–1915 Atjeh (Tapa Tuan); in the Netherlands study of Arabic; 1917–1921 consul at Djeddah; 1922 in the East Indies interim adviser Inlandsche Zaken (Native Affairs); 1923–1926 assistant resident Central Celebes (Poso); 1926–1937 adviser Inlandsche Zaken; 1949–1950 journey through Indonesia for the purpose of reorganization of education there; in the Netherlands i.a. charged with the publication of the *Ambtelijke adviezen* by C. Snouck Hurgronje. (*Jaarboek van de Maatschappij der Nederlandsche Letterkunde;* 1953/55, p. 59/*RAvNI*).

The collection consists mainly of papers drawn up and received when he was (interim) adviser Inlandsche Zaken, together with papers of before 1923 and after 1937. For the names and subjects mentioned in the shelflist see the index.

Entry: Shelflist.

Size: *C.* 0.50 running metre (H 1085, 69 items).

**J. Hageman JCz, September 30th 1817 The Hague – November 30th 1871 Pasuruan** [628]

*C.* 1835 as soldier to the Dutch East Indies; 1836–1844 clerical officer (1836–1837 Batavia, 1837–1838 Besuki, 1838–1844 Surabaja); 1842, 1845 journeys to Europe and America; 1846–1860 several secretaryships, including *c.* 1852–1860 clerk to the landraad (district council) Surabaja; 1860–1871 retired in the East Indies; published many scientific papers particularly about Java. (Supplements from the *Nederlandsche Spectator.* 1872, p. 42).

Apart from preliminary papers from *c.* 1650 and papers concerning America the collection mainly consists of papers after 1816, viz. reports, statistics, articles, correspondence etc. concerning administration, British interregnum, population, Chinese, Javanese (outside Java), trade, products, woods, finance, army, navy, slavery, piracy, sciences, history, chronology, fables, legends and Hindu-antiquities in the following districts and places: Balambangan, Bali, Bankalan, Batu-Litjin, Bawean, Borneo (East and West), Grisee, Java (mainly East), Kangean, Kutei, Lamongan, Landak, Laut-Pulau, Lumadjang, Madjalengka, Madura, Mampawa, Modjokerto, Modjopait, Pamakasan, Pontianak, Sedaju, Sukoredjo, Sumenep, Sunda area, Surabaja and the Principalities, written, except by himself, also by: Endrokoesoemo, Sjarief Abdula Habashi, A. von Humboldt, KITLV, G.A. Lange, H. Mac Gillavry, P. Melvill van Carnbee, J.A.C. Oudemans, . . . Versteeg, J. van Vollenhoven and H. von Dewall.

Size: *C.* 0.80 running metre (numbered H 14–21, H 24–28, H 32–44, H 83, H 84, H 99, H 102, H 108, H 118, H 137, H 214, H 255–258, H 343, H 459–461).

See also the following publications written by Hageman

*Handleiding tot de kennis der geschiedenis, aardrijkskunde, fabelleer en tijdrekenkunde van Java.* Batavia, 1852. 2 vol.

Geschiedenis van het Bataafsch en Hollandsch gouvernement op Java, 1802–1810. In *TBG*, vol. 4(1855), p. 333–382.

Geschiedenis van het Hollandsch gouvernement op Java. In *TBG*, vol. 5(1856), p. 164–284.

Bijdrage tot de geschiedenis van Borneo. In *TBG*, vol. 6(1857), p. 220–246.

De Engelschen op Java. In *TBG*, vol. 6(1857), p. 290–457.

Bijdragen tot de kennis van de residentie Madoera. In *TNI*, vol. 20(1858), I, p. 323–352, II, p. 1–25.

Bijdragen tot de kennis van de residentie Soerabaja. In *TNI*, vol. 20(1858), II, p. 85–104, vol. 21(1859), I, p. 17–34, 105–126, II, p. 129–164.

Geschiedenis der Soenda-landen. In *TBG*, vol. 16(1867), p. 193–251, vol. 17(1868), p. 178–257, vol. 19(1870), p. 201–269.

**G.A.J. Hazeu, August 22nd 1876 Amsterdam – December 1st 1929 Wassenaar** [629]

Study of Sanskrit and Indonesian languages; 1898–1904 teacher Javanese language at the Gymnasium Koning Willem III (Grammar School Koning Willem III) at Batavia; 1904–1906 official East Indies languages, attached to the adviser Inlandsche en Arabische Zaken (Native and Arab Affairs) C. Snouck Hurgronje; 1906 succeeds Snouck Hurgronje, till 1914 adviser Inlandsche Zaken, from 1911 also director Onderwijs en Eerediensten (Education and Religious Affairs); 1914–1916 in the Netherlands charged with the reorganization of East Indies education; 1916–1919 government commissioner for Inlandsche en Arabische Zaken; 1920–1928 professor of Javanese language at Leyden; worked on the Gajo-Dutch dictionary.

The collection consists almost completely of papers drawn up and received when he was adviser Inlandsche Zaken 1904–1919, together with papers about education and a draft of the Gajo dictionary. For the names and subjects mentioned in the shelflist see the index.

Entry: Shelflist.

Size: *C.* 2 running metres (H 1083, 81 items).

**O.L. Helfrich, October 6th 1860 Serang – April 29th 1958 Voorburg** [630]

1881–1882 agricultural school Buitenzorg; 1882–1883 aspirant comptroller Besuki; 1883–1892 (aspirant) comptroller Benkulen (1883 Kroe, 1885 Benkulen, 1888 Mana); 1892–1893 comptroller Atjeh (Sabang, Kota Radja); 1893 West Coast of Sumatra; 1893–1897 Tapanuli (Mandailing, Batang Natal); 1898–1901 Palembang; 1901–1909 (assistant) resident Djambi; 1909 till he retired in 1912 resident Benkulen; in the Netherlands 1916–1919 archivist Zuid-Sumatra-Instituut (South

Sumatra Institute); 1919–1921 governor Curaçao; member of the Commissie voor het Adatrecht (Committee for Customary Law) (from 1943 honorary chairman); member of the board and honorary member (1950) of the KITLV. (Supplements from *BKI;* vol. 114(1958), p. 249).

The collection consists of records of service; reports of retiring officials, maps, articles etc. concerning Benkulen and Djambi; letters 1881–1959 from J. Brandes, H. Kern, C. Snouck Hurgronje, H.N. van der Tuuk and others; papers about the Besemahers with correspondence 1939–1948; papers 1918 by L.J.M. Swaab about Native education in South Sumatra; papers 1807–1861 concerning the legal system in Benkulen.

Size: *C.* 0.80 running metre (H 665, H 799, H 800, H 813).

**J.D. van Herwerden, January 18th 1806 – September 9th 1879** [631]

In 1825 arrived in the East Indies as a sailor; then official Algemene Secretarie (General Secretariat) at the bureau Maleische Taal (Malay Language); comptroller Landelijke Inkomsten en Cultures (Public Revenue and Cultures) Kediri (Ngrowo); 1834–1840 assistant resident Kediri (Ngrowo); 1841–1844 Probolinggo; 1844–1848 resident Madiun; in 1848 to the Netherlands where he set up the *Haagsch Dagblad;* known as an antagonist of W.R. van Hoëvell. (*Indische Gids*. 1879, II, p. 928/*RAvNI*).

In addition to a number of preliminary papers from *c.* 1650, the collection consists of reports, statistics, accounts, articles, letters, notes, etc., after 1816 concerning the Principalities 1812–1859 (letting of land, the Getas lands, Tembelongan, Seloegoer, Gagattan, Ploembon, Mlamblong and Ampel, Javanese jurisdiction Djojobojo (by J.W. Winter), genealogies, titles, payoongs); [J.] Olivier [Schilperoort] at Palembang 1822; Riouw 1825–1827 by Mohamad Zien Seid (about Riouw, Lingga and Billiton), Ch. van Angelbeek and C.L. van Ranzow (about Riouw and Djohor); piracy 1830 by Said Hasan Habasi; Dajaks in West Borneo, Malays, Arabs 1830; Ngrowo 1833, 1835; cinnamon culture 1836; Madiun 1836 by L. Launy; Raad van Indië (East Indies Council) 1837 by J.P. Cornets de Groot van Krayenburg; pilgrims Mecca 1838 by R.K.F. Krämer; Madura 1839 by D.F.W. Pietermaat, with complaints of the sultan of Sumenep about C.G. Goldman; Bagelen 1840; Native population and languages 1840; landed property, woods, fish ponds, nipah in Java 1847 by D. Lange; memorandum East Indies possessions 1864.

Size: *C.* 0.15 running metre (H 86, H 261, H 369, H 477–499).

**J. van Hinte, July 13th 1889 Muiden – March 1st 1948 Amsterdam** [632]

1912–1915 teacher of geography and history Leeuwarden; 1915–1917 Den Helder; 1917–1919 Amsterdam; from 1919 Openbare Handelsschool (Public Commercial School) Amsterdam; 1926 thesis *Nederlanders in Amerika;* studied colonization problem and for that purpose made tours through the United States (1921), Australia (1933 in the course of which he also visited the Dutch East Indies and

South Africa) and a second tour through the Dutch East Indies (1938); he was i.a. member of the board of the KNAG (from 1923) and the Nederlandsche Zuid-Afrika Vereeniging (Dutch South Africa Association). (*TKNAG;* 2nd run, vol. 55(1948), p. 289).

The collection consists of the manuscript *Nederlanders in Amerika,* 1928; and articles, texts of lectures, letters and notes 1920–1948 concerning the United States, Surinam, South Africa, New Guinea; concerning New Guinea i.a. papers 1930–1938 by and/or about the Nederlandsche Emigratie Vereeniging Nieuw-Guinea (Dutch Emigration Association New Guinea), Nieuw-Guinea Comité (New Guinea Committee), Stichting Immigratie en Kolonisatie "Nieuw-Guinea" (Foundation Immigration and Colonization "New Guinea"), De Vaderlandsche Club (The Fatherland Club), Vereeniging "Kolonisatie Nieuw-Guinea" (Association "Colonization New Guinea"), W.C. Klein, J. van der Weide, J. Zwierzycki, photographs.

Size: *C.* 0.35 running metre (H 784, H 786).

**F.W. Junghuhn, October 26th 1809 Mansfeld – April 24th 1864 Lembang** [633]

1830 qualified as a physician at Berlin; 1831 sentenced to imprisonment of 10 years for a duel; 1833 escaped and as a soldier in the foreign legion to Africa till 1834; 1835–1845 army surgeon at Batavia; 1838 drafted to, and 1845–1848 member of the Natuurkundige Commissie in Nederlandsch-Indië (Physical Committee in the Dutch East Indies); 1848–1855 scientific activities in the Netherlands; 1855–1857 inspector Natuurkundige Onderzoekingen in Nederlandsch-Indië (Physical Research in the Dutch East Indies); 1857–1864 charged with experimentations in cultivation of cinchona in Java.

In the collection records of service 1849–1862; diplomas 1837–1858; catalogues of his library and instruments; letters 1846–1864 from and/or to . . . Boursse Wils, A. de Codrika, Prince Frederik, A. Henley, J.D. van Herwerden (about the Semeru), W. Narsan Lees, Ch.F. Pahud, F. Toetterlen, J.E. de Vrij (about i.a. cinchona cultivation); papers 1918 concerning the publication by L. van Vuuren of Junghuhn's *Die Battaländer auf Sumatra;* papers 1865–1967 about Junghuhn by Délilah, K. Fens, I. Groneman, W.H. Hoogland, H. Rochussen, C.Ph.C.E. Steinmetz, W.H. de Vrieze, A. Wichmann; photographs, pictures concerning the Junghuhn and De Vrij families, . . . Gerstäcker, Buitenzorg, Gunung Lamongan, Lembang, Gunung Malabar, Gunung Salak, Sindanglaja, Tangkuban Prahu, cinchona experimental gardens; papers concerning his father W.F. Junghuhn and his wife J.L.F. Junghuhn-Koch.

Entry: *Inventaris van de collectie Junghuhn;* [door F.G.P. Jaquet]. 's-Gravenhage, 1965. With supplement inventory.

Size: *C.* 0.15 running metre (H 886, 96 inventory items).

See also C.Ph.C.E. Steinmetz. Culturele gegevens uit familiepapieren; Dr. Franz Wilhelm Junghuhn (1809–1864). In *Cultureel Indië;* 4th vol. 1942, p. 60–72, 173–182.

**Raden Adjeng Kartini, April 21st 1879 Majong – September 17th 1904 Rembang** [634]

Plays a part in the field of the Javanese women's emancipation; from 1903 till her death she managed a small school, at first in Japara, later in Rembang; she became known owing to the publication of her letters *Door duisternis tot licht*, edited by J.H. Abendanon in 1911.

The collection consists of papers originating from H.Th. de Booy, including photocopies of letters 1900–1965 from Kartini and her sisters Kardinah and Roekmini (for the letters from C.J.G. de Booy see H 954); papers 1913–1965 collected by R. Nieuwenhuys; papers 1902–1964 originating from C. Vreede-de Stuers; photographs 1893–1903. For the names and subjects mentioned in the inventory see the index.

Entry: Inventory.

Size: *C.* 0.10 running metre (H 897, 31 inventory items).

**R.A. Kern, September 26th 1875 Leyden – March 23rd 1958 Leyden** [635]

1879–1906 (aspirant) comptroller Preanger Regencies (1901 Trogong (Garut), 1902 Bodjong Lopang); 1906–1908 Madiun (Patjitan); 1908–1914 Preanger (Bandung, also municipal councillor (1909) and member of the branch bank Volkscredietwezen (People's Credit System) 1909–1911); 1914–1917 assistant resident Pekalongan (Brebes, also president branch bank Volkscredietwezen); 1917–1920 Surabaja (Modjokerto, also president branch bank Volkscredietwezen); 1920–1926 adviser Inlandsche Zaken (Native Affairs) at Batavia, also member of the board of the Bataviaasch Genootschap voor Kunsten en Wetenschappen (Batavian Society of Arts and Sciences) 1920–1926, member central audit committee Volkscredietwezen 1921, member Commissie van Toezicht van de Rechtshogeschool (Supervisory Committee of the University of Law) 1921–1926, member of the board PKB 1921–1926; 1926–1942 lecturer of Sundanese State University Leyden, taught also Buginese and Makasar languages and for a short time Javanese and Islam; from 1926 secretary, subsequently treasurer Instituut Kern; 1927–1937 secretary Oosters Instituut (Oriental Institute); 1927–1951 member of the board KITLV (1955 honorary member); 1936–1955 president Islam-Stichting (Islam Foundation). (*BKI;* vol. 114 (1958), p. 345/*RAvNI*).

The collection consists of papers drawn up and received in his functions from 1897, with as the centre of gravity his function of adviser Inlandsche Zaken. For the names and subjects mentioned in the inventory see the index.

Entry: Y.S. Hidayat. *Inventaris van de papieren van R.A. Kern*. Leiden, 1977.

Size: *C.* 1.25 running metres (H 797, 498 inventory items).

**A.H.W. de Kock, March 24th 1808 Surabaja – June 17th 1891 The Hague** [636]

Son of H.M. de Kock; 1824 second lieutenant, 1824–1828 lieutenant adjutant to the lieutenant governor general, 1825–1830 Java war; 1830 adjutant to the governor general; 1832–1836 captain adjutant, 1833 expedition on the West Coast of Sumatra; 1838 major general staff West Coast of Sumatra, 1841 expedition there; 1841 lieutenant colonel, interim resident Palembang; 1846 on leave; 1848 government commander Bali, second commanding officer Bali war, interim resident Besuki, discharge from military service; 1848 resident Djokjakarta; 1851 discharge; 1852 interim resident Madiun; 1854 resident Batavia; 1855 member of the Raad van Indië (Indies Council), 1855–1856 president PKB; 1857 on leave; 1859 till he retired in 1860 vice president of the Raad van Indië. (*RAvNI*/Stamboek Oost-Indische ambtenaren).

The collection contains concerning the Dutch East Indies "memorie van overgave" (report retiring official) 1819 of H.M. de Kock, governor of the Moluccas; reports, accounts etc. 1821–1848 about the residency Palembang by J.W. Broers, H.T.O. Buschkens, A.H.W. de Kock, H.M. de Kock, C.F.E. Praetorius and J.C. Reynst; report 1830 by H.M. de Kock about Diponegoro; notes by A.H.W. de Kock about the history of Djambi and Palembang; concerning Siam "De Siamsche histories" by Abdullah bin Mohamad el Mazrie 1823; concerning Japan papers 1830 by G.F. Meylan about Nagasaki.

Size: *C.* 0.20 running metre (H 71, H 270–274, H 287–297, H 299, H 304, H 307, H 309, H 340).

See also Legenden van Djambi. In *TNI;* vol. 8(1846), part 4, p. 33–56.

See also no. 219.

**V.E. Korn, July 1st 1892 The Hague – June 25th 1969 The Hague** [637]

1910 study of indology at Leyden; 1913 BB official at Parepare; 1914 aspirant comptroller Celebes (Parepare, Makasar, Sopeng); 1914 comptroller Ternate; 1915 Batjan; 1917 Bali and Lombok (Tabanan, Denpasar); 1921 on leave in the Netherlands, graduates in 1924 on the thesis *Het adatrecht van Bali;* 1925 comptroller Denpasar; 1926 Singaradja; 1927 secretary of the residency of Bali and Lombok; 1927 assistant resident Bali and Lombok; 1928 Southern and Eastern District of Borneo (Samarinda); 1929 Bali and Lombok; 1930 on leave in the Netherlands, revised edition of his thesis; 1933 East Coast of Sumatra (Asahan); 1934 Solok (Sawahlunto); 1936 resident Djambi; 1936 Tapanuli; 1939 retired; 1939 extraordinary professor of adatrecht (customary law) of the Dutch East Indies at Leyden; 1945 ordinary professor; 1958 retired.

The collection consists of official papers and papers which Korn collected as expert in customary law, i.a. about Celebes, Bali, Borneo, East and West Coast of Sumatra, Djambi and Tapanuli, and also a number of private papers. For the names and subjects mentioned in the list see the index.

Entry: F.G.P. Jaquet. *Lijst van stukken deel uitmakend van de collectie Korn.* Leiden, 1973.

Size: *C.* 7.50 running metres (Or 435, 515 items).

**B.J. Lambers, July 7th 1901 Lunteren – January 19th 1966 Heemstede** [638]

1926–1927 official Raad van Justitie (Council of Justice) at Batavia; 1927–1928 clerk Council of Justice Padang; 1928–1929 landrechter (police magistrate) East Coast of Sumatra (Kisaran); 1929–1932 substituut officier (deputy prosecutor) Council of Justice Batavia; 1934–1935 deputy prosecutor Council of Justice Makasar; 1936–1939 solicitor general Court of Justice Paramaribo; 1939–1941 deputy prosecutor Council of Justice Batavia; 1941–1942 public prosecutor Council of Justice Surabaja; after the war till 1948 chief officer at the disposal of the attorney general; resigns and becomes head of the department of housing of Sociale Zaken (Social Welfare) at Batavia till 1949; in the Netherlands from 1950 solicitor's office in Haarlem. *(RAvNI).*

The collection includes articles about Java by I. Groneman in the *Nieuws van den Dag* 1888–1890; documents of the case concerning the Lettinga and Boom affairs 1931; newspaper articles 1931–1932 about various legal cases; reports when he was chief officer at the disposal of the attorney general concerning Japanese war crimes, military operations in South Celebes, the Ratulangi c.s. affair 1945–1947; papers concerning the rehabilitation of Indies NSB followers 1940–1954.

Size: *C.* 0.20 running metre (H 973).

**M.J.Th. Lievegoed-van Buren Schele** [639]

Wife of A.J. Lievegoed (1878–1946), journalist, editor of the *Sumatra Post* (1901–1914) in Deli, editor in chief of *De Locomotief* (1917–1925) at Semarang in the Netherlands i.a. on the staff of the *NRC.* Mrs. Lievegoed took an active part in all kinds of women's movements, she was engaged in the organization of the exhibition "De Vrouw 1813–1913" (Women 1813–1913); in the Netherlands she was from 1926 member of the board of the "Vereeniging tot Oprichting van Jonge-Vrouwenscholen in Nederlandsch-Indië" (Association for the Establishment of Schools for Young Women in the Dutch East Indies). (*Jaarboek van de Maatschappij der Nederlandsche Letterkunde;* 1946/47, p. 87).

The collection includes papers concerning welfare work in the East Indies 1895–1913, women's vote 1911–1932, the exhibition "De Vrouw 1813–1913"; letters 1908–1913 from A.H. Jacobs, F.I. Kasteleyn-Beyerinck, M.C. Kooy-van Zeggelen, C.E. Rahdervoet, J. van der Schuyt and others; minutes of the Vereeniging tot Oprichting van Jonge-Vrouwenscholen in Nederlandsch-Indië with annual reports, rules etc. of the branches of Semarang, Malang and Ambarawa 1911–1938; articles etc. on the death of C. van Vollenhoven (1933) collected by A.J. Lievegoed.

Size: *C.* 0.25 running metre (H 974).

**A. Ligtvoet, ? – 1879 Balang Nipa** [640]

1865–1866 official at the disposal of the government of Celebes (Makasar); 1866–1868 clerk Inlandsch Bureau (Native Bureau) Celebes (Makasar); 1868–1878 secretary Inlandsche Zaken (Native Affairs) Celebes (Makasar), from 1869 also member of the school committee there; 1878 till his death in 1879 assistant resident Eastern Districts of Celebes (Balang Nipa). *(RAvNI).*

The papers in this collection are almost all pertaining to Celebes and dependencies, viz. Adjataparang, Barru, Boni, Buton, Gowa, Kajeli, Laikang, Laiwu, Luwu, Mandar, Masanrebulu, Sopeng, Sulu Islands, Sumbawa, Tanette, Tontoli, Wadjo, the administration (i.a. during the time when J.A. Bakkers was governor), the princes there; moreover historical notes concerning Makasar and Mandar and historical dictionary South West Celebes, with Luwu, Toradja and Saleijer.

Size: *C.* 0.45 running metre (H 226, H 606–609).

See also A. Ligtvoet. Transcriptie van het dagboek der vorsten van Gowa en Tello; met vertaling en aanteekeningen. In *BKI;* 4th run, vol. 4(1880), p. 1–259.

**J.M. van Lijf, ? – c. 1957** [641]

1935–1942 government official in Celebes, viz. 1935–1936 aspirant comptroller Sungguminasa (Djeneponto), 1936–1937 Luwu (Makale), 1937–1939 comptroller Bonthain (Saleijer, 1939 Sindjai), 1939–1942 Luwu (Masamba). *(RAvNI).*

The collection consists of a memorandum of H.T. Lanting about Makale Rantepao 1936 and material collected by Van Lijf for a work about Luwu 1930–1950.

Size: *C.* 0.45 running metre (H 785, H 789).

**H.L.Ch. te Mechelen, October 5th 1841 Rembang – May 30th 1917 Bandung** [642]

1864–1872 (aspirant) comptroller for the Landelijke Inkomsten en Cultures (Public Revenue and Cultures) Rembang; 1872–1875 Preanger; 1875–1879 teacher of Javanese and Malay at the gymnasium Koning Willem III (grammar school) at Batavia; 1882–1889 assistant resident specially charged with the suppressing of opium smuggling (1882–1885 Japara, 1885–1886 Semarang, 1886–1889 Rembang); 1889–1893 chief inspector opium affairs; 1893–1906 in the Netherlands, resigns in 1898, returns to Java in 1906 where he dies in 1917; he was a well-known hunter and an authority on wajang. (*Nederlandsch Indië Oud en Nieuw;* 2nd vol., 1917/18, p. 305).

The collection consists of Javanese and Malay texts and papers concerning opium smuggling by himself, W.P. Groeneveldt, L.W.C. Keuchenius (objections against the acquisition of steam ships) and the residents of Rembang [A.C. Uljee] and Japara [W.C.J. Castens] 1875–1894.

Size: *C.* 0.25 running metre (H 422, H 423).

**P. Melvill van Carnbee, May 20th 1816 The Hague – October 24th 1856 Batavia** [643]

1831 naval cadet; 1835 first voyage to the East Indies; 1837 second voyage to the East Indies; 1838 naval lieutenant; 1839 for the third time to the East Indies, official with hydrographic service of the Naval Department at Batavia for the purpose of the composition of sea charts; 1845 to the Netherlands; 1850 to the East Indies as adjutant to vice admiral E.B. van den Bosch; 1853 discharge from naval service, employed as cartographer with the Indisch sedentaire zeewezen

(East Indies sedentary maritime matters); 1856 naval captain lieutenant, director of the Marine-etablissement (naval yard) in Onrust.

The collection consists mainly of papers concerning the Dutch East Indies, viz. memorandums, statistic material, notes, letters, drawings etc. 1810–1854 about hydrography, cartography, geology, volcanology (i.a. West Coast Sumatra, Gunung Talang), nautical knowledge, ethnography and economy of himself, L.W. Beijerinck and L. von Buch; gold and coal exploitation in Celebes (Gorontalo) 1828–1844; essays by F.A. Schuring about the Batu Islands; reports about Sumatra 1843–1849; travel journals 1850 about i.a. piracy in the Riouw archipelago.

Size: *C.* 0.25 running metre (H 143, H 145, H 150, H 161, H 162, H 190–196, H 222–225, H 508, H 582).

**J.F.W. van Nes, born in Utrecht** [644]

Up to 1819 law study at Utrecht; after functions with the judiciary at Zutphen and Arnhem in 1823 to the East Indies, where he was employed till 1826 at the Hooggerechtshof (High Court of Justice) at Batavia; 1826 assistant resident Surabaja; 1827 resident Djokjakarta; 1831 Pasuruan; 1839 to the Netherlands; 1842 resident Semarang; 1843 member of the Raad van Indië (East Indies Council); 1847 to the Netherlands; 1851 vice president of the Raad van Indië; 1853 to the Netherlands. (*RAvNI/Moniteur des Indes Orientales et Occidentales;* II (1847/48), p. 93).

Besides preliminary papers from 1705 the collection consists of contracts 1749–1830 with ministers of the court of Djokjakarta; papers 1773–1803 concerning the court of Djokjakarta and Hamengkubuwono II by i.a. W.H. van IJsseldijk and I.G. van den Berg; political papers 1788–1846 concerning Surakarta; proclamations 1799 of the Hooge Regering (Supreme Government); papers of W.H. van IJsseldijk concerning the journey of G.W. van Imhoff 1799; papers 1818–1830 concerning the Principalities, i.a. concerning landed property, Tan Tjoe Hong, a translation by C.F. Winter of the Angger Ageng, a report by P.F.H. Chevalier; papers 1825 concerning trade, economy and politics in Java, i.a. by G.A.G.Ph. van der Capellen and H.G. Nahuys van Burgst pertaining to Diponegoro; memorandum L.P.J. du Bus de Gisignies for C.Th. Elout about colonization 1827; papers 1828 concerning sale of land in Cheribon with papers by J.B. van Linge and J.I. van Sevenhoven; statistical, topographical and economical notes 1829 by R.M. Schabbing about Banda and the South Eastern and South Western Islands; memorandum 1830 by Van Nes about Islamic customs concerning marriage and division of property; memorandum and annual report 1832–1836 by Van Nes i.a. about the sugar cultures at Pasuruan with a letter from Kertoadinegoro.

Size: *C.* 0.25 running metre (H 29, H 90, H 94–97, H 104, H 107, H 112, H 230, H 391, H 395, H 397).

See also D.C. Steyn Parvé. *Het koloniale monopoliestelsel getoetst aan geschiedenis en staathuishoudkunde;* nader toegelicht. Zalt-Bommel, 1851, p. 1–154 "memorie

Du Bus". J.I. van Sevenhoven. "Project-instructie voor de Commissie, belast met het verkoopen aan de Javanen op aanvrage van de erven of pakarangans, geleegen in eene dessa". In *BKI;* new series, vol. 4(1861), p. 317–341.

**A. Neytzell de Wilde, August 8th 1877 Utrecht – 1939** [645]

1903–1905 deputy registrar High Court of Justice and High Court Martial; 1905 vice president landraad (district council) Kediri and Ngandjuk; 1906–1908 president landraad Bangil; 1908–1910 teacher of law at the Opleidingsschool voor Inlandsche Ambtenaren (Training College for Native Officials) at Probolinggo; 1910–1916 teacher, teacher-director and adjutant director of the Opleidingsschool voor Inlandsche Rechtskundigen (Training College for Native Jurists) at Batavia; 1916–1920 at the disposal of the attorney general; 1920–1923 with the High Court of Justice and the High Court Martial; 1923–1924 on leave; 1924–1925 raadsheer (counsel) with the High Court of Justice; 1925–1929 president of the Volksraad (People's Council); 1929–1935 head department of legal and League of Nations affairs of the Ministry of Colonial Affairs; 1935–? professor of colonial agronomy at Wageningen.

The collection consists mainly of papers of the Volksraad 1925–1928, concerning politics, finance, oil development, hereditary lease, ground tax, death penalty in the East Indies, Chinese.

Size: *C.* 0.15 running metre (H 768).

**R. Nieuwenhuys, June 30th 1908 Semarang –** [646]

1935–1940 teacher of the Dutch linguistics and literature Surabaja, Djokjakarta, Semarang, Batavia; 1940–1942 teacher Literaire Faculteit (Faculty of Literature) of the university at Batavia; 1940–1941 editor of *De Fakkel;* 1947–1952 head Culturele Zaken, Departement van Onderwijs (Cultural Affairs, Department of Education), Kementerian PPK; 1948–1952 editor secretary *Oriëntatie;* 1953 teacher Amsterdam; 1964–1973 also on the scientific staff of the KITLV.

The collection consists of documentation connected with Nieuwenhuys' researches relating to the East Indies belles-lettres and the European community in the Dutch East Indies, viz. notes concerning A.M. Courier dit Dubekart, G. Lebret, T. Ottolander, S.E.W. Roorda van Eijsinga and G.J.P. de la Valette; photocopies of letters 1868–1871 from J. Heemskerk Azn. to J.J. Hasselman concerning C. Busken Huet; "Aspecten van de koloniale problematiek in de Zuid-Nederlandse roman" by J. Geraerts 1966; photocopy of the diary by lieutenant Knoerle about Diponegoro 1830; letters and photographs 1883–1908 of the families of Wieseman and Broese van Groenau at Djokjakarta; "De Indo en Tong Tong; een onderzoek naar de Indo-Europeaan vroeger en nu" by R.J. Kuiper and H.G. Surie 1967; correspondence 1967–1968 with A.D. van Mook-Maureau; letters, notes, articles 1865–1965 concerning I. Groneman for the purpose of the handbook *Oost-Indische spiegel* by Nieuwenhuys; letters, notes, articles 1960–1965 concerning H.N. van der Tuuk for the purpose of *Tussen twee vaderlanden* and *De pen in gal gedoopt* by

Nieuwenhuys, with letters from Van der Tuuk to i.a. R. d'Ablaing van Giessenburg, J.L.A. Brandes, NBG, P.J. Veth *c.* 1880; correspondence 1956–1962 with J.W. Meyer Ranneft about the Binnenlands Bestuur (East Indies Civil Service) and other colonial subjects; photocopies of papers 1856–1857 concerning W.J. Leyds; essay concerning Hotel De Boer at Medan by H.L. Burghardt-de Boer; documentation about J. Olivier and Tjalie Robinson.

Size: *C.* 0.50 running metre (H 669, H 766, H 772, H 914, H 928, H 930–932, H 940–942, H 960–962, H 984).

**A. Palmer van den Broek** [647]

The collection consists of papers collected by A. Palmer van den Broek and originating from and/or written by W. Palmer van den Broek, principal Inlandsche Kweekschool (Native Teachers Training College) at Surakarta, "Beschrijving van Nederlandsch Oost-Indië ten dienste van den Inlander" 1880, translations of the Babad Dipanegara, Wiwaha Djarwa and the Brata Soenoe 1832–1870; Ch.L.J. Palmer van den Broek, government official (papers 1891–1893 concerning Nias, papers 1894–1897 concerning North Achinese chiefs (including genealogical trees), memorandums, reports, letters etc. 1908–1910 concerning Simelungun and the Karo area, papers 1890 concerning himself with outline maps of New Guinea); A. Meursinge, teacher Koninklijke Academie (Royal Academy) of Delft, lecture notes about geography and ethnology of the Dutch East Indies 1849.

Size: *C.* 0.15 running metre (H 555, H 568, H 587, H 589, H 592, H 593, H 595, H 728).

See also W. Palmer van den Broek. *Nederlandsch Oost-Indië;* of, Handleiding tot de kennis van Nederlandsch-Indië ten dienste van de Javanen opgesteld en in het Javaansch vertaald door Mantrie Goeroe Soerono te Soerakarta. S.l. and s.a.

**Van Rees** [648]

This collection consists of the papers concerning and/or originating from O. van Rees and his son D.F.W. van Rees.

a. *O. van Rees, January 4th 1823 Culemborg – March 10th 1892 Arnhem*
1837–1851 official Algemene Secretarie (General Secretariat) (1837 clerk, 1850 referendary, 1851 assistant secretary); 1851–1853 inspector of Finance; after his leave in 1856 resident Kedu; 1857–1860 Bagelen; 1860–1864 Surabaja; 1864–1868 member of the Raad van Indië (East Indies Council); 1868 discharge and to the Netherlands; 1869–1872 second time member of the Raad van Indië; 1872–1873 member of the Second Chamber; 1873–1878 vice president of the Raad van Indië; 1879 minister of Colonial Affairs; 1879–1881 member of the Second Chamber; 1881–1884 president of the Second Chamber; 1884–1888 governor general of the Dutch East Indies; from 1888 retired in the Netherlands.

Among the papers of O. van Rees were included commentary 1864 by Van Rees on the regulations for undertakings of agriculture and industry, correspondence mainly with his daughter S.A. Mijer-van Rees and A.J.M. van Genderen Stort-van Rees, A. Mijer, D.F.W. van Rees and O. van Rees jr; papers concerning O. van Rees (22 inventory items).

b. *D.F.W. van Rees, March 11th 1863 Surabaja – October 31st 1934 Montreux*
1882 aspirant comptroller Departement van Binnenlands Bestuur (Department of East Indies Civil Service); 1883–1888 idem residency of Batavia; 1888–1896 official Algemene Secretarie (General Secretariat), 1888 comptroller, 1889 principal clerk, 1891 referendary; 1896 secretary of the Government and on leave to the Netherlands; back in the East Indies from 1898–1900, together with H.J.W. Lawick van Pabst, member of the Commissie Naleving van Voorschriften betreffende de Verhuring van Gronden door Inlanders aan Niet-Inlanders (Committee Observance of Regulations concerning the Lease of Land by Natives to Non-Natives) in Java and Madura; 1900–1903 secretary of the Government; 1903–1905 director Binnenlands Bestuur; 1905–1906 general secretary of the Government; 1906–1908 member of the Raad van Indië (East Indies Council); 1908–1914 vice president of the Raad van Indië, 1914–1921 in the Netherlands member and secretary of the Koloniaal Instituut (Colonial Institute); 1921–1934 member and vice president of the Permanente Mandaten-Commissie van de Volkenbond (Permanent Mandates Commission of the League of Nations) in Switzerland.

Originating from D.F.W. van Rees are service records; papers 1899–1914 as East Indies official; papers 1914–1921 as member of the board of the Koloniaal Instituut; papers 1922–1934 as member of the Permanente Mandaten-Commissie (68 inventory items). For the names and subjects mentioned in the inventory see the index.

Entry: *Inventaris van de collectie Van Rees;* [door F.G.P. Jaquet]. 's-Gravenhage, 1965.

Size: *C.* 0.25 running metre (H 895, 91 inventory items).

**G.P. Rouffaer, July 7th 1860 Kampen – January 8th 1928 The Hague** [649]

1877 study Polytechnic University of Delft, 1879 broken off, travels to Italy, Germany, France, England and Spain; 1885–1890 first journey through the East Indies; 1898–1908 assistant secretary KITLV, also secretary of the Commissie van Bijstand ten bate van het Oudheidkundig Onderzoek op Java en Madoera (Advisory Committee for the Purpose of Archaeological Research in Java and Madura), editor *TKNAG,* founder of the Linschoten Vereeniging (Linschoten Association) and the NV Boeatan (Boeatan Ltd); 1909–1911 second journey through the East Indies; 1911 honorary member KITLV; 1913–*c.* 1925 member of the board and president Linschoten Vereeniging; 1921 honorary degree of doctor of literature and philosophy State University of Leyden. (*BKI;* vol. 84 (1928), p. 163).

The collection includes notes, diaries, reports etc. about his two journeys through Asia, viz. concerning Further India, British India, Bali, Flores, Java, Malaya, Philippines, Sumba, Sumbawa, Timor (i.a. memorandum L. van Vuuren); reports, essays, notes etc. about Java and the Principalities concerning archaeology, history, art, batik, kreeses, Javanese law, chronology, genealogies of Javanese princes, Kalang Kiwa-tengen, opium; essays, notes, letters and other papers concerning the Principalities, i.a. copies from the archives of Djokjakarta and Surakarta 1646–1889, papers 1816–1889 concerning i.a. princes, kratons (Padjang and Kartasura) with drawings, Mangkunegara, agriculture, opium, by J. Crawford, W.F. Engelbert van Bevervoorde, R.G. Kilian, H.G. Nahuys van Burgst, Rouffaer, J.W. Winter and others; essay about vessels in the East Indies, 1904–1907 with a letter from N. Adriani; essay by J. Habbema about sirih 1900; notes about taxation in Atjeh; the Islam in North Sumatra; petroleum in Perlak; letters from Portuguese Timor 1735–1785 with letters 1914 from T.L. van Veeren and C. Lulofs and a memorandum 1911; correspondence with E.C. Ecker, F.A. Liefrinck, W.B. Tholen and others 1879, 1899; notes and incomplete biography of Multatuli.

Size: *C.* 0.80 running metre (H 396, H 473, H 673–675, H 677–679, H 682–700, H 712, H 716, H 721–723, H 754, H 756, H 757, H 767, H 798, H 875).

See also the articles written by Rouffaer in the *Encyclopaedie van Nederlandsch-Indië* about "Tijdrekenkunde" and "Vaartuigen".

Encyclopaedie-artikelen van Dr. G.P. Rouffaer. In *BKI;* vol. 86(1930), p. 191–215.

R. Nieuwenhuys. Rouffaer en Multatuli. In *BKI;* vol. 116(1960), p. 408–423.

G.P. Rouffaer. Tentoonstelling van Oost- en West-Indisch bewerkt metaal. In *TNI;* 1900, p. 543–546.

G.P. Rouffaer. Waar kwamen de raadselachtige Moetisalah's (aggrikralen) in de Timor-groep oorspronkelijk vandaan? In *BKI;* vol. 50(1899), p. 409–675.

J.W. Winter. Beknopte beschrijving van het hof Soerakarta in 1824; met voorwoord en eenige noten van G.P. Rouffaer. In *BKI;* vol. 54(1902), p. 15–172.

**L.J. Schippers, August 8th 1870 – 1951 The Hague** [650]

1894–1897 (aspirant) comptroller Bantam (Rangkasbitung); 1897–1908 Semarang (1897 Ambarawa, 1905 secretary of the resident Semarang); 1908–1910 in the Netherlands for a training at the Bestuursacademie (Academy for Administration) at The Hague; 1910–1912 comptroller Surabaja; 1912–1919 assistant resident for the police of Surabaja; 1919–1920 resident Surabaja; 1920–1922 director of the Binnenlands Bestuur (East Indies Civil Service); 1922–1923 member of the board of the Protestantsche Kerken (Protestant Churches) in the Dutch East Indies.

The collection includes service records, conduct sheets etc.; diplomas, contracts of marriage etc.; "Volksraadstukken" (papers of the People's Council) 1920–1921 concerning the Binnenlands Bestuur; reception books; family papers (diary of his parents 1870, letters when he was a child 1883, diary F. Schippers about the voyage from the East Indies to the Netherlands 1916, verse album, photographs).

Size: *C.* 0.15 running metre (H 953).

**C.Ph.C.E. Steinmetz, May 23rd 1884 – June 1953** [651]

1910–1914 official Algemene Secretarie (General Secretariat), also from 1913 member of the board of the Buitenzorgsche Kunstkring (Buitenzorg Art Society); 1914–1916 official at the disposal of the Department of Justice; 1916–1920 deputy prosecutor Raad van Justitie (Council of Justice) Batavia; 1922–1930 official at the Department of Finance (1926–1930 official at the generale thesaurie (general treasury)); 1930–1935 official with the service Opium- en Zoutregie (Monopoly of Opium and Salt) at Batavia (1930 inspector, 1931 head); in the Netherlands i.a. official of the archives of the KIT 1941–1949 and a contributor to the periodical *Cultureel Indië;* vol. 3(1941)–8(1946). *(RAvNI).*

The collection consists of the papers, photographs and drawings which Steinmetz collected when he was official of the archives of the KIT and which have been partly embodied in his articles in *Cultureel Indië*. For the names and subjects mentioned in the inventory see the index.

Entry: Inventory.

Size: *C.* 0.15 running metre (H 898, 53 inventory items).

See also the articles under the heading "Culturele gegevens uit familiepapieren", which were mainly written by Steinmetz.

Nederland en Japan; eenige gelegenheidsgedichten van P. Knuttel. In *Cultureel Indië;* vol. 3(1941), p. 169–190.

An article about Junghuhn, see no. 633.

Van trouwen en sterven onder de compagnie. In *Cultureel Indië;* vol. 4(1942), p. 208–222; vol. 5(1943), p. 31–46.

Zilveren gedenkborden. In *Cultureel Indië;* vol. 5(1943), p. 128–137.

Twee ongepubliceerde portretten van Mr. H.W. Daendels, maarschalk van Holland, gouverneur-generaal van Nederlandsch-Indië. In *Cultureel Indië;* vol. 5(1943), p. 185–190.

De schipbreuk der familie [F.V.H.A.] de Stuers op het Lucipararif, 1837. In *Cultureel Indië;* vol. 5(1943), p. 271–285. (The collection contains no papers on this).

Gouverneur-generaal D.J. de Eerens en zijn adjudant C.Ph.C. Steinmetz. In *Cultureel Indië;* vol. 6(1944), p. 34–47.

Het journaal van George Webb (1787–1837); of, De kracht der middelmaat. In *Cultureel Indië;* vol. 6(1944), p. 207–222.

A.M. Lubberhuizen-van Gelder. Hendrik Veeckens, een ambtenaar van den ouden stempel. In *Cultureel Indië;* vol. 7(1945), p. 89–106.

Indische familiedocumenten. In *Cultureel Indië;* vol. 7(1945), p. 215–216.

Een artistieke gouverneur-generaal; Mr. J.W. van Lansberge (1830–1905); naar gegevens uit het familie-archief van Van Lansberge. In *Cultureel Indië;* vol. 8(1946), p. 143–151.

See also the following articles written by Steinmetz. Book review of J.L. van den Berg van Saparoea. De tragedie op het eiland Saparoea; herinneringen mijner jeugd. In *Cultureel Indië;* vol. 5(1943), p. 237–238.

Oud-Hollandsche buitenplaatsen van Batavia. In *Cultureel Indië;* vol. 7(1945), p. 171–181.

**W.L. de Sturler, June 15th 1802 Tiel – November 24th 1879 The Hague** [652]

*C.* 1819 as second lieutenant to the East Indies; 1821–1825 as officer at Palembang (1821 expedition against the sultan, 1824 first lieutenant engineer); 1826–1830 officer in the Java war; 1830–1837 as captain of the Engineering Corps the chief engineer on duty for the inspection of the fortresses at Djokjakarta and Klaten; 1837–1840 drafted to the Department of War; 1840 pensioned as major, in the Netherlands a member of the Maatschappij der Nederlandsche Letterkunde (Society of Dutch Literature). (*Indische Gids;* 1881, I, p. 191).

The collection contains mainly papers about Palembang from 1655; treaties and other official papers by H.W. Muntinghe, Mahometh Badaroeding, Achmat Naeyam Oedien and Kramadjaja 1818, 1823; papers of the assistant resident J.E. de Sturler, memorandums, accounts, reports etc. concerning the population, trade, journal of a journey to Ulu Musi, complaints about J.C. Reynst as a result of an amuck 1824–1829; studies and notes 1842–1870 by W.L. de Sturler about adat (i.a. at Kikim), history and princely houses of Palembang, mission work in Java, trade in rattan, mining matters, report 1843 by L. Blankenburg of an excursion to the doeson Krong (Gumai Lembar); among this collection was also included "Het eiland Banka en zijne uitwendige gesteldheid" by L. Ullmann (originating from A.L. de Sturler); report 1826 by . . . Gibson about his voyage to Japan.

Size: *C.* 0.25 running metre (H 85, H 170–176, H 183–186, H 188–189, H 243, H 371, H 400, H 561, H 583).

See also L.W.C. van den Berg. Rechtsbronnen van Zuid-Sumatra; uitgegeven, vertaald en toegelicht. In *BKI;* vol. 43(1894), p. 1–352.

J.W.J. Wellan. Een tournée in het Palembangsche vóór honderd jaren. In *Tijdschrift voor economische geographie;* vol. 16(1925), p. 101–108, 148–156.

**B.C.C.M.M. van Suchtelen, September 7th 1885 The Hague – ?** [653]

1907–1910 KNIL officer in Java and Bali; 1910–1918 government official Flores; 1918–1922 Djambi; 1922–1925 Atjeh; 1925–1928 assistant resident Borneo (also president of the municipal council Bandjermasin); 1929–1930 Atjeh; 1931–1933 resident Southern and Eastern District Borneo; 1933–1936 governor East Coast of Sumatra.

The collection consists of a "memorie van overgave" (report left for his successor) 1931–1933 of resident of the Southern and Eastern District of Borneo with photographs; the Van Suchtelen dossier concerning the Bijzonder Uitvoerrecht (Special Export Duty) *c.* 1935.

Size: *C.* 0.15 running metre (H 522, H 610).

**Van Swieten** [654]

For the history of Asia are relevant in the first place the papers originating from J. van Swieten; besides the collection contains some papers of other members of the family.

a. *J. van Swieten, May 28th 1807 Mainz – September 9th 1888 The Hague*
   1824 second lieutenant; 1827–1830 Java war; 1830 in the Netherlands 1st lieutenant; 1835 returns to Java, captain; 1842 major; 1844 lieutenant colonel West Coast of Sumatra; 1848, 1849 expeditions in Bali; 1849–1858 military governor West Coast of Sumatra, 1853 major general, 1856 aide-de-camp to the King, 1857 treaty with Atjeh; 1858 lieutenant general, commander of the Dutch East Indies army, expedition against Boni; 1862 retired; 1864 in the Netherlands staatsraad (councillor of state) in extraordinary service; 1864–1866 member of the Second Chamber; 1867–1872 chairman of the Indisch Genootschap (East Indies Society); 1873 expedition to Atjeh, conquers in 1874 the kraton of Kuta Radja; 1874 returns to the Netherlands, in retirement publicist. This part of the collection contains, besides biographical and genealogical papers, official papers from 1824.
b. *Other members of the family*
   For the history of Asia are relevant the papers originating from F.H.L. van Swieten, L.S.C. Straeter-van Swieten, V.A. van Swieten, E.Chr. van Swieten, J. van Swieten jr.

For the names and subjects mentioned in the inventory see the index.

Entry: H.H. Scholtens. *Inventaris van de papieren van het geslacht Van Swieten.* Leiden, 1975.

Size: *C.* 1.50 running metres (H 997, 229 inventory items).

NB. The collection has been given to the Institute in loan.

**G.L. Tichelman, January 31st 1893 Palembang – January 3rd 1962 Overveen** [655]

Training colonial agriculture in Wageningen and Deventer; 1913–1916 planter in South Sumatra and Tapanuli; 1916–1923 (aspirant) gezaghebber (district commissioner) in Amboina (1917 Amboina, 1917 Ceram (Piru), 1918 Ceram (Amahai)); 1923–1929 gezaghebber Southern and Eastern District of Borneo (1923 South Coast (Tanahbumbu), 1926 Hulu Sungei (Barabai)); 1931–1933 comptroller Atjeh (1931 Meureudu, 1932 Lhokseumawé); 1933–1937 comptroller East Coast of Sumatra (1933 Asahan (Batu Bara and Labuhan Batu), 1934 Simalungun and Karo area (Pematang Siantar)); 1936 titular assistant resident; in the Netherlands i.a. on the scientific staff of the Koninklijk Instituut voor de Tropen (Royal Institute for the Tropics), member of the board of the Vereeniging Oost en West (Association East and West) branch Haarlem, honorary member of the Vereniging van Indische Verlofgangers en Gerepatrieerden (Association of East Indies Persons on Leave and Repatriates), editor of the periodicals *Indië Roept* (1948) and *Wordend Indië* (1948).

The collection consists mainly of papers 1901–1960 collected and written as publicist and a member of the staff of several institutes, such as the Koninklijk Instituut voor de Tropen and Oost en West. For the names and subjects mentioned in the shelflist see the index.

Entry: Shelflist.

Size: *C.* 5 running metres (H 814, 122 items).

See also no. 258.

**C.J. van der Vlis, ? – October 23rd 1842 Surakarta** [656]

Theologian, Javanist; 1839 in the service of the NBG to Surakarta, where he was occupied with the Javanese bible translation till he died in 1842.

The collection consists of a diary 1840–1842 and papers concerning ethnography, archaeology and Hindu antiquities in Java (Djeto, Sung-Kuh, Tampir, Tjandi Krikil, Tjandi Lumbung, Tjandi Peta, Tjandi Sewu), Javanese linguistics, papers of A. Simons and . . . Winter, papers concerning Pangeran Adipati Mangkunegoro and the susuhunan Pakubuwono IV.

Entry: Inventory of H 389.

Size: *C.* 0.10 running metre (H 341, H 389).

**C. van Vollenhoven, May 8th 1874 Dordrecht – April 29th 1933 Leyden** [657]

Professor at Leyden of the adatrecht (customary law) of the Dutch East Indies and constitutional and administrative law of Surinam and Curaçao; founder of the Adatrechtstichting (Customary Law Foundation), co-founder of the Oosters Instituut (Oriental Institute).

The collection consists completely of notes and documentation for his lectures and publications. For the names and subjects mentioned in the shelflist see the index.

Entry: Shelflist.

Size: *C.* 4.50 running metres (H 1086, 68 items).

See also the following publications by Van Vollenhoven.

*De eendracht van het land.* 's-Gravenhage, 1913.

*De drie treden van het volkenrecht.* 's-Gravenhage, 1918.

*Het adatrecht van Nederlandsch-Indië.* Leiden, 1918–1933. 3 vol.

*De Indonesiër en zijn grond.* Leiden, 1919.

*De ontdekking van het adatrecht.* Leiden, 1928.
*Verspreide geschriften;* bezorgd door F.M. van Asbeck. Haarlem, 1934–1935. 3 vol.

**J.W.J. Wellan** [658]

1893–1894 aspirant comptroller at the disposal of the Department of Binnenlands Bestuur (East Indies Civil Service); 1895 at the disposal of the assistant resident of Palembang; 1895 Menado (Ratakan); 1898 comptroller Southern and Eastern District of Borneo (Tanah Laut); 1900–1903, 1904–1907 Palembang (Musi Ilir); 1907 secretary of the regional administration Palembang, also municipal councillor 1909–1910; 1910 official at the disposal of the Department of Binnenlands Bestuur; 1911–1912 assistant resident East Coast of Sumatra (Medan); 1915 Palembang, also president of the Palembangsche Volksbank (Palembang People's Bank); 1917 official at the Encyclopaedisch Bureau (Encyclopaedic Bureau) at Batavia; in the Netherlands i.a. secretary/archivist of the Zuid-Sumatra-Instituut (South Sumatra Institute) and contributor to *Cultureel Indië. (RAvNI).*

Besides preliminary papers from *c.* 1730 the collection consists of papers 1825–1917, collected by Wellan mainly in connection with the history of Palembang and Djambi, viz. "memories van overgave" (reports left for his successor), "registers van handelingen en besluiten" (indexes of proceedings and resolutions), statistics, travel reports, letters and notes concerning the Aru Islands, Benkulen, Danau, Djambi, British administration in Sumatra, financial affairs, Javasche Bank (Java Bank), coffee, Lais, Lingga, Madiun, Moluccas, NHM, Padang, Padangse Bovenlanden, Palembang, pepper cultivation, Ponorogo, Priaman, Raad van Indië (East Indies Council), Riouw, West Coast of Sumatra, Ternate, Tiko, Timor; by and/or about A.J. Bik, J. van den Bosch, J. de Clercq Zubli, J.C. Daansen, C.P.J. Elout, E. Francis (for other papers – all concerning Borneo – and service record of Francis see H 247, H 250–253, H 358), H.J. van de Graaff, Th.A.L. Heyting, N.L. Koops, H. Raders, J.C. Reynst, J.J. Rochussen, G.D. Schlegel, D.B. Schuurman, V. de Serière, H.J.J.L. de Stuers, J.E. de Sturler, the sultan of Lingga, E. Verbeke, B.C. Verploegh, J. Walland (for correspondence and other papers of Walland see H 399 and H 580) and J.W.J. Wellan.

Size: *C.* 0.45 running metre (H 166, H 507, H 531–549, H 551–554, H 556–558, H 563, H 581, H 590).

See also E. Francis. *Herinneringen uit den levensloop van een Indisch ambtenaar van 1815–1851;* medegedeeld in brieven. Batavia, 1856, 1859. 3 parts in 2 vol.

E.B. Kielstra. Sumatra's Westkust van 1826–1832. In *BKI;* vol. 37(1888), p. 216–380.

E.B. Kielstra. Sumatra's Westkust van 1833–1835. In *BKI;* vol. 38(1889), p. 161–249, 312–379, 467–514.

P.H. van der Kemp. Eene bijdrage tot E.B. Kielstra's opstellen over Sumatra's Westkust. In *BKI;* vol. 44(1894), p. 257–320, 525–615.

W.M.F. Mansvelt. *Geschiedenis van de Nederlandsche Handel-Maatschappij.* Haarlem, [1925, 1926]. 2 vol.

A.J. Bik. *Dagverhaal eener reis, gedurende het jaar 1824;* tot nader verkenning der eilanden Kefing, Goram, Groot- en Klein-Kei en de Aroe-eilanden; uitgegeven door A.J.E.A. Bik. Leiden, 1928.

**L.C. Westenenk, February 3rd 1872 Penawangan – May 2nd 1930 Wassenaar** [659]

1892–1893 government official Bandung; 1893–1897 (aspirant) comptroller Western District Borneo (1893 Tajan, 1894 Sanggan and Sedakan, 1895 Putus Sibau); 1897–1900 comptroller West Coast Sumatra (1897 Pajakumbo, 1898 Sawah Lunto); 1900–1903 Atjeh (1900 Olehleh, 1901 Idi, 1901 Kota Radja, 1902 Olehleh); 1905–1909 comptroller West Coast Sumatra (Fort de Kock); 1909–1913 assistant resident West Coast Sumatra (1909 Fort van der Capellen, 1912 Fort de Kock); 1914–1915 inspector general for the reform of Turkish Armenia at Istanbul; 1915–1920 resident Benkulen; 1918 member of the Volksraad (People's Council); 1920–1921 resident Palembang; 1921–1924 governor East Coast Sumatra (Medan); 1924–1925 member of the Raad van Indië (East Indies Council); 1925–1929 in the Netherlands adviser for Indonesian students; 1927 teacher of Minangkabau at the Instituut Kern (Kern Institute); 1928 government commissioner of the Koninklijke Nederlandsch-Indische Luchtvaartmij (Royal Dutch East Indies Air Lines) and the KPM (L.C. Westenenk. *Het rijk van Bittertong;* met een biografie van de schrijver; door H.T. Damsté. 's-Gravenhage, 1932).

The collection consists of papers about Java, Sumatra, Borneo and Armenia, with some preliminary papers from 1825, and of correspondence with his family *c.* 1904–1926. For the names and subjects mentioned in the shelflist see the index.

Entry: Shelflist.

Size: *C.* 0.40 running metre (H 963, 19 items).

## Miscellaneous papers [660]

A large portion of the collection of the western manuscripts consists of miscellaneous papers, subdivided into official papers, non-official papers and written estates of deceased persons. The papers have been further subdivided according to geography and subject. Of the papers after 1795 *c.* 110 manuscript numbers concern the whole of the Dutch East Indies/Indonesia, *c.* 110 numbers Java and Madura, *c.* 70 numbers Sumatra, *c.* 20 numbers Borneo and Celebes, *c.* 30 numbers the Moluccas and New Guinea, *c.* 15 numbers the Lesser Sunda Islands and *c.* 10 numbers South East Asia, mainly Japan. For the names and subjects listed in the catalogue sub no. 614(2a) see the index.

## Oriental manuscripts [661]

This collection consists of miscellaneous papers in the following languages:

Arabic, Achinese, Balinese, Batak, Javanese, Lampong, Madurese, Malay, Minangkabau, Sundanese and other Indonesian languages. Several papers have been removed from collections found in the western manuscripts.

Entries: P.S. van Ronkel. Catalogus der Maleische handschriften van het KITLV. In *BKI;* vol. 60(1908), p. 181–248.

H.H. Juynboll. Catalogus der Javaansche, Balineesche en Madureesche handschriften van het KITLV. In *BKI;* vol. 69(1914), p. 386–418.

Ph.S. van Ronkel. Aanvulling der beschrijving van de Maleische en Minangkabausche handschriften, benevens een Atjehsch handschrift, in het bezit van het KITLV. In *BKI;* vol. 103(1946), p. 555–606.

Beschrijving der [Javaanse] handschriften van het KITLV; opgemaakt in 1964 door Th.G.Th. Pigeaud, naar aanleiding van zijn boek *Literature of Java* (see List of bibliographical works no. 186); met supplement en de beschrijving van enige Javaanse teksten aangetroffen in de verzameling "Westerse Handschriften".

The numbering of the first three catalogues are out of date. *C.* 1960 the oriental manuscripts have been renumbered, of which a manuscript list has been drawn up.

Size: *C.* 20 running metres (505 items).

NB. See also the Korn collection (no. 637), which, though belonging to the oriental manuscripts collection, for this guide has been considered to be a western collection, because of the great number of papers written in Dutch.

**Maps and marine charts** [662]

The collection consists of *c.* 2000 catalogue numbers, of which *c.* 1600 concern the Dutch East Indies/Indonesia, *c.* 100 of these are marine charts. A large part of the maps are concerning Java (*c.* 600 numbers) and Sumatra (*c.* 400 numbers). The Institute possesses of the whole of the Dutch East Indies *c.* 100 numbers, Borneo *c.* 70 numbers, Celebes *c.* 80 numbers, Moluccas and New Guinea *c.* 100 numbers, Lesser Sunda Islands *c.* 80 numbers. Besides there are *c.* 125 maps concerning Asia and Oceania, subdivided into: Asia (*c.* 75 catalogue numbers), Persia, India, Oceania, Australia, Polynesia, Melanesia, Philippines.

The collection of maps mainly consists of topographical atlases and maps, and in addition political, military, agricultural, climatic, traffic, Roman Catholic and Protestant mission and population maps, mainly made by the topographical services at Batavia and The Hague.

Entries: [G.P. Rouffaer]. *Catalogus der land- en zeekaarten toebehorende aan het KITLV.* 's-Gravenhage, 1898.

W.C. Muller. *Catalogus der land- en zeekaarten van en uitgegeven door het KITLV.* 's-Gravenhage, 1913.

Catalogue on index cards, geographically arranged.

The numbering of the first two catalogues is out of date. For the names of places and cartographers of the manuscript maps see the index.

NB. Associated with the Centrale Catalogus voor Kartografie (Central Catalogue for Cartography).

**Drawings, paintings etc.** [663]

The collection consists of *c.* 2000 drawings and about ten paintings. About 1700 are pertaining to the Indonesian archipelago, *c.* 50 to Ceylon and 4 to Japan. For the names of persons and places mentioned in the catalogue see the index.

Entry: J.H. Maronier. *Pictures of the tropics;* a catalogue of drawings, water-colours, paintings and sculptures in the collection of the Royal Institute of Linguistics and Anthropology in Leiden. The Hague, 1967.

**Photographs, picture postcards** [664]

*C.* 25,000 photographs, 500 albums, 3000 negatives on glass, 5000 picture post-cards, *c.* 22,000 of these concern the Indonesian archipelago, *c.* 2000 the rest of Asia. The collection includes photographs of the following photographers: Cephas, Charls en Van Es, I. van Kinsbergen, O. Kurkdjian, J.A. Meessen, H.G. Rime-stad, H. Salzwedel, Woodbury and Page, and in addition a series of photographs made by i.a. Th. van Erp and J.J. de Vink of the Oudheidkundige Dienst (Archaeological Service).

Based on the catalogue on index cards drawn up in 1940 (comprising *c.* 15,000 photographs) the following subdivision can be made: Java *c.* 12,000 photographs, of which West Java *c.* 4500 (i.a. Batavia 2000, Buitenzorg 750, Preanger 500), Central Java 3000 (i.a. Kedu 700, Djokjakarta 1200, Surakarta 350), East Ja-va 1700 (i.a. Surabaja 500, Kediri 300, Malang 300), the rest of the photographs concerns landscapes, villages etc. which could not be exactly localized, and in addition photographs of persons, objects etc. Sumatra *c.* 5000 photographs, of which Atjeh, *c.* 750, East Coast Sumatra 800, Tapanuli 500, West Coast Suma-tra 1000, Palembang 500, Lampong Districts 250, the rest of the photographs have not been localized. Borneo *c.* 750 photographs, of which Southern and Eastern District *c.* 500, West Borneo 250, Celebes *c.* 1500 photographs, of which Govern-ment Celebes *c.* 1000 (i.a. Makasar 300), Menado 500, Moluccas, New Guinea *c.* 1500 photographs, of which Amboina *c.* 300, Ternate 300, New Guinea 900, Lesser Sunda Islands *c.* 1500 photographs, of which Timor *c.* 200, Flores 200, Sumba and Sumbawa 100, Bali 800, Lombok 200. Asia *c.* 2000 photographs, subdivided into: Asia, Malaysia, Singapore, Malaya, North Borneo, Labuan, Brunai, Serawak, Cocos and Christmas Islands, British India, Ceylon, Burma, Portuguese Timor, East Asia, Japan, China, Shanghai, Hongkong, Annam, Cambodia, Siam, Philippines, Central Asia, Australia, Australian New Guinea.

Entries: A geographically classified catalogue on index cards and a subject catalogue. The headings of the subject catalogue are folk types; personalities; houses, interiors; buildings and public institutions; wards, villages, towns; land-scapes; traffic; art, manufacture, industry, cultures, fishery; archaeology; prehisto-rical antiquities; Polynesian antiquities; Hindu-Buddhist monuments; Hindu art; Hindu inscriptions; Islamic antiquities; western antiquities; antique objects.

NB. A revised catalogue based on the UDC is forthcoming.

## Rijksherbarium (National Herbarium), 1829

6 Schelpenkade, 2313 ZT Leyden.
Admission by appointment.
Research finished in 1979.

**Archives** [665]

The Rijksherbarium was established in 1829; activities directed to phytography and related fields.

The archives contain official and private correspondence (from 1829–1871 very fragmentary), diaries, lists of plants given to the museum; for the history of Asia may be relevant the papers by and/or about: Amdjah, S. Binnendijk, C.L. Blume, H.A.B. Bünnemeyer, D.W. Burger, E.A. Forsten, J. Elbert (Sumatra expedition), J.G. Hallier (0.50 running metre), J.K. Hasskarl, J.C. van Hasselt, F.W. Junghuhn (cinchona), S.H. Koorders, G.J.A. Kollmann, P.W. Korthals (including a diary of his activities for the Natuurkundige Commissie voor Nederlandsch-Indië (Physical Committee for the Dutch East Indies), in total 1 running metre of papers), H. Kuhl, H.J. Lam, 's-Lands Plantentuin (Government Botanical Garden) at Buitenzorg, A. Loher, J.P. Lotsy, A.C. Maingay, H. Raap, C.G.C. Reinwardt, Ph.F. von Siebold, J.J. Smith, J.B. Spanoghe, F.L. Splitgerber, W.F.R. Suringar (mainly Surinam), J.E. Teysmann, M. Treub, W.H. de Vriese, A. Zippelius.

Entry: Part of the archives have been arranged alphabetically according to author.

Size: *C.* 130 running metres.

## Rijksmuseum voor Natuurlijke Historie (National Museum for Natural History), 1820

2 Raamsteeg, 2311 PL Leyden.
Admission by appointment.
Research finished in 1979.

**Archives** [666]

The Museum was established in 1820 at an initiative of the minister of Public Education of 1819; the zoological collections of the Leyden University (including those of the king), the collection of 's-Lands Kabinet van Natuurlijke Historie (National Gallery of Natural History) and the C.J. Temminck collection were the basis; the object was to put together a collection for the benefit of scientific research, of secondary importance were the activities in the field of education; the Museum maintained contacts with the Natuurkundige Commissie voor Nederlandsch-Indië (Physical Committee for the Dutch East Indies) partly established under the auspices of the Museum, official collectors (with a Government instruction) such as Ph.F. von Siebold and H. Bürger, amateur collectors such as J.A. Kruyt (consul in Djeddah), E. Jacobson and P. Buitendijk.

The archives consist of a chronologically arranged correspondence with the Ministry of Education and the board of governors of the Leyden University; an alphabetically to correspondent arranged correspondence with other organizations and persons; papers outside the province of correspondence, such as reports and diaries of the members of the Natuurkundige Commissie voor Nederlandsch-Indië. For the history of Asia and Oceania may be relevant the papers of i.a.: F. Adèr-Verver, M.E.G. Bartels, L.J.W. Benschop, J.C.B. Bernelot Moens, H.A. Bernstein, C.L. Blume, H. Boie, M. Boissevain, J.F.R.S. van den Bossche, H. Brummund, J.A. Buddingh, P. Buitendijk, H. Bürger, . . . Colffs, [H.] Croockewit [Hzn], K.W. Dammerman (about the Krakatau, 's-Lands Plantentuin (Government Botanical Garden), Lampong, Sumba, Idjen, Sumatra, Celebes, Lesser Sunda Islands, Moluccas, New Guinea; 1 running metre), P. Diard, P. van Eecke, F.H.O. Finsch, E.A. Forsten (Menado), Genootschap "Natura Artis Magistra" (Society "Natura Artis Magistra"), W. de Haan, B. Hagen, J.C. van Hasselt, J.A. Herklots, W.C. van Heurn (Simalur), J. Hewitt, Y. Hirase, . . . Hodgson (Nepal), D.J. Hoedt, L. Hörner, A. Horst (Ternate), F.W. Junghuhn, C.J. Kaiser, P.N. van Kampen, . . . Köhn, P.W. Korthals, J.A. Kruyt, H. Kuhl, J.W. van Lansberge, H.A. Lorentz, E.W.A. Ludeking, H.C. Macklot, D.G. Müller, S. Müller, museums at Batavia, Serawak and Sydney, S.C.J.W. van Musschenbroek, P.G. Neeb, A.W. Nieuwenhuis (Borneo), H. van Oordt van Lauwenrecht, P. van Oort, B.N. Overdijk, Pacific Science Association (H. Boschma), . . . Palmer van den Broek, Pan Indian Ocean Science Association, W.J. Pompe van Meerdervoort (Japan), G. van Raalten, C.G.C. Reinwardt, L.D.H.A. Renesse van Duivenbode, A.C. Ricord, J.G.F. Riedel, H. Rolle, W.F.H. Rosenberg, C.B.H. von Rosenberg, M. Schaalje, G. Schlegel, C. Schneider, C.A.L.M. Schwaner, J. Semmelink, Ph.F. von Siebold, W. Spies, J.E. Teysmann, M. Treub, Q.M.R. Ver Huell, A.G. Vorderman, W.H. de Vriese, A.R. Wallace, M. Weber, G.F. Westerman, G.F. Wienecke. The archives contain also photographs of persons. The archives have been made up to the present.

Entries: Indexes to the official correspondence with the Ministry of Education and the board of governors; up to 1915: A. Gijzen. *'s Rijks Museum van Natuurlijke Historie, 1820–1915*. Rotterdam, 1938.

Size: *C.* 50 running metres: official correspondence *c.* 13 running metres, the rest of the correspondence *c.* 20 running metres, papers outside the province of correspondence *c.* 17 running metres.

**Drawings** [667]

The collection consists of some hundreds of drawings and pictures, mainly representing animals, i.a. of Java and Japan.

# CHAPTER VIII

# MISSIONARY ORGANIZATIONS

## Zendingsbureau van de Christelijk Gereformeerde Kerken (Mission Bureau of the Christian Calvinist Churches), 1928

144 Stevinweg, 1222 SV Hilversum.
Admission by appointment.
Research finished in 1979.

**Mission archives of the Christelijk Gereformeerde Kerken (Christian Calvinist Churches), 1928 –** [668]

From 1928 the Christelijk Gereformeerde Kerken (Christian Calvinist Churches) were active in mission in Celebes, with Mamasa as centre, the first missionary was A. Bikker, in 1930 succeeded by M. Geleijnse, the main activities were directed to education.

The archives, which have not been arranged yet, contain the residue of the activities from 1928.

Size: *C.* 6 running metres.

## Zendingscentrum van de Gereformeerde Kerken in Nederland (Missionary Centre of the Calvinist Churches in the Netherlands), 1946

18 De Beaufortweg, 3833 AG Leusden.
Admission free. No lending of archival material.
Research finished in 1974.

**Archives** [669]

The Zendingscentrum (Missionary Centre) keeps the archives of the generale deputaten (general delegates) of the mission of the Gereformeerde Kerken (Calvinist Churches) see no. 670 and of the Zendende Kerk van Amsterdam (Missionary Church of Amsterdam) see no. 671. Included are here also the descriptions of the archives of the other missionary churches of the Gereformeerde Kerken see no. 672–677. The addresses of the institutions where these papers arc kept are mentioned after each description of the archives.

For literature concerning the mission of the Gereformeerde Kerken see *De Zending in Oost en West;* verleden en heden; bewerkt en samengesteld door H.D.J. Boissevain; vol. 2. 's-Gravenhage, 1943.

**Zending der Gereformeerde Kerken (Mission of the Calvinist Churches)** [670]

These archives are divided into two subdivisions, viz. the archives of the Nederlandsche Gereformeerde Zendingsvereeniging (Dutch Calvinist Missionary Association) and the archives of the Christelijk Gereformeerde Kerk/Gereformeerde Kerken (Christian Calvinist Church/Calvinist Churches).

a. *Nederlandsche Gereformeerde Zendingsvereeniging (Dutch Calvinist Missionary Association), 1859–1894*
In May 1859 the Nederlandsche Gereformeerde Zendingsvereeniging was established, with Java as missionary field (areas: Tegal from 1859, Banjumas from 1867, Bagelen from 1869, Sumba from 1880, Kedu from 1888 and Djokjakarta from 1889). In 1894 the Vereeniging turned its activities over to the Gereformeerde Kerken, although formally it continued to exist.

This part of the archives consists of minutes 1859–1896; copy letter books 1862–1893, partly with alphabetic index; cash books 1873–1893; papers concerning the establishment and the history of the Vereeniging; personal files with letters and other papers of J.J. van Alphen, Ph. Bieger, H. Guillaume (only about his training), M. Hildering, R.J. Horstman, P. Jansz, F. Lion Cachet (journey 1890/91), R.M. Tengku Nataadmata (only about his training), J.G. Scheurer, H. Stoové, H.F.M. Uhlenbusch, A. Vermeer (the first missionary, from 1861 in Tegal), J. Wilhelm (diaries 1881–1892, cash books, Javanese sermons), J.P. Zuidema.

b. *Christelijk Gereformeerde Kerk/Gereformeerde Kerken (Christian Calvinist Church/Calvinist Churches), 1861–1892, 1892–1949*
The Christelijk Gereformeerde Kerk started in 1861 with mission work. From 1873 it was active in Batavia, in 1889 it took over Sumba from the Gereformeerde Vereeniging. In 1892 the Christelijk Gereformeerde Kerk associated with the Nederlandsch Gereformeerde Kerk (Dutch Calvinist Church), which together became the Gereformeerde Kerken in Nederland (Calvinist Churches in the Netherlands). In 1894 they took over the work of the Nederlandsche Gereformeerde Zendingsvereeniging. In 1896 the Generale Synode (General Synod) decided to let a local church to take care of the missionary areas. Sumba was attended to by the churches in the provinces of Groningen, Drenthe and Overijssel and also by eastern Friesland. Java was subdivided as follows:

Purbolinggo (Banjumas) by Rotterdam with southern South Holland;

Purworedjo (South Bagelen) by Utrecht and Gelderland;

Surakarta (up to 1922) and Djokjakarta by Amsterdam with North Holland;

Kebumen and Wonosobo (Kedu since 1930) by Heeg with Friesland;

Magelang (Kedu) by Middelburg with Zeeland and North Brabant;

Wonosobo (up to 1930) and Surakarta (since 1922) by Delft with northern South Holland.

This part of the archives also includes the papers originating from the Christelijk Gereformeerde Kerk. There are minutes of the Algemene Zendingscommissie der Christelijk Gereformeerde Kerk (General Mission Committee of the Christian Calvinist Church) 1861–1892, with alphabetic registers; minutes of the Deputaten Synodi van de Gereformeerde Kerken in Nederland voor de Zending onder de Heidenen en Mohammedanen (Delegates of the Synod of the Calvinist Churches in the Netherlands for the Mission among the Heathen and Mohammedans) 1892–1949; correspondence 1868–1927; copy letter books 1899–1940, with alphabetic indexes; general chest up to 1940 with registers; papers Gereformeerde Kerk at Batavia, Bandung, Medan and Surabaja; papers L.Adriaanse 1894–1897, J.J. van Alphen 1903–1928, zendende kerken (missionary churches) Amsterdam, Delft, Heeg, Middelburg, Rotterdam and Utrecht 1899–1930, C. de Best 1865–1867, mission among Chinese, Keucheniusschool, Hollandsch-Inlandsche Kweekschool (Dutch-Native Training College for Teachers), Zendingsconsulaat (Missionary Consulate), J.G. Scheurer, S.J.P. Goossens, J.P. Zuidema, Sumba, Wonosobo, K. van Dijk, Petronella Hospital.

Entry: Alphabetic index.

Size: *C.* 9 running metres.

**Zendende Kerk van Amsterdam (Missionary Church of Amsterdam), 1887–1945** [671]

Missionary areas: Surakarta (up to 1922), Djokjakarta.

The archives consist of minutes 1898–1900; minutes commissie van advies (advisory committee) 1901–1905, 1919–1945; minutes deputaten medische zending (delegates medical mission) 1896–1918; papers schools and mission 1916–1945; records of letters received 1887–1922 and letters despatched 1900–1922; correspondence 1897–1945; accounting papers 1905–1945; files of the areas of Surakarta and Djokjakarta; papers medical mission, Petronella Hospital, hospital at Surakarta, Vereeniging Dr. Scheurers Hospitaal (Association Dr. Scheurer's Hospital), several physicians and nurses; personal files of D.J.B. Allaart, H.A. van Andel, D. Bakker, F.L. Bakker, S.J.D. Esser, K.P. Groot, S. van der Ley, F.C. van Marle, M.G. Muller, J. Offringa, A. Pos, C. van Proosdy, H.S. Pruys, G.J. van Reenen, J.C. Rutgers, J.G. Scheurer, D. Verhagen, W.L. Vogelzang, C. Zwaan; photographs.

Entry: Alphabetic index, with summary of the archives.

Size: *C.* 2 running metres.

**Sumba Zending (Sumba Mission), 1896–** [672]

The archives consist of minutes 1896–1932, 1939–1943, 1947– ; accounting papers 1896–1926; copy books of letters despatched 1896–1926; personal files of J.S. Bedet, J. Berg, P.G. van Berge, H. Bergema, C. de Bruyn, Miss H. Cleveringa, J.F. Colenbrander, T. van Dijk, W. van Dijk, J.L. Erkelens, S.J.P. Goossens, D.H. Krijger, L.P. Krijger, P.L. Lambooy, M. Lock, P.J. Luyendijk, T. Mobach, L. Onvlee, W. Pos, F. Veldstra, J.A. Vellinga, D.K. Wielenga, G. Zigterman.

Size: *C.* 21 running metres.

The archives are deposited at 2 Marnixkade, Groningen.

**Zendende Kerk van Rotterdam (Missionary Church of Rotterdam), 1936–** [673]

These archives have been destroyed by fire, only a minutebook 1936 up to now has been preserved. Letters from G.J. Ruyssenaers are published in A. van Veelo. *In memoriam;* ter herinnering aan den arbeid van G.J. Ruyssenaers, missionaire dienaar der Kerk van Rotterdam in de residentie Banjoemas op Java. Rotterdam, 1908.

Letters from B.J. Esser are published in the *Kerkbode van Rotterdam.*

**Zendende Kerk van Utrecht (Missionary Church of Utrecht), 1896–** [674]

Missionary area: Purworedjo.

The archives consist of minutes 1896– ; accounting papers from 1942; letters received and despatched from 1896; papers of L. Adriaanse, B. Alkema, G.J. Dreckmeyer, J.C. Flach, H.P. Ingwersen, A.F. Keuchenius, J. van der Linden, H. Nieboer, H. Slettenaar, N. Strijker, F. Veldstra; besides some papers of B. Allaart, G.A. Allaart-Scheurer, H.A. van Andel, D. Bakker, F. Bast, J.H. Bavinck, H. Bergema, C.W.Th. van Boetzelaer van Asperen en Dubbeldam, G.A. Both, K. van Dijk, B.J. Esser, P.H. van Eijk, L. Götzen, K.P. Groot, W.S. de Haas, W.J. Hartman, G. Haspers, R. Hooisma, A.W.F. Idenburg, T. Kuypers, S. van der Ley, F.G. van Marle, A. Merkelijn, A.M. Netelenbos, L. Netelenbos, C. Nieuwenhuyze, J. Offringa, H. Pluyter, A. Pos, J.A.C. Rullman, J.G. Scheurer, M.M.C. Verleur, C. Versteeg, A.H. Vonk, D.J.B. Weyers, A.J. van Zanten, P.H. Zuidema, C. Zwaan; photographs.

Size: *C.* 3 running metres.

The archives are deposited in the Bethelkerk, 1 Burg. Norbruislaan, Utrecht.

**Zendende Kerk van Heeg (Missionary Church of Heeg); 1899–1957** [675]

Missionary areas: Kebumen and Wonosobo (since 1930).

The archives consist of minutes from 1906; letters received and copybooks of letters despatched from 1899; papers of A. Antheunisse, D. Bakker, J. Berg, H. Bergema, M.O. Bolt, A. Brummelkamp, G.J. Dreckmeyer, K. van Dijk, G.B. Wiggers, C.L. de Graaf, H.M.R. Hangelbroek, H.F. ten Katen, E. Klein,

G.H. van Koeverden Brouwer, L.W. Korvinus, E. Meinders, H.C. Muys, J.W. Oosterhuis, W. Peddemore, T. Pel, W.C. Rinck, H. Sanders, T.T. Sybenga, D. Verveen, A.H. Vonk; photographs.

Entry: Annual indexes.

Size: *C.* 10 running metres.

The archives are deposited at the Zendingsburo Gereformeerde Kerken van Friesland, 28 Stationsweg, Leeuwarden.

**Zendende Kerk van Middelburg (Missionary Church of Middelburg), 1910–1959** [676]

Missionary area: Magelang.

The archives consist of minutes from 1931; accounting papers from 1910; letters received and despatched; papers of G.J. Dreckmeyer, A. Merkelijn, C. van Nes, H. Schut; photographs.

Size: *C.* 10 running metres.

The archives are deposited in the Hofpleinkerk, Middelburg.

**Zendende Kerk van Delft (Missionary Church of Delft), 1900–1954** [677]

Missionary areas: Wonosobo (up to 1930), Surakarta (since 1922).

The archives consist of minutes from 1900; accounting papers; letters received from 1900; papers of B. Allaart, H.A. van Andel, D. Bakker, J.H. Bavinck, J. Berg, S.J. Bos, H.G. Brink, C. Bronkhorst, G.J. Dreckmeyer, B.J. Esser, P.H. van Eijk, C.A. Gehrels, L. Goemaat, G.J. Hoekstra, G.D. Kuiper, C.P. van Leersum, A. Merkelijn, H.S. Pruys, H. te Velde, D. Verhagen, J.S. Wiersema, S.U. Zuidema; photographs.

Entry: *Inventaris van het zendingsarchief van de Gereformeerde Kerk van Delft als zendende kerk en de zamenwerkende kerken.* [Delft], 1931.

Size: Up to 1930 *c.* 1 running metre.

The archives up to 1930 are deposited at the Gemeente Archief Delft (Municipal Archives of Delft), 169 Oude Delft, Delft. The archives after 1930 are deposited in the Immanuel Kerk, 118 De Colignystraat, Delft.

## Raad voor de Zending der Nederlandse Hervormde Kerk (Mission Council of the Dutch Reformed Church), 1951

11 Leidsestraatweg, 2341 GR Oegstgeest.
Admission free. Lending (except archival material).
Photocopying equipment available.
Research finished in 1970.

**Library** [678]

30,000 books; *c.* 300 periodicals, of which 100 currently received; card index of periodical articles; literature surveys.

Directed to mission, theology and religious sciences, linguistics, geography and ethnology (up to 1940 mainly concerning the Dutch East Indies/Indonesia).

The library is also the library of the Nederlandse Zendingshogeschool (Dutch Missionary College) and is composed of the libraries of the corporations, which were amalgamated in 1951 (see no. 679–694).

For literature concerning the different corporations and organizations see i.a.

*De zending in Oost en West;* verleden en heden; bewerkt en samengesteld door H.D.J. Boissevain. 's-Gravenhage, 1934, 1943. 2 vol.

## Archives and archival material [679–695]

### Introduction to the archives and collections [679]

The mission of the Reformed Dutch Church was mainly directed to the Dutch East Indies/Indonesia. Only after 1949 there arose a new orientation towards Africa, although New Guinea/Irian Barat remained an important missionary field up to 1963. Consequently the missionary archives contain, besides papers concerning the organization in the Netherlands, mainly papers concerning the mission in the East Indies. Besides the archives of the Gereformeerde Zendingsbond (Calvinist Missionary Union) (117 Utrechtseweg, Zeist) all the Reformed missionary archives are deposited in the depot at Oegstgeest.

The archives may be subdivided into the following periods: 1797–1927 (see no. 680–689), 1928–1952 (see no. 690–692) and 1953– (see no. 693–694). Moreover the Mission Bureau keeps the missionary archives sent over from New Guinea (no. 695) and a few collections (no. 696–698).

The archives 1797–1927 are arranged according to personal names. That means that besides series such as minutes, journals and drafts of outgoing letters, the papers concerning and/or originating from a certain person were kept together in a "dossier" (file) bearing the name of that person. These bundles were called "persoonsdossiers" (personal files). After 1927 one switched to an arrangement based upon the UDC.

The repository contains a "financieel archief" (financial archives) of *c.* 20 running metres kept separately, containing ledgers, journals, budgets, accounts, cashbooks, financial reports etc. up to 1946 of the NZG, NZV, UZV, JC, S en TC, NZS and ZB. These papers are mentioned with the description of the separate corporations.

Entries: Some general entries have been made to the archives, viz.: Index 1797–1927; indexes "zendingsvelden" (missionary fields) 1928–1952, 1953–1962; indexes "binnenland" (interior) 1946–1948, 1948–1950, 1951–1962. The indexes are lists with catchwords and personal names alphabetically arranged, referring to the location in the archives.

NB. Moreover, separate catchword registers have been made to the archives of all the corporations, mentioned as "entry" with the description of the separate archives.

The archives repository also contains a name index, including biographical data of missionaries in the service of the Doopsgezinde Zending (Mennonite Mission), Evangelische Missionsgesellschaft zu Basel (Basel Mission), Java-Comité (Java Committee), Nederlandsch Zendelinggenootschap (Dutch Missionary Society), Nederlandsche Zendingsvereeniging (Dutch Mission Association), Salatiga-Zending (Salatiga Mission), Sangi- en Talaud-Comité (Sangi and Talaud Committee) and Utrechtsche Zendingsvereeniging (Utrecht Mission Association).

**Nederlandsch Zendelinggenootschap (Dutch Missionary Society), NZG 1797** [680]

The NZG was established December 19th 1797, being the first Reformed missionary corporation. Missionary fields were the Moluccas 1797–1864, Timor 1797–1854, 1873– *c.* 1900, Zuid-Wester Eilanden 1797–1841, Celebes (Minahasa from 1822, Bolaang Mongondow from *c.* 1825, Poso from 1887, Makasar and Gorontalo from 1851), Eastern Java from 1848, Deli (Karo-Batak mission) from 1890.

The archives include "actenboeken" (instrument books) and extract instruments 1797–1952; copybooks (copies of letters despatched) 1798–1920; accounting series and papers 1798–1946; minutes with extracts of several Dutch departments 1828–1929; minutes (up to 1928) of several conferences of missionaries i.a. in the fields Deli, East Java and Poso; correspondence (up to 1928) with i.a. the Ministry of Colonial Affairs, NBG, Zendingsconsulaat (Missionary Consulate), departments and members of the committee in the Netherlands; reports, statistics, documentation etc. concerning the Van Deventerstichting (Van Deventer Foundation), the Vrouwen Hulpvereeniging (Women's Assistance Association), training of missionaries, missionary work and history, education (i.a. the school for girls at Tomohon), medical work, adat etc. in the following territories: Moluccas (Amboina, Saparua, Ceram, Buru, Ternate), Celebes (Minahasa, Poso), Sangihe- and Talaud Islands, Batak area, East Java, Timor; personal files, for the names see the index.

In the archives of the NZG is deposited the collection of:

*J.W. Gunning, June 15th 1861 Utrecht – November 25th 1923 Noordwijkerhout*
1885 minister Eerbeek; 1897 director NZG, later till his death of the gezamenlijke zending (cooperating mission); travelled through the East Indies archipelago in 1900 and 1916–1918; promoted the establishment of the Nederlandsche Zendingsschool (Dutch Missionary Training School) (1905, 1917 at Oegstgeest), the Zendingsconsulaat (Missionary Consulate) (1906) and the Zendingsbureau (Mission Bureau) (1908). (D. Crommelin. In memoriam J.W. Gunning. In *De Opwekker.* 1923, p. 457/Necrology in *Nederlandsch Zendingsblad.* 1924, p. 3).

The collection consists of sermons 1885–1908, 1913, 1922; correspondence 1902–1920; drafts of outgoing letters; memorial book 1898–1923; diary; papers concerning the school question in the Minahasa.

Entry NZG archives: Alphabetic catchword index.

Size: *C.* 45 running metres.

**Nederlandsche Zendingsvereeniging (Dutch Missionary Association), NZV 1858** [681]

The NZV was established December 2nd 1858, as a corporation separated from the NZG. The missionary fields were West Java (Batavia, Cheribon, Bandung, the Preanger) from 1862, South East Celebes from 1915.

The archives include minutes 1858–1912; copybooks (letters despatched) 1859–1912; accounting series and papers 1859–1946; instructions for missionaries 1859–1914, 1940–1944; applications; minutes of the Indramaju Committee 1916–1941 and the conferences of missionaries in West Java 1907–1912, 1915–1921, 1928, 1932; correspondence with the Ministry of Colonial Affairs and the Zendingsconsulaat (Missionary Consulate); reports, statistics, documentation etc. concerning the history of the NZV, training of missionaries, education, medical work, missionary stations in West Java, Sundanese and Chinese mission in West Java, mission in Celebes, schools at Bandung, the flood at Indramaju, the Sangihe and Talaud Islands; personal files, for the names see the index.

Entry: Alphabetic catchword index.

Size: *C.* 11 running metres.

**Utrechtsche Zendingsvereeniging (Utrecht Mission Association), UZV 1859** [682]

The UZV was the second separation from the NZG. The corporation was established April 13th 1859. The missionary fields were New Guinea from 1862, Halmahera from 1866, Buru from 1885, the Sulu Islands from 1913 (only a few years), Bali 1866–1881, South Celebes 1895–1905.

The archives include minutes and proceedings 1859–1928; accounting series and papers up to 1946 (incomplete); drafts of outgoing letters 1859–1929; minutes of various Dutch departments; minutes and annual reports of the missionary conferences at Halmahera 1900–1926, New Guinea 1902–1927, Buru 1916–1927; papers of the Zendingsconsulaat (Missionary Consulate) 1906–1927 about New Guinea, Halmahera and Buru and of the Vrouwen Hulpvereeniging (Women's Assistance Association) 1864–1906 about New Guinea, Halmahera and Sangihe; reports, statistics, documentation etc. concerning the history of the UZV, missionary work in New Guinea, Halmahera (i.a. adat and cocopalm plantations), Buru; personal files, for the names see the index.

Entry: Alphabetic catchword index.

Size: *C.* 9 running metres.

**Java-Comité (Java Committee), JC 1855** [683]

The JC is a committee that on March 24th 1855 became independent from the Genootschap voor In- en Uitwendige Zending (Society for Internal and External Mission) at Batavia. At first the JC was a money collecting auxiliary missionary association, after 1864 the Committee started to do missionary work independently. Missionary fields were Sumatra (Angkola) from 1864, East Java (Madura, Besuki) from 1879.

The archives include minutes 1855–1933; rules 1923–1933; accounting series and papers 1855–1946; drafts of outgoing letters 1886–1893, letters received 1868–1906; correspondence with the Zendingsconsulaat (Missionary Consulate); personal files of H.W. van den Berg, J. Dammerboer, O. Dedecker, J.H. Duyverman, J.J. Eggink, W.H.Th. van Hasselt, H. Hendriks, F. Schelfhorst, J. Warneck.

Entry: Alphabetic catchword index.

Size: *C.* 3 running metres.

**Sangi- en Talaud-Comité (Sangihe and Talaud Committee), S en TC 1887** [684]

The S en TC has been established in 1887 from the JC as an independently acting committee. Fields of operations were the Sangihe and Talaud Islands.

The archives include minutes 1903–1929; annual reports and statistics 1896–1934; accounting series and papers 1900–1946; drafts of outgoing letters and bundles of correspondence 1903–1921; letters, documentation and other papers 1859–1896 originating from the Zendeling-Werkman (Missionary-Worker); reports, documentation etc. concerning the history of the Committee, education, cocopalm plantations, adat in the Sangihe and Talaud Islands; personal files, for the names see the index.

Entry: Alphabetic catchwords index.

Size: *C.* 4.50 running metres.

**Centraal-Comité voor de Oprichting en de Instandhouding van een Seminarie nabij Batavia (Depok) (Central Committee for the Establishment and Maintenance of a Seminary near to Batavia (Depok)), Centraal-Comité Depok *c.* 1870** [685]

The object of the Committee was to establish, after having collected the necessary money, a seminary for Indonesians who wanted to become a minister or evangelist. Finally this seminary was opened in 1878.

The archives include minutes 1871–1907; cashbooks etc. 1874–1916, 1922–1927; drafts of outgoing letters 1873–1876, 1894–1922; papers concerning the establishment of the seminary; proceedings, annual reports etc. of the seminary 1878–1927; reports of the director of the seminary 1878–1905; rolls of the pupils 1878–1926; correspondence of the Committee 1905–1926 with the seminary, the Zendingsconsulaat (Missionary Consulate), the inspection of schools and various missionaries;

letters from pupils 1920–1926; personal files of A. Bierens de Haan, C. Dijck, C. van Engelen, C. de Graaf, N. Heertjes, H. van Hogerzeil; photographs of pupils.

Entries: Survey of the archives made by the Committee; alphabetic catchwords index.

Size: *C.* 2 running metres.

**Nederlandsche Zendingsschool (Dutch Missionary Training School), NZS 1905** [686]

The NZS was established in 1905 by the NZG and the UZV. The object was the common training of missionaries of both corporations. After the S en TC, the Doopsgezinde Zendingsvereeniging (Mennonite Missionary Association), the Gereformeerde Zendingsbond (Calvinist Missionary Union) and the Zeister Broedergemeente (Moravian Brotherhood of Zeist) had joined, the NZV decided in 1919 to have its missionaries also trained at the NZS. From 1905–1917 the school was established at Rotterdam, from 1917 at Oegstgeest.

The archives include minutes with rules and regulations 1905–1946; accounting series and papers 1905–1952; reports and correspondence concerning the training of missionaries and missionary wives and concerning medical mission; correspondence with J. Stuiving concerning the building of the school at Oegstgeest 1913–1918; correspondence with the Ministry of Colonial Affairs; correspondence with the Zendelingenbond (Union of Missionaries) with the newspapers *Rottebode* and *Soerat Saudara* 1903–1933; personal notes etc. of pupils 1850–1878, 1896, 1904; namebook of the pupils at Rotterdam 1905–1917; papers concerning the installations of A.M. Brouwer and A. van Velthuizen; files concerning the cases of Burger and Van der Wielen; newspaper cuttings concerning education 1915, political parties 1916, Indië Weerbaar 1917, Sarekat Islam 1917; personal files of F. Fokkens, J.W. Gunning (tour of inspection 1917/1918), S. van der Linde, J.H. Schelfhorst.

Entry: Alphabetic catchwords index.

Size: *C.* 5 running metres.

**Samenwerkende Zendings-Corporaties (Co-operating Missionary Corporations), SZC 1908** [687]

The co-operation in the training of missionaries brought on also co-operation in the field of administration and organization. The SZC came into being *c.* 1908. The directors were appointed by the associated corporations. Associated were the initiators NZG and UZV and later S en TC 1913, NZV 1923, JC 1931, Doopsgezinde Zendingsvereeniging (Mennonite Missionary Association) *c.* 1931, whereas the Rheinische Missionsgesellschaft at Barmen in 1913 and the Evangelische Missionsgesellschaft at Basel in 1925 had their representatives in the SZC.

The archives consist, besides a series of minutes 1914–1924 and accounting papers 1920–1923, mainly of correspondence and files, i.a. letters received and despatched 1913–1923; correspondence with the International Mission Council, American missionary corporations, the German mission, the Rheinische Missionsgesellschaft, the Evangelische Missionsgesellschaft, the Zendingsconsulaat (Missionary Consulate), UZV, S en TC, Evangelische Broedergemeente (Moravians), NBG, various clerical boards, NHK; statistics NZG, UZV, S en TC, NZV; reports world mission conference 1910; reports mission conferences Poso, East Java, New Guinea, Buru, Halmahera, Deli; reports of the official tours of J.W. Gunning 1906–1908, 1916–1918 and K.J. Brouwer 1927–1928; reports etc. of and about the mission in West and Central Java 1939–1948, the Rheinische Missionsgesellschaft 1832–1855, 1917–1924, Evangelische Missionsgesellschaft, Berliner Missionsgesellschaft 1915–1923; reports of the advisory committee i.a. concerning education and medical work 1916–1927; reports H. Kraemer concerning Sangihe; memorandum G.B. Tiekstra concerning the Minahasa; papers about adat and mission in the Batak area i.a. by J.H. Meerwald; papers J.W. Gunning and J.G. Kielstra concerning the legal status of Indonesian Christians; documentation concerning education, medical work, the liquor problem, the Roman Catholic mission, the Pinkstergemeente (Pentecostals), the first world war, Buru, Halmahera, Poso, Minahasa, Sangihe and Talaud, New Guinea, Java, Bali; personal files of K.J. Brouwer, D. Crommelin, J.W. Gunning, M.A. Kelling, M. Lindeborn; photographs.

Entry: Alphabetic catchwords index.

Size: *C.* 5.50 running metres.

**Zendingsbureau (Mission Bureau), ZB 1908** [688]

The ZB may be considered to be the secretariat of the SZG. However, the papers have been kept separate from those of the SZG.

The archives consist mainly of accounting series and papers up to 1946 (*c.* 4.50 running metres). Moreover correspondence with the Doopsgezinde Zendingsvereeniging (Mennonite Missionary Association) 1907–1927, the Gereformeerde Zendingsbond (Calvinist Missionary Union) 1910–1924, the JC 1910–1926, the Zending der Gereformeerde Kerken (Mission of the Calvinist Churches) 1912–1924, the Evangelische Broedergemeente (Moravians) 1912–1927, the Nederlandsche Christelijke Studentenvereeniging (Dutch Christian Students Association) 1915–1926, the Nederlandsch Luthers Genootschap (Dutch Lutheran Society) 1919–1923, the Salatiga Mission 1921–1924, the Committee Tasikmalaja 1924, the NHK 1925, J.W. Gunning and W. Oettli 1919–1920; papers of the Nederlandsch-Indische Zendingsbond (Dutch East Indies Missionary Union) 1882–1910, Zendingsstudieraad (Mission Study Council) and J. Rauws.

Entry: Alphabetic catchwords index.

Size: *C.* 5 running metres.

**Miscellaneous papers** [689]

Of a number of bundles and papers it was impossible to decide to which archives they belonged. These include:

a. Minutes of the Christelijk Nationaal Zendingsfeest (Christian National Mission Festival) 1873–1902 with papers 1924–1925.

b. Minutes of the Algemeene Nederlandsche Zendingsconferentie (General Dutch Mission Conference), ANZC 1877–1940 with circulars 1889–1916 and reports 1912–1913.

c. Minutes of the Commissie van Advies voor Kerkelijke Zending (Advisory Committee for Church Mission) 1912–1913.

d. Minutes of the Centraal Zendingscomité (Central Mission Committee) at Haarlem 1917–1928.

e. Minutes of the Christiana-fonds (Christiana Fund) 1919–1936.

f. Minutes of the interior action meetings 1935–1939.

g. Index of letters received and despatched of the mission in Halmahera 1912–1915, New Guinea 1913–1922.

h. Correspondence of the Christelijke Werkman/Zendeling-Werkman (Christian Workman/Missionary Workman) 1850–1890 i.a. with . . . Burgers, E. Michaelis and F.A. Gützlaff.

i. Correspondence with the missionary consul C.W.Th. van Boetzelaer van Asperen en Dubbeldam 1906–1915 with index of letters received and despatched 1913–1918.

j. Correspondence of the Centraal Lectuur-Bureau (Central Literature Bureau) 1929–1932.

k. Correspondence "at home and abroad" 1946–1948.

l. Reports of official journeys 1900–1901.

m. Index of letters received and despatched of J.W. Gunning during his official journey 1917–1918 with papers about education in Minahasa and Sangihe 1903–1917.

n. Report of the journey of J.B. Mott 1926–1927.

o. Papers of the "Indische Kerk" (Protestant church in Dutch East Indies) 1905–1927, i.a. concerning Sumba and the Minahasa and correspondence with S.D. Jansz, A.J. Kelling, G. van Trehn-Boon, Native representatives, persons in the Minahasa, sermons.

p. Papers concerning international mission conferences, Edinburg 1910, Continuation Committee of World Missionary Conference 1912–1920, Wereldraad van Kerken (World Council of Churches) 1922–1923 of C. Boeke.

q. Papers of the Nederlandsche Zendingsstudieraad (Dutch Mission Study Council) 1912–1925, the Nederlandsche Zendingsraad (Dutch Mission Council) and the NBG 1915–1917.

r. Papers concerning the Ermelo Zending (Ermelo Mission), UZV, Salatigazending (Salatiga Mission) 1910 of P.A. Gericke, Kisar-zending (Kisar Mission) 1927, the Elisabethvereeniging (Elisabeth Association) at Utrecht 1923–1925.

s. Papers non provided education 1909–1915.

t. Papers NBG by J.F.C. Gericke.

u. Papers of the Ministry of Colonial Affairs and the East Indies government.

v. Malay Press Reviews 1920.

w. Papers concerning Menado 1825, Minahasa, South Celebes, Sangihe, West Java, the Principalities 1826 of H. Mac Gillavry, East Java, Salatiga.

x. Papers concerning military homes.

y. Papers N. Adriani 1892–1893, N.E. Breeman 1926, R. Grundeman 1868–1873, H.C. Klinkert 1868–1889, O.J. van Limburg Stirum 1885–1894, B.F. Matthes 1859–1892, E. Nijland 1888–1892, L.J. van Rhijn 1849–1877.

z. Cyclostyles by C.W. Nortier concerning his mission review 1940–1945.

Entry: Alphabetic catchwords index.

Size: *C.* 2.50 running metres.

**Archief 1928–1952 "Zendingsvelden" (Archives 1928–1952 Missionary Fields)** [690]

Partly owing to the co-operation between the several corporations in 1928 it was decided to make the archives of the associated corporations of the SZC (see no. 687) into one. The archives 1928–1952 are subdivided into 10 headings. The headings are split up into *c.* 1300 subjects. The subjects are classified according to corporation. The whole classification is based upon the UDC. After 1946 the papers concerning the Netherlands were kept separately and the archives "Binnenland" (Interior) were made side by side with the archives "Zendingsvelden". Some subjects in the archives "Zendingsvelden" 1928–1952 concerning the Dutch East Indies/Indonesia are: minutes, reports, rules, statistics (i.a. NZG, UZV, NZV, S en TC, JC, Vereeniging Nederlandsche Zending (Association Dutch Mission)), minutes and reports of various conferences (i.a. Bolaang Mongondow, Deli, East Java, Poso, JC, West Java, South East Celebes, New Guinea, Halmahera, Sangihe and Talaud), missionary work general (Bali, Batak area, Bolaang Mongondow, Buru, Central Celebes, South East Celebes, South West Celebes, Deli, Halmahera, East Java, West Java, Minahasa, New Guinea (entirely, from 1948), North New Guinea, West New Guinea (Bentuni, Berau, Kapauer), Sangihe and Talaud Islands, Java Committee (1931–1936), Zendingsbureau (Mission Bureau), Zendingsconsulaat (Missionary Consulate), transfer to other corporations, women's missionary work, missionary work among the Chinese in East and West Java, evangelization, liturgy etc.), social service work, education, schools, higher schools of divinity, orphanages, medical work, hospitals, propaganda, cocopalm plantations (Halmahera, Buru, Sangihe), financial affairs, personnel affairs (leave, pension, salary, equipment, applications etc.), missionary workers (alphabetically arranged to name).

Entries: Index 1928–1952 "Zendingsvelden" (Missionary Fields); alphabetic indexes for letters received and despatched.

Size: *C.* 20 running metres.

**Vrouwen-Zendingsbond/Vrouwen-Zendingswerk (Women's Mission Union/ Women's Missionary Work), VZB 1929–** [691]

The VZB, later called Vrouwen-Zendingswerk, was established in 1929 with the object to assist the missionary corporations with the training of female missionary workers and with missionary work among women and girls in Java. In 1938 the VZB placed itself as an independent organization at the disposal of the SZC. The archives of the VZB, however, have always been kept separately from the archives 1928–the present.

The archives include minutes 1928–1946 with papers concerning the establishment in June 1929; cash books 1929–1937; rules, surveys 1928–1937; lists of members; correspondence 1947–1955; papers concerning conferences, meetings etc. from 1939; lectures; minutes training committee 1939–1946; papers concerning the training 1930–1941; papers concerning medical work 1948–1950; minutes and other papers of the Nederlandse Zendingshogeschool (Dutch Missionary College); correspondence and other papers concerning New Guinea (i.a. Papoese Huishoudschool (Papuan Housekeeping School)), Minahasa, Celebes, Halmahera; correspondence C.H.C. Gunning.

Size: *C.* 2.50 running metres.

**Zendingspensioenfonds (Mission Pension Fund), 1937–** [692]

The archives of the Zendingspensioenfonds (Mission Pension Fund) too have been kept separately from the archives 1928–the present.

The archives consist of papers 1937–1964, including bank books, journals, ledgers, income and expenditure accounts, correspondence, contracts.

Entry: Alphabetic index.

Size: *C.* 3.50 running metres.

**Archief 1953–1962 "Zendingsvelden", 1946–1962 "Binnenland" (Archives 1953–1962 "Missionary Fields", 1946–1962 "Interior")** [693]

Although this part of the archives, just like the following parts (see no. 694, 695), on the whole is not within the period 1796–1949, nevertheless it is mentioned here, because it is closely connected with the preceding archives (see no. 680–692). In 1951 the Reformed missionary corporations amalgamated, after, in 1942, a more close co-operation had started with the foundation of the Hervormde Zendingsraad (Reformed Missionary Council). June 28th 1951 the Raad voor de Zending der Nederlandse Hervormde Kerk (Council for the Mission of the Reformed Dutch Church) was established. The co-operation, naturally, left its residue in the archives. The subjects were no longer divided according to corporation (see no. 690). In 1953 was decided to limit the number of subjects to *c.* 550. The two parts "Zendingsvelden" and "Binnenland" remained. The part "Binnenland" contains mainly papers concerning the organization in the Netherlands. The classification is in principle the same as the classification of the "Zendingsvelden", with the exception of the period 1948–1950, when the papers were arranged alphabetically according to catchword.

Some subjects from the part "Zendingsvelden" are: rules, minutes, reports etc., personnel, financial affairs, churches (Bolaang Mongondow, Central Celebes, East Celebes, West Celebes, Halmahera, East Java, West Java, Karo-Batak, Minahasa, the Moluccas, Sangihe and Talaud), transfer of missionary work to other corporations or churches (the same districts as above and also Bali, Deli, New Guinea, Sumatra). Kept separately are a series of "personeels- en salarisdossiers 1947–1963" (personnel and salary files 1947–1963) concerning New Guinea, alphabetically arranged to personal names (3 running metres) and a series of correspondence with and about "Papoese jongens en meisjes 1956–1963" (Papuan boys and girls 1956–1963), mainly about possibilities for their training in the Netherlands.

Entries: Index "Zendingsvelden" 1953–1962; indexes "Binnenland" 1946–1948, 1949–1950, 1951–1962; alphabetic index for letters received and despatched.

Size: "Zendingsvelden" *c.* 12 running metres; "Binnenland" *c.* 14.50 running metres.

**Archief 1963–1972 (Archives 1963–1972)** [694]

These archives, the "working archives", are, for the time being, arranged per annum according to subject. It is intended to join the subjects together every ten years (after having destroyed the duplicates) and to transfer the archives from the administrative office to the archives repository.

**Archief Nederlands Nieuw-Guinea (Archives Dutch New Guinea)** [695]

These archives were transferred from New Guinea to the Netherlands in 1963. They include papers 1946–1963 of the Reformed mission and consist of:

a. The archives of the Nederlands Sprekende Gemeente van de Evangelisch Christelijke Kerk (Dutch Speaking Parish of the Evangelical Christian Church), ECK 1950–1963, including: baptismal registers, registers of marriages, registers of church members, minutes of meetings of the clerical boards (Manokwari, Fak Fak, Hollandia, Sarmi, Serui, Sorong, Ifar, Sentani) with accounts of various meetings and with reports.

b. Part of the archives of the Kerkekantoor Hollandia (Church Office Hollandia) 1946–1962, including annual reports, papers of the Raad voor Volksopvoeding (Council for Popular Education), papers concerning politics, adat, agriculture, mission, education, ASB, the South Pacific Conference 1958, correspondence, annual reports 1953–1961 of the "terreinleiding" (local direction).

c. Papers 1948–1963 originating from the "terreinleiding" including reports of tours, papers concerning financial matters, church building at Hollandia, mission, medical work, the Stichting Protestants Christelijk Onderwijs (Foundation Protestant Christian Education) with minutes, decrees and salary papers, the ASB, language policy, the NHM at Makasar, correspondence with the governor, the Zendingsbureau (Mission Bureau), the Raad van Zending (Council for the Mission), H. van Arkel, D.J. Baars, J. Baars, L. Beck, H.J. Brinkman, G.J. Clay, E. Dürkstra, . . . den Dulk, O. Ewoldt, E. Gijsbers, D.A. ten Haaf, A. van Heerden, H.A. van Herwaarden, F.C. Kamma, D. Kapteijn, W.C. Klein, C. Knijf, I.S. Kijne, R.E.H. Marcus, P. Messie, A.M. Middag, E. Osok, H. de Ridder, A. Rigters, F.J.S. Rumainum, L.C. Swaan, H.J. Teutscher.

Entry: Alphabetic catchwords index.

Size: *C.* 2 running metres.

## Private persons [696–698]

### H. Kraemer, May 17th 1888 Amsterdam – November 11th 1965 Driebergen [696]

1906–1910 training at the Nederlandsche Zendingsschool (Dutch Missionary Training School) at Rotterdam; 1911–1921 study of Indonesian literature at Leyden; 1921–1927 in the service of the NBG in Java, travels from Djokjakarta, i.a. to Amboina, Minahasa, Sangihe; 1928 conference Jerusalem; 1929 to India; 1930–1935 in the service of the NBG for the second time to Java (Surakarta), travels to the Batak area, Bali and Borneo; studies the Islam in relation with Christianity, the Sarekat Islam, Indonesian nationalism; in the Netherlands he received the honorary degree of doctor of theology at Utrecht 1936; 1937–1947 professor of the history of the modern non-Christian religions at Leyden; 1938 conference at Tambaram in Ceylon, from there he travels to i.a. Sumatra, Java, South Celebes and Bali in 1939; 1947–1955 director of the Oecumenisch Instituut (Oecumenical Institute) and of the Wereldraad van Kerken (World Council of Churches) (Geneva); 1956–1957 visiting professor Union Theological Seminary at New York. (A.Th. van Leeuwen. *Hendrik Kraemer;* dienaar der wereldkerk. Amsterdam, 1959./C.F. Hallencreutz. *Kraemer towards Tambaram;* a study in Hendrik Kraemer's missionary approach. Upsala, 1966./J. van der Hucht. *Prof. Dr. H. Kraemer, 17 mei 1888 – 11 november 1965;* bibliografie. [Katwijk aan Zee, 1966].)

The collection, also called the "archief NBG" (NBG archives) because of Kraemer's activities with the NBG, includes: deeds of appointment, minutes of the

NBG, Zendingsconsulaat (Missionary Consulate), Nederlandsche Zendingsraad (Dutch Missionary Council), Instituut Kern (Kern Institute); annual reports and accounts GZB (Rantepao mission), West Java mission, East Java mission, travel reports 1928 Amboina, Minahasa, Sangihe, 1935 Borneo (by him and by J.W. Gottin), 1939 Batak area, papers NBG 1923–1942 i.a. concerning his translation of the bible, H.N. van der Tuuk; papers 1930–1940 concerning the Hoge Theologische School (Theological University) at Batavia; the Indische Kerk (East Indies Church) i.a. Makasar, Saleijer; Deli mission (Karo-Batak); Doopsgezinde zending (Mennonite mission) in Java and Sumatra; Java i.a. history, religion, mission, mission among Chinese, agricultural law, Surakarta, Tenger reports 1933, 1934, word lists; Bali i.a. mission, translation of the bible, education of Balinese, desa regulations, minutes, accounts, reports, correspondence; Timor i.a. Timor church, Rotinese; papers Islam and Christianity; Protestant and Roman Catholic mission; education; medical work; nationalism; Indonesian press; adat i.a. Java, Bali, Timor, Amboina, Leitimor, New Guinea; papers concerning J. Held and H.C.H. Hildering; correspondence *c.* 1910–1935 with i.a. the Nederlandse Hervormde Kerk (Reformed Dutch Church), C.W.Th. van Boetzelaer, C. Crommelin, H.C.H. Hildering, . . . Jaffray, A.C. Kruyt, A. Nortier, E. René van Ouwenaller, J. Ritsema, B. Schuurman, D.J. van Wijk; reports etc. concerning the Tambaram conference i.a. of C.J. Marle Davis; papers Indonesia 1945–1949 i.a. nationalism, RTC; university lecture notes; articles by himself; Javanese, Arabic, Malay, Sundanese and Balinese manuscripts; photographs concerning the mission and H. Hiebink Rooker.

Entry: Alphabetic catchwords index.

Size: *C.* 5 running metres.

**Kruyt** [697]

This collection contains papers originating from A.C. Kruyt and his son J. Kruyt jr. The collection is also called "NZG archief" (NZG archives) because both were active for this corporation.

*A.C. Kruyt, October 10th 1869 Modjowarno – January 19th 1949 The Hague*
1890 as missionary for the NZG to Central Celebes (Poso), where he was head of the Normaalschool voor Zendingsonderwijzers (Training College for Missionary Teachers) at Pendolo from 1913–1929; travelled through the archipelago to study animism from 1905–1906 and 1920–1921; returns to the Netherlands after he retired in 1932; 1913 honorary degree of doctor of theology of the university of Utrecht; correspondent of the Koninklijke Akademie van Wetenschappen (Royal Academy of Sciences) in the East Indies.

*J. Kruyt jr.*
1917 as missionary for the NZG to Central Celebes (Kora Walelo); 1924 Pendolo, 1925 Kora Walelo; after his leave head of the Normaalschool voor Zendingsonderwijzers at Pendolo from 1930–1939, Tentena 1940–1942; he was also chairman-secretary of the Conferentie van Zendelingen (Conference of Missionaries). (Supplements *RAvNI*).

The collection consists of diaries of A.C. Kruyt 1914–1922, reports and correspondence of his journeys through the Batak area and Central Celebes (s.a., 1924, 1930), papers concerning the [P.] Schuyt-Kruyt affair; accounts, reports, minutes, statistical data, bills etc. 1831–1953 originating from A.C. Kruyt and J. Kruyt jr i.a. concerning schools (education, gurus, HIS, training of missionaries etc.), hospitals, language matters (i.a. Baree), adat, relations with Binnenlands Bestuur (East Indies Civil Service), Handelsonderneming Midden-Celebes (Trading Concern Central Celebes), Native Christians, history of the period of occupation 1940–1945, the provinces Luwuk, Makasar, Menado, Poso, Tentena, conferences Central Celebes and East Java; letters 1890–1940 from and/or to the central committee, Zendingsconsulaat (Missionary Consulate), Leger des Heils (Salvation Army), GZB, N. Adriani, M.L. Adriani-Gunning, [C.] Albers, [M.J.] van Baärda, [H.B.] de la Bassecourt Caen, N. van den Bent, . . . van den Berg, . . . Beukers, C.W.Th. van Boetzelaer, J. Brokken, . . . Diekerhof, E. Dijkhuis, [W.G.] van Eelen, . . . Engelberg, S.J. Esser, A. van der Flier, [J.] Fortgens, E. Fries (at Nias), [E.] Gobée, J.F. van Hasselt, F.W. Hering, J. Held, [R.] Hofman, [A.] Hueting, R. Klodic, A.C. Kooistra, H. Kraemer, A.C. Kruyt, J. Kruyt sr. (at Modjowarno), J. Kruyt jr, J.H. Kruyt-Moulijn, J. Kruyt-Oosterlee, . . . Kubler, [A.A.] van de Loosdrecht, G. Maan, . . . Maree, [J.H.] Meerwaldt, . . . Najoan, [J.H.] Neumann, A. Nortier, . . . van Ossenbruggen, H. Perdok, K. Riedel, J. Ritseman, . . . Sarasin, J. Schut, [P.] Schuyt, [C.] Snouck Hurgronje, H.J. Spreeuwenburg, . . . van der Torre, R. Veldhuis, J. Warneck, J.H. Wesseldijk, . . . Winkler, J. Woensdrecht, H. Zuppinger.

Entry: Survey of the archives.

Size: *C.* 3.20 running metres.

**I.S. Kijne, May 1st 1899 Vlaardingen – March 12th 1970 Leyden** [698]

As missionary for the UZV to North New Guinea, from 1923–1926 at Mansinam and from 1926–1958 at Miei, where he was head of the Normaalschool voor Zendingsonderwijzers (Training College for Missionary Teachers); during the war he was interned in Sumatra. (Supplements *RAvNI*).

The collection, also called UZV archives, consists of the papers drawn up by Kijne in connection with his work in New Guinea. His notes and studies concerning New Guinea languages have been given to the KITLV.

The collection has no inventory yet.

Size: *C.* 2 running metres.

## Photographs [699]

Besides the photographs and maps, which the several archives contain and which are to be found by means of the indexes 1797–the present, a separate collection of *c.* 12,000 photographs is present at the Zendingsbureau (Mission Bureau). Of these *c.* 11,000 are pertaining to the Dutch East Indies/Indonesia. They have been subdivided into the following headings:

a. Portraits, *c.* 1200;

b. School work (general, Bali, Halmahera, Central Java, East Java, New Guinea, Sangihe), *c.* 1800;

c. Parish work (Amboina, Bali, Batu Islands, Bolaang Mongondow, Borneo, Buru, Central Celebes, West Celebes, South and East Celebes, Deli, Halmahera, Central Java, East Java, West Java, Kangean, Minahasa, Moluccas, New Guinea, Sangihe- and Talaud Islands, Tapanuli, Timor), *c.* 3000;

d. Medical work (general, Bali, East Java, West Java, New Guinea), *c.* 1200;

e. Geography and Anthropology (Amboina, Bali, Batu Islands, Borneo, Buru, Celebes, Deli, Halmahera, Java (general), Central Java, Madura, New Guinea, Sangihe, Sumatra, Sumba, Lesser Sunda Islands, Timor), *c.* 3500.

Entry: Alphabetic catchwords index.

## Provincialaat Paters der HH Harten (Religious Order of the Fathers of the Holy Hearts), 1800

26 Seminarieweg, 4854 PB Bavel.
Admission by appointment.
Research finished in 1979.

**Archives of the Paters der HH Harten (Fathers of the Holy Hearts), 1800–** [700]

1800 foundation of the congregation at Paris; 1827 first missionaries to Hawaii; 1834 mission in East Oceania: Marquesas Islands, Tahiti (Gambier Islands); 1849 mission in the Tuamotu archipelago; 1863 Easter Island; 1893 Cook Islands; 1896 first house in the Netherlands (Grave); 1923 establishment of a Dutch province of the Paters der HH Harten, mission in Banka, Billiton and the Riouw archipelago, first missionaries: Th. Herkenrath, M. van Soest, I. Meyer, G. Jeanson and A. Bruyns; 1940 Cook Islands entrusted to the Dutch province; 1948 mission in New Zealand.

The archives are the residue of the activities of the congregation both in the Netherlands and in the mission territories mainly from 1914, though there is also documentation present of the previous period. The archives also include a collection of photographs (*c.* 5 running metres) and a number of periodicals and works of members of the congregation (*c.* 25 running metres). For the history of Asia and Oceania are relevant, besides the more general papers and series such as regulations, circular letters etc., especially the correspondence and other papers pertaining to the mission territories outside Europe, which have been kept separate, viz. Indonesia/Singapore (correspondence with V. Bouma, M. van Soest, G. van der Westen, E. Tromp, W. van Soest, J. van Berkom, W. de Bruin, B. Renckes, C. Molenkamp, H.J. Peters, I. Meyer). Banka, where were active i.a. V. Bouma, Th. Mul, B. Bakker, A. Nijssen, A. Mars, D. van Gorp, L. van

Gelder, P. van der Knaap, P. Nieuwe Weme, P. Heuver; Hawaii, Cook Islands, New Zealand and Australia. Moreover the archives contain papers about Brazil and Mozambique, documentation about Pater Damiaan (Molokai), correspondence of the Indische Missie-vereeniging (East Indies Mission Association) from 1940 and papers about mission exhibitions.

Size: The total size is *c.* 100 running metres, on Indonesia *c.* 0.50 running metre, on Oceania 0.20 running metre and on Australia and New Zealand *c.* 0.10 running metre.

Permission is necessary for consultation of the papers.

**Photographs** [701]

Though the collection of photographs is part of the archives, it is described here as a separate collection. The collection consists of geographically arranged photographs about Europe, Cook Islands, Singapore, New Zealand, Chatham Islands, Brazil, Belgian Congo, Banka (Belinju, Dhoen Thong, Muntok, Pangkal Pinang, Sambong, Sungeiliat), Billiton (Tandjung Pandjang, Manggar), Riouw and Lingga (Moro, Tandjung Balei, Tandjung Pinang), Mozambique, Oceania (Hawaii, Gambier, Marquesas, Easter Island, Tahiti, Morea, Tuamotu, Tubuai), United States, albums of Th. Mul, M. van Soest, W.M. van der Brugt and C. Molenkamp.

Size: *C.* 5 running metres.

## Congregatie der Priesters van het Heilig Hart van Jesus (Congregation of the Priests of the Holy Heart of Jesus), SCJ 1878

54 Duivelsbruglaan, 4835 JH Breda.
Admission by appointment.
Research finished in 1979.

**Archives, 1878–** [702]

In 1878 the SCJ was established by father L. Dehon. In 1900 the first settlement in the Netherlands was started. Besides in Europe and North America, in Zaïre and Kamerun, in Chile, Argentina, Uruguay and Brazil the congregation has been active in South Sumatra (dioceses of Palembang and Tandjung Sakti).

The archives are the residue of the activities within and outside Europe. They include surveys, reports and correspondence about Palembang and surrounding districts. Archival documents about the missionary activities of the congregation are also to be found in the Missieprocuur (mission office) at Cadier en Keer.

Size: *C.* 120 running metres.

Permission is necessary for consultation of the papers.

Owing to the preparation of an inventory, the archives are not open for the public for some time.

See also *Het Rijk van het Heilig Hart van Jesus*. 1901–1968.

*Missiestemmen*. 1920–1938.

*De Missiepost*. 1925–1950.

*Missiealmanak*. 1927–1959.

*Katholiek leven bisdom Palembang*. 1927–1940.

*Oogstvreugd*. 1933–1960.

*Annalen SCJ;* Nederlandse Provincie. 1945–1967.

## Provincialaat Missionarissen van de H Familie (Religious Order Missionaries of the H Family), MSF 1895

209 Tilburgseweg, 5051 AG Goirle.
Admission by appointment.
Research finished in 1979.

**Archives of the Kongregatie van de Missionarissen van de H Familie (Congregation of the Missionaries of the H Family), 1895–** [703]

1895 foundation of the congregation by J.P. Berthier with the object of establishing Apostolic schools for late vocations of boys of 14 till 24 years old to train them for priests missionaries; later schools in Belgium, France and Germany; 1910 mission in Brazil; 1926 taking over of the mission territory of the Capuchins in the Southern and Eastern Districts of Borneo, mission province in Laham (Mahakam), from 1930 also mission stations at Samarinda, Balikpapan, Banjarmasin and Tarakan; 1938 Apostolische Prefectuur (Apostolic Prefecture) Banjarmasin, first prefect J. Kusters; 1949 Apostolic Prefecture erected into a Apostolisch Vicariaat (Apostolic Vicariate), J. Groen first vicar; 1955 divided into the Apostolic Vicariates Banjarmasin and Samarinda; 1961 the vicariates Banjarmasin and Samarinda erected into a diocese. From 1932 mission in Semarang in co-operation with the Jesuits. Apart from Indonesia the missionaries of the Dutch province of the H Family have also been active in Chile and Argentina.

The archives are the residue of the activities from 1895 within and outside the Netherlands. As the archives are still in course of preparation, it is impossible to find out which series and subjects they include.

Size unknown for the time being.

Owing to the preparation of the archives there is no possibility of consultation for some time.

See also *Bode van de H Familie.* 1907–1966.

*Jaarboek voor het Katholiek huisgezin.* 1949–1959.

F. Nolte. *Historische Skizze der Kongregation der Missionäre von der Heilige Familie. C.* 1950. 5 vol.

Forthcoming. M. Coomans. *Adat, kultuurverandering en evangelisatie;* geschiedenis van het bisdom Samarinda en onderzoek naar de verhouding tussen verkondiging en de socio-kulturele veranderingen onder de Daya-stammen van Oost Kalimantan na 1900. Thesis.

## Bisdom Haarlem (Diocese of Haarlem)

80 Nieuwe Gracht, 2011 NJ Haarlem.
Admission by appointment.
Research finished in 1979.

**Archives** [704]

The archives of the diocese contain correspondence with Indonesia 1855–1921, 1953, 1973; moreover Herderlijke brieven van de Apostolisch Vicaris van Batavia (Pastoral letters from the Apostolic Vicar of Batavia) 1894–1930.

## Abdij van Berne (Abbey of Berne), 1134

49 Abdijstraat, 5473 AD Heeswijk-Dinther.
Admission by appointment.
Research finished in 1979.

**Norbertijnen (Praemonstratensen) van de Abdij van Berne (Premonstratensians of the Abbey of Berne), 1920–** [705]

The community was established at Berne in 1134, not active in public after the reformation; 1857 restoration of abbey life at Heeswijk; from 1893 mission in North America; from 1920 unsuccessful attempts at mission work in the Dutch East Indies; from 1929 finally mission work in India, viz. in Madras (Phirangipuram, Guntur, Jabalpur), first bishop from 1933 was C. Dubbelman.

The archives consist of correspondence, biographies of the *c.* 50 missionaries who worked in India and photographs.

Size: The mission part is *c.* 0.50 running metre.

# Orde der Minderbroeders Kapucijnen (Order of the Franciscan Friars Capuchins), OFM CAP 1609

12 Van der Does de Willeboissingel, 5211 CB 's-Hertogenbosch.
Admission by appointment.
Research finished in 1979.

**"Archivum Capuccinorum Hollandiae", 1882–** [706]

The Minderbroeders-Kapucijnen established themselves in the Netherlands in 1609; 1882 independent province of the order; 1905 apostolic prefecture established in Borneo (taken over from the Jesuits), erected into the apostolic vicariate Pontianak in 1918, 1926 Southern and Eastern District handed over to the missionaries of the H Family, 1939 Sintang handed over to the Montfortanen, 1946 Ketapang handed over to the Passionisten (Passionists), 1961 Pontianak archbishopric; 1911 apostolic prefecture established in Sumatra (taken over from the Jesuits), 1923 prefecture Padang for the Capuchins, Banka, Billiton and Riouw for the Paters van de HH Harten (Fathers of the H Hearts), Benkulen (later Palembang) for the Priesters van het H Hart (Priests of the H Heart), 1932 Padang erected into a vicariate, 1941 transfer of the see of the vicariate from Padang to Medan, 1952 Padang transferred as a separate prefecture to the Italian Societeit van Franciscus Xaverius (Society of Franciscus Xaverius) at Parma, 1959 South Tapanuli and Nias (capital Sibolga) transferred as a separate prefecture to the German Capuchins, 1961 Medan archbishopric; 1936–1965 active with the Centraal Missie Bureau (Central Mission Bureau), from 1966 the Capuchins are in charge of a parish in Jakarta.

The archives consist of the residue of the activities within and out of the Netherlands from 1882, to be distinguished into: generalia (general administration, *c.* 2 running metres); provincialia (administration province, study, training, activities within and out of the Netherlands (except mission), *c.* 35 running metres); conventualia (papers pertaining to the 21 monasteries in the Netherlands, *c.* 27 running metres); missionalia (general 1882–1954, to and from the mission 1905–1974, visitations 1911–1971, agreements with other congregations 1906–1957, apportion of goods between Superior Regularis and Apostolic Vicar (Borneo 1934–1953, Sumatra 1938–1951), foundation Domus religiosae (Borneo 1936–1953, Sumatra 1938–1952), papers Borneo (correspondence 1905–1969, reports, statistics etc. 1909–1960, special works (junior technical schools 1931–1955, education 1934, 1947–1952, klein seminarie (preparatory seminary) 1932–1957, hospitals 1953, 1977), Stations of the Cross 1905–1961, maps 1930–1951, photographs, chronicle 1899–1951, war reports 1941–1946), papers Sumatra (correspondence 1909–1968, reports, statistics etc. 1913–1972, special works (junior technical school 1938-1950, education 1929–1950, klein- en groot seminarie (preparatory seminary and seminary) 1948–1952, Maria legion 1953), Stations of the Cross 1912–1954, maps 1940–1950, photographs, chronicle Sumatra mission 1912–1950, chronicle Banka-Billiton

1849, 1912, 1924, war reports 1941–1946), publicity (i.a. *Borneo-almanak* 1905–1960 and Sint Fidelis Missiebond (Saint Fidelis Mission Association) 1906–1960, economy 1905–1967, mission course and mission activities 1918–1964, Centraal Missie Bureau (Central Mission Bureau) 1931–1963, Indische Missie Vereniging (East Indies Mission Association) 1908–1966, general mission institutions (reproduction, Memisa, Raptim etc.), Gedelegeerd Missie Comité (Delegate Mission Committee) 1944–1946, Comité Uitzending Missionarissen (Committee Sending Out of Missionaries) 1932–1946, Commissie van Adviseurs inzake Missiebelangen bij de Souvereiniteitsoverdracht (Advisory Committee concerning Mission Interests at the Transfer of Sovereignty) 1946–1950, size missionalia *c.* 8 running metres); personalia (papers concerning i.a. the 178 Dutch Capuchins who were active in Indonesia, in total *c.* 55 running metres).

Size: In total *c.* 135 running metres.

Permission is necessary for consultation of the papers.

See also *Oud Massaia.* 1938–1966. (Included are many letters from missionaries).

*Zondagsblad voor het Katholieke huisgezin.* 1898– . From 1922 continued under the title: "De Missie-illustratie". From 1929 continued under the title: "Christus Koning". From 1945 continued under the title: "Katholiek volk". From 1947 continued under the title: "Volksweekblad". From 1960–1963 continued under the title: "Rotonde".

*Borneo-almanak.* 's-Hertogenbosch, 1911–1961.

*Sint Fidelisklokje;* gratis orgaan voor de leden van het Sepharijns werk der heilige missen (St. Fidelis Missiebond), Velp, 1922–1960.

*Met kap en koord;* orgaan van de Nederlandse Minderbroeders Kapucijnen. Eindhoven, 1961– .

Gentilis Aster. *Morgen is mijn eiland rijk;* vijftig jaar Capucijnenmissie op Borneo. Tilburg, 1955.

Gentilis Aster. *God kapt zich een weg door de bossen.* Voorhout, 1957.

Gentilis Aster. *Een volk ontdekt Christus;* de Katholieke missie onder de Bataks op Sumatra. Voorhout, 1959.

Gentilis Aster. *Wij hebben andere belangen;* vijftig jaar Kapucijnenmissie op Sumatra. Tilburg, 1961.

**Other papers** [707]

With the Missieprocuur (Mission office) of the Capuchins (165 Korvelseweg, Tilburg) are deposited collections of maps, drawings, photographs and slides i.a. about Borneo, Sumatra, Tanzania and Chile. The objects collected by the Capuchins from 1905 have been given in loan to the Volkenkundig Museum (Ethnographical Museum), in Tilburg, 24 Kloosterstraat.

# Archiefbewaarplaats van de Nederlandse Provincie der Jezuieten (Archives Repository of the Dutch Province of the Jesuits), SJ 1849

2 Houtlaan, 6500 GV Nijmegen.
Admission by appointment.
Research finished in 1979.

## Introduction [708]

In the archives repository are deposited, besides the archives of the actual Province, also the archives of abolished houses and organizations (i.a. the archives of the weekly magazine *De Linie*) of the Jesuits. In addition a collection of documents has been started. As the Provincialaat (Office of the Provincial) functions as archives centre for the Jesuits, an inventory of the archives of the Sint Claverbond (Saint Claver League) is included below, see no. 711.

See also J. Barten. Het archief van de Nederlandse Provincie der Jezuïeten. In *Jaarboek van het Katholiek Documentatie Centrum, Nijmegen*, 1973, p. 180–184.

## Archives and archival material [709–711]

### Archives of the Nederlandse Provincie der Jezuïeten (Dutch Province of the Jesuits), 1849– [709]

1592–1773 mission of Dutch Jesuits; 1805 activities in the Netherlands resumed; 1849 foundation of the Nederlandse Provincie der Jezuïeten (Dutch Province of Jesuits); from the 16th century forward missionary activities in Asia; 1808 first apostolic prefecture in Batavia; 1859 first Jesuits appointed for the East Indies Mission; the Jesuits had ecclesiastical places in Java (Ambarawa, Bandung, Bara, Batavia, Buitenzorg, Cheribon, Djokjakarta, Klaten, Madiun, Magelang, Malang, Mendut, Muntilan, Ungaran, Salatiga, Sedaju, Semarang, Sukabumi, Surabaja, Surakarta), Sumatra (Kuta Radja, Medan, Padang, Sungeislan, Tandjung Sakti), Borneo (Singkawang, Smitau), Celebes (Makasar, Menado, Tomohon, Woloan, Kendari Baai), Flores (Koting, Larantuka, Lela, Maumere, Sika), Sumba (Laora), Timor (Atapupu, Lahurus) and the Moluccas (Kei, Bomfia, Kapaur, Kessewooi). (*Chronologisch overzicht van de werkzaamheid der Jezuïeten in de missie van N. O–I.;* bij den 75sten verjaardag van hun aankomst in de nieuwe missie, 1859–9 Juli 1934; samengesteld door A.I. van Aernsbergen. Bandung etc., 1934).

The archives are the residue of the activities from 1849. They consist of 6 headings, viz. actual administration (with preliminary papers from 1847), files of special subjects, administrative forms, separate houses, litterae annuae (historia domus, necrologia), overseas missions. The last heading includes travel reports of missionaries, letters, reports and other papers by and about F.C. Heijnen, M. van den Elzen, P.M. Vrancken, J. Lijnen, A.C. Claessens, G. Metz, J. de Vries,

G. Wenneker, P.J. van Santen, C.J.F. le Cocq d'Armandville, Zusters van Tilburg (Sisters of Tilburg) at Padang, Broeders van St. Louis (Brothers of St. Louis) at Surabaja, Indische Missie-Vereeniging (East Indies Mission Association), J. Mertens, F. Fleerakkers, J. Mulder, A.Th. van Hoof, J.G.A. van Baal, J.F.M. Sträter, J.J. van Rijckevorsel, E.E.A. Engbers, J. Hoeberechts, G.A.M. Schmedding, A.J.A. van Kalken, C.J.E. de Quay, Canisius College at Batavia, school statistics (1926–1940), Shanghai, Australia, KNIL chaplains (1945–1946), Sint Claverbond (St. Claver League), data about various buildings, I. Vogels, P. Willekens, C. Kolfschoten, E.S. Luypen, A.P.F. van Velsen, Zambesi mission, Philippines, Comité Uitzending Missionarissen (CUM, Committee for Sending out of Missionaries), Gedelegeerd Missie Comité (GMC, Deputed Mission Committee), Vereeniging Missionarissen (Association Missionaries), Centraal Katholiek Koloniaal Bureau (Central Catholic Colonial Bureau), Centraal Missiecommissariaat (Central Mission Commissariat).

Entry: Inventory in typescript.

Size: *C.* 20 running metres.

Permission is necessary for consultation of the papers.

**Collection of documents** [710]

This collection consists of papers collected as documentation for the activities of the Provincialaat. The collection is subdivided into headings. For the history of Asia and Oceania are relevant the headings AD (overseas missions, except Indonesia) and I (Indonesian mission).

a. Overseas missions except Indonesia. The greater part of this heading consists of papers before 1796 and concern mission in China, Siam, Japan, Agram (India), South America, the Cape of Good Hope, Curaçao, Surinam, North America, Zambesi, Madagascar, Celebes, Philippines and Syria.

b. Indonesian mission. This heading consists of printed and manuscript sources. Archival documents are present concerning the expulsion of J. Grooff from the East Indies in 1846, travel accounts and correspondence (J.D. Escherich 1845, A.N. Heuvels 1844, 1845), "officiële stukken" (official papers) (M. van den Elzen, J.B. Palinckx, J.F. van der Hagen, G. Metz, A.G. Terwindt, J. Meijer, A. Geelen, B. Mutsaers, C.W.J. Boelen, J. Hendrichs, C.W.J. Wenneker, J. Kraaijvanger, W.J. Staal, H. van Rijckevorsel, P.J. van Santen, J.J.H. Roebroek), treaty with Japan 1856, Timor 1859, Atjeh 1824–1873, J.H. Muller, school in Tomohon 1903, building of the cathedral in Batavia, training of missionaries, Dutch-Native teachers' seminaries, conference in Muntilan 1929, Centraal Missiebureau (Central Mission Bureau) 1933–1939, Flores 1873, J. de Vries SJ, A.J. Asselbergs, F.G.J. van Lith, J.B. Prennthaler, papers originating from G.D.A. Jonckbloet (photographs and correspondence of Noto Soeroto, H. Thans, Multatuli), H.C. Verbraak (Padang, . . . de Wolf at Medan, Atjeh), A. van der Velden 1882–1918, F.C. Heijnen, J.W.M. Timmers, correspondence F.H. de Bruyn (Ambarawa, Surabaja, Semarang, Salatiga 1865–1894), J.B. Palinckx (Djokja, Surabaja 1890–1894), Th.J. Peters (Batavia, Sindanglaja 1870), J. Hendrichs (Batavia, diary 1871–1879), J. van Meurs (Batavia, Padang, Djokja, Surabaja 1871–1873), C.W.J. Wenneker (Deli, Surabaja, Weltevreden 1880–1896), A. Kortenhorst (Batavia, Weltevreden, Muntok 1890–1895), letters from Batavia (G.D.A. Jonckbloet, A.P.M. Dijkmans, J. Lancée 1890–1891), letters from Semarang (H. Fisscher, P. Neyboer, J. Keller, J. Hoevenaars, L. Hebrans, Th. van Hout 1891–1896), W. van den Heuvel (Batavia, Cheribon, Bandung, Lahat 1891–1896), J.D. Opdenkamp (Semarang, Padang 1892–1896), Chr. Schets (Cheribon, Weltevreden 1894–1896), P.H. Diederen (Surabaja, Semarang 1895–1896), F. Voogel (Magelang, Bandjar Negara 1890–1895), A.P. Smit (Padang 1873, 1890–1892), H.C. Verbraak (Padang, Pantai Perak 1872–1895), Th. van Swieten (Padang, Atapupu 1891–1892), J. van Meurs (Tandjung Sakti 1888–1891), W.L. Jennissen (Tandjung Sakti 1890–1891), H. Schräder (Singkawang, Pontianak, Tandjung Padang 1893–1896), H. Looymans (Smitau, Singkawang, Udjung Aru Sedjiram with photographs 1890–1896), H. Mulder (Sebruang Smitau, Nanga Sedjiram, Larantuka 1893–1896), J. van Meurs (Surabaja, Menado 1880–1884), F.C. Heijnen (Minahasa 1881), A.H.J.H. Bolsius (Tomohon 1891–1893), B. Mutsaers (Menado 1892), J. Onel (Menado 1896), documentation (C.J.F. le Cocq d'Armandville, G. de Groot, P.J. van Santen, J. de Vries, P.M. Bonnike, T. Heslinga, J. Kraaijvanger, W.J. Staal, J.B. Palinckx, E.S. Luypen, A.P.F. van Velsen, L.C.M. Weve), report on the Japanese invasion 1942, comments on Tao Teh King (Lao Tsze), travel report of J. de Vries to Flores and Timor, letters from Th. Kolfschoten i.a. about M. Cohen as deserter executed in Atjeh, G. Creutz Lechleitner (Saint Melania Work, Muntilan, communism, Javanese language, art, culture, dancing, Raden Mas Waloejo), schools in Klaten, Djokja, Batavia, Muntilan, travel account by W.R. van Hoëvell in East Java 1847 with drawings and photographs (copy), papers about the internment camp at Bandung, G. Vriens, J.C.J. Burgers, J. Creijghton, D.C. Wammes, C.H.J.C.M. Doumen, L.A.C. Wevers, A.E. Soegijapranata, P. Willekens, J.J. ten Berge.

Entry: Inventory in typescript.

Size: *C.* 50 running metres (heading AD *c.* 0.60 running metre, heading I *c.* 3 running metres).

Permission is necessary for consultation of the papers.

**Sint Claverbond (Saint Claver League), 1889–** [711]

The archives are deposited at the foundation, address 7 Graafseweg, Nijmegen.

The Sint Claverbond (Saint Claver League) originated from the Herencongregatie van het Jezuïetencollege (Gentlemen's Congregation of the Jesuit College) of Maastricht; 1880 independent section called "Roomsch Katholieke Missievereeniging" (Roman Catholic Mission Association), with the object of collecting contributions for the Jesuit mission in the East Indies; 1889 establishment of the national association Sint Claverbond belonging to the Provincialaat (office of the

Provincial) of the Jesuits, publication of the periodical *Berichten uit Nederlandsch Oost-Indië;* 1925 reorganization of the association into Stichting 'Sint Claverbond' (Foundation 'Saint Claver League'), meanwhile expansion of the activities: publication *Algemeene Missiekalender* and collections for the benefit of the Java mission called *Het Liefdewerk de Javabus.*

The archives are the residue of the activities from 1889 and have been subdivided into 6 headings, viz. general (organization, rules, correspondence with the Holy See, the Generalaat (Office of the General) and the Provincialaat (Office of the Provincial), A.J.A. van Kalken, C.J.E. de Quay, F.G.J. van Lith, J. Grooff, J.Th.F.M. Noijons, A.E. Soegijapranata, P.Ch. Soetapanitra, L.M. van Rijckevorsel, P. Willekens, documentation about i.a. the Chinese and mission), papers propaganda (correspondence, lectures, Nijmeegsch Missie-Comité (Nijmegen Mission Committee), Java-club), papers of the editorial and publishing offices of the periodical, papers about various projects and actions (mission museum, Pantjasila-beweging (Pantjasila movement)), Indonesian students, Sanata Dharma, Loyola College Semarang, schools in Java, All Asian Seminar Madras, Propaganda Fide, Priester Missie-Bond (Priests Mission League), miscellanies (i.a. geography and ethnology, Javanese songs, statistics), financial papers.

Entry: Inventory in typescript.

Size: *C.* 8 running metres.

Permission is necessary for consultation of the papers.

### Photographs [712]

The collection consists of an unknown number of photographs about the activities of the Jesuits, including portraits of missionaries.

## Paters Montfortanen (Fathers Montfortans), SMM 1705

12 Montfortlaan, 5688 NM Oirschot.
Admission by appointment.
Research finished in 1979.

### Archives Paters Montfortanen (Fathers Montfortans), 1705– [713]

The congregation was established in 1705; 1905 Dutch province under French provincial; 1922 first Dutch provincial; in addition to Haiti (1871), Denmark (1901), Iceland (1903), Malawi (1903), Columbia (1904), Mozambique (1922), Zaire (1933), Brazil (1967), the Montfortans were active in mission from 1939 in Sintang (West Borneo), after having made a first effort to this end in 1924.

The archives are the residue of the activities of the Montfortans within and out of the Netherlands from 1881 and consist of official and private correspondence, reports and other papers, maps, photographs, printed matter.

Size: *C.* 25 running metres, of which the part about Sintang *c.* 0.50 running metre.

Permission is necessary for consultation of the papers.

See also *Onze Missionarissen.* 1909– . From 1935 continued under the title: "Scepter en Kruis". From 1947 continued under the title: "Alles voor allen". From 1965 continued under the title: "Orante". *Montfortaanse Almanak.* 1936–1957.

## St. Joseph's Missionary Society of Mill Hill, 1866

15 Boerhaavelaan, 4708 AE Roosendaal.
Admission by appointment.
Research finished in 1979.

**St. Joseph's Missionary Society of Mill Hill, 1866–** [714]

The congregation was established by H. Vaughan in Mill Hill in 1866 with the sole object of doing missionary work; 1890 establishment of the Mission House in Roosendaal; from 1875 mission in India (Madras 1875, Nellore 1928); from 1887 Kashmir and Kafirstan (Rawalpindi – "Pakistan"); from 1881 Labuan and North Borneo (Serawak); from 1905 Philippines; from 1886 New Zealand (Maoris); also from the end of the 19th century mission in Uganda, Kenia, Belgian Kongo ("Zaire"), Camerun, Sudan, Falkland Islands, Peru, Chile, Brazil; after 1950 also mission in Malaysia. (*Emigrants for God;* the story of the Mill Hill Fathers. London, 1962).

The archives are the residue of the activities within and out of the Netherlands, and include reports, correspondence, scientific publications, articles, diaries and personal notes. Data about all the territories mentioned above are present in the archives.

Inventory is forthcoming.

Size: *C.* 10 running metres.

Permission is necessary for consultation of the papers.

See also *Annalen van het Missiehuis te Roosendaal;* vol. 1–74. 1890–1967. From 1921(vol. 32) continued under the title: "Annalen der St. Joseph's Congregatie van Mill Hill". From 1936(vol. 47) continued under the title: "Mill Hill; (maandblad van de St. Joseph's Congregatie)".

**F.J.M. Westerwoudt, March 14th 1861 Amsterdam – April 13th 1898 Kuching** [715]

1881 study of theology at Warmond; 1883 received holy orders, admitted into the St. Joseph's Missionary Society; 1885–1898 missionary in Serawak, mission among the Dajaks. (E. v[an] R[ijckevorsel-] W[esterwoudt]. *Een Nederlandsch Missionaris;* Father Felix Westerwoudt; 3rd ed. Tilburg, 1921).

The collection consists of private papers and possessions collected in connection with plans for his canonization.

Size: *C.* 1 running metre.

Permission is necessary for consultation of the papers.

**Photographs** [716]

The collection has been geographically arranged and consists of material about Africa, New Zealand (3 albums), South America, India (16 albums), Rawalpindi (8 albums), Philippines (7 albums), Borneo (9 albums), Asia general (1 album), mission general (13 albums), in addition a collection of portraits of members of the congregation and about forty films about i.a. Kashmir, Tahiti, Nellore, Nove Ziena, Guntur, China, India, Islamic marriages, Madras, Bombay, Pakistan, Rawalpindi, Islam in Africa and Asia.

## Congregatie van de Fraters van Onze Lieve Vrouw, Moeder van Barmhartigheid Tilburg (Congregation of the Brothers of our Lady, Mother of Charity Tilburg), 1844

54 Gasthuisring, 5041 DT Tilburg.
Admission by appointment.
Research finished in 1979.

**Library** [717]

90,000 books and an unknown number of periodicals.

Directed to history and pedagogics.

**Publications** [718]

*Mededeelingen*. 1932– . From 1952 continued under the title: "Congregatie-nieuws". From 1958 continued under the title: "Mededelingen". From 1968 continued under the title: "Korte berichten". From 1976 continued under the title: "Berichten".

*Verleden en heden*. 1937– . From 1958 continued under the title: "Ontmoetingen".

**Archives, 1844–** [719]

The congregatie van de Fraters van Tilburg (Congregation of the Brothers of Tilburg) was established in 1844 and approved by the Pope as a religious institution in 1861; 1886 first missionary station in the Antilles; 1902 Surinam; 1924 Indonesia (1924 Menado and Tomohon, 1923–31 dec. 1976 Padang, 1927 Medan, 1934 Makasar, 1950 Balige, 1968 Sitoli, 1977 Langgur); 1958 Kenya, 1958–1975 Zaire, 1959 South West Africa; 1960 Brazil.

The archives are the residue of the activities within and out of the Netherlands; the following headings are to be found in the archives: generalia, regionalia, localia, personalia, photograph archives, remaining files; the headings have been subdivided, if possible, into personnel lists, chronicle, photographs and papers about the history; for the history of Asia are relevant the papers about the activities in Celebes and Sumatra.

Size: *C.* 250 running metres, of which *c.* 6 running metres especially concerning Indonesia, *c.* 2 running metres regionalia and *c.* 4 running metres localia.

Permission is necessary for consultation of the papers.

## Missiehuis Tilburg (Roman Catholic Mission House Tilburg), 1882

204 Bredase Weg, 5038 NK Tilburg.
Admission and consultation by appointment. Lending out (except of archival material).
Research finished in 1970.

**Library** [720]

25,000 books; an unknown number of periodicals, mainly concerning the mission, about ten concerning the Dutch East Indies/Indonesia. Directed to humanistic studies, practical theology, mission.

**Publications** [721]

*Annalen van Onze Lieve Vrouw van het Heilig Hart.* 1882–1958. From 1958–1968 continued under the title: "Rerum ecclesia; MSC-annalen van Onze Lieve Vrouw van het Heilig Hart". The publication of this periodical was discontinued in 1969.

**Archief van de Missionarissen van het Heilig Hart (Archives of the Missionaries of the Holy Heart), MSC 1882–** [722]

In 1882 the missionaries settled in Tilburg, December 22nd 1902 the apostolische prefectuur Nederlands Nieuw-Guinea (apostolic prefecture Dutch New Guinea) was established (New Guinea, Kei-, Tanimbar-, Aru-Islands, Ceram, Banda, Amboina, Halmahera and the neighbouring islands), which area was taken over from the Jesuits. February 13th 1903 M. Neijens was appointed to be the first prefect, with H. Geurtjens as fellow worker. The same year they established themselves in the Kei Islands. Langgur became the central and head station.

In 1905 the missionaries settled in South New Guinea, where Merauke became the head station. From 1910 the MSC was active in the Tanimbar Islands. The second apostolic prefecture of the MSC in the Dutch East Indies was Celebes. In 1919 they took this area over from the Jesuits. The head stations there included

Menado from 1919 and Tomohon from 1924. In 1924 the MSC took over from the Jesuits the area Central Java, where in 1932 the prefecture Purwokerto was established. And finally from 1931 the MSC did mission work among the Chinese in Batavia. Besides in mission work the MSC was also engaged in i.a. technical education (carpentry, building of houses etc. by missionaries), guidance to the female population (by mission sisters) and medical work (also by mission sisters). From 1908 the MSC was active in the Philippines, respectively in Surigao, Lipa 1914, Manilla 1918, Mactan (Opon and Cordoba) 1929, Agusan 1935, Nueva-Ecija 1937. From Germany missionaries were sent to Rabaul from 1889. The MSC has been active in Australia from 1885. (*Een kwart-eeuw apostolaat;* gedenkboek bij het zilveren jubileum van de missie in Nederlandsch-Nieuw-Guinea en de Molukken, 1903–1928. Tilburg, [1928]./*MSC over de wereld, 1854–1954.* Tilburg, 1954).

The archives of the MSC have been arranged as follows:

a. "Zaken-archief" (archives of matters), subdivided into: Official documents and correspondence of the provincial administration with the general administration and with ecclesiastical and secular authorities.

   Papers concerning and correspondence with the several mission areas. Each area arranged seperately according to year. Besides papers concerning Australia (Sydney), Brazil, Oceania (Rabaul) and the Philippines (Mindanao) this heading includes concerning the Dutch East Indies/Indonesia papers about the following organizations, territories and subjects: Centraal Missie-Bureau Djakarta (Central Mission Bureau Djakarta); Indische Missie-Vereeniging (East Indies Mission Association) IVM; native clergy; Kei Islands (i.a. civil administration, reports, correspondence); New Guinea (i.a. arrangement mission areas, Nieuw-Guinea-Comité (New Guinea Committee), mission and outsiders, civil administration, reports, correspondence, accounts, vicars, prefects, superiors (Merauke), geography, miscellanies); Tanimbar Islands (i.a. civil administration, reports, correspondence); Celebes, head station Menado (i.a. reports, correspondence, accounts); Central Java (i.a. Java mission and the government, reports, correspondence, accounts).

   Papers concerning and correspondence with the houses and works in the Netherlands and other ecclesiastical provinces.

   Size of the "Zaken-archief": *C.* 50 running metres.

b. "Personen-archief" (archives of persons), including personal notes of the members of the Dutch province of the MSC. (*C.* 7 running metres).

c. "Archief-bibliotheek" (library for the archives), consisting of documentation pertaining to the work of the MSC, including books, pamphlets etc. written by members of other provinces of the congregation; congregational expences; manuscript-essays, -articles etc. mainly of members of the Dutch province (i.a. concerning geography, ethnology and languages of the various mission areas); periodicals, which are being published by the Mission House or to which members are or were contributors; congregational periodicals; publications of and about the various mission areas; documentation consisting of newspaper cuttings, alphabetic according to personal names. (*C.* 70 running metres).

d. "Foto-archief" (archives of photographs). See no. 725. The total size of the archives: *C.* 130 running metres.

**Maps** [723]

The collection has been arranged as follows:

a. New Guinea/Irian Barat, 31 maps, of which 8 manuscript: 2 outline maps, linguistic areas of New Guinea by P. Drabbe, mission map of West New Guinea (Vogelkop), mission map of the peninsula Onin (West New Guinea, Gulf of Maccluer), Mappi territory 1954, South New Guinea with Frederik Hendrik Eiland (Kolepom; Dolak), mission map of South New Guinea. Moreover 13 photocopies of manuscript maps concerning mission areas in South New Guinea, viz. outline map, Merauke, Tanah Merah, Kimaam (Frederik Hendrik Eiland), Kepi and Arare, Bapai Rivier, Merauke-Okaba, Mappi Rivier, Aworkat, Bade, Agats, Aboge and Wildeman Rivier, Mindiptana, Getentiri.

b. Kei Islands, Tanimbar, South Moluccas, 11 maps of which 5 manuscript, viz. Kei, Tanimbar and Jamdena, small mission map Jamdena, Tanimbar, Aru.

c. Celebes/Sulawesi Utara, 5 maps of which 1 manuscript, viz. Sangihe and Talaud.

d. Philippines, 1 atlas and 13 maps of which 1 manuscript.

e. Brazil, 1 atlas and 11 maps of which 4 manuscript.

f. Other maps, 5 maps of which 1 manuscript, viz. map of the work of the MSC all over the world.

Total amount: 2 atlases and 71 maps.

**Drawings, paintings** [724]

Besides about ten paintings, the Mission House possesses a collection of *c.* 80 coloured drawings by P. Vertenten concerning different population groups in New Guinea, the musical instruments used by them and garments and objects used for ceremonial dances, *c.* 1930.

**Photographs** [725]

The collection consists of some thousands of photographs concerning all the areas where the MSC was active. In principle the arrangement is as follows:

a. Photographs of persons (alphabetic to name) with memorial pictures.

b. Photographs of various houses.

c. Photographs of various mission areas. Concerning the Dutch East Indies/Indonesia i.a. population, residential districts, buildings, events in i.a. Kei, New Guinea, Tanimbar, North Celebes and Central Java.

Entry: The collection is only partly arranged.

Size: *C.* 10 running metres.

## Minderbroeders Nederland (Minors in the Netherlands), OFM 1853

13 Deken Roesstraat, 3581 RX Utrecht.
Admission by appointment.
Research finished in 1979.

**Orde der Minderbroeders Franciscanen (Minors Franciscan Friars), 1853–** [726]

The order was founded in 1210; from a Low German province, founded in 1217–1239, a Dutch speaking province was branched off in 1529; 1853 restoration of the Dutch province; 1870 mission in China; 1889 Brazil; 1931 West Java; 1935 British India (Karachi and Bangalore); 1936 New Guinea (Hollandia).

The archives consist mainly of the residue of the activities of the Franciscans after 1910 within and out of the Netherlands, and in addition of deposited archives. For the history of Asia may be relevant the "missie-archief" (mission archives) consisting of correspondence of the mission secretary with regular superiors overseas, local bishops overseas, individual Friars and a limited number of documents of the management; moreover the archives contain the deposited papers of the bishop of Luanfu (Shansi) and finally the personal inheritances of the Friars and reports of the meetings of the board may contain passages about mission.

Entry: The arrangement and inventory of the archives is still in course of preparation, with the result that part of the archives are not, or on a very limited scale, accessible at the moment.

Size: Total size *c.* 150 running metres, "missie-archief" *c.* 6 running metres, deposited "Chinese archivalia" (Chinese archival material) *c.* 5 running metres.

Permission is necessary for consultation of the papers.

See also J.A. de Kok. Het archief van de Nederlandse Provincie van de Orde der Minderbroeders Franciscanen. In *Jaarboek van het Katholiek Documentatie Centrum*. 1973, p. 185-189.

*Sint Anthonius*. 1895–*c*. 1927.

*Neerlandia Seraphica*. 1927–1977.

**Photographs** [727]

Separate photograph archives are in an early stage of building up.

# CHAPTER IX

# LIST OF SOME OTHER RELEVANT INSTITUTIONS

In this list are included the addresses of institutions which in principle do not keep any archival material, though there is a possibility that one may find a few documents and manuscripts there. However, the institutions owe their value in the first place to collections of a different kind, such as collections of objects, drawings, pictures, paintings, maps and documentation.

For the compilation of the list were used:

*Bibliotheek- en documentatiegids voor Nederland, Suriname en de Nederlandse Antillen;* 3rd ed. 's-Gravenhage, 1971.
*De Nederlandse musea.* 5th ed. 's-Gravenhage, 1970.

After the addresses below, the nature of the collections is indicated in catchwords between brackets. The addresses are arranged alphabetically in order of place names.

**Institutions** [728]

*Amsterdam*

a. Nederlands Filmmuseum (Dutch Film Museum). 13 Paulus Potterstraat, 1071 CX. (Film, photographs)

b. Rijksmuseum (National Museum); departments of "Aziatische Kunst" (Asian Art) and "Nederlandse Geschiedenis" (Dutch History). 21 Hobbemastraat, 1071 XZ. (Objects, paintings, drawings, pictures)

c. Rijksmuseum Nederlands Scheepvaartmuseum (National Museum Dutch Maritime Museum). Lands Zeemagazijn, 1 Kattenburgerplein, 1018 KK. (Objects, maps, drawings, pictures)

d. Rijksprentenkabinet (National Gallery of Prints). 1 a Jan Luykenstraat, 1071 CJ. (Drawings, pictures)

e. Stichting Cultuurgeschiedenis van Nederlanders Overzee (Foundation Cultural History of Dutchmen Overseas). 21 Hobbemastraat, 1071 XZ. (Documentation)

*Breda*

f. Volkenkundig Museum "Justinus van Nassau" (Ethnographical Museum "Justinus van Nassau"). 13 Kasteelplein, 4811 XC. (Objects)

*Delft*

g. Nederlands Legermuseum "Generaal Hoefer" (Dutch Army Museum "Generaal Hoefer"). 1 Korte Geer, 2611 CA. (Objects, drawings, pictures)

h. Topografische Dienst (Ordnance Survey). 9 Westvest, 2611 AX. (Maps)

i. Volkenkundig Museum "Nusantara" (Ethnographical Museum "Nusantara"). 4 St. Agathaplein, 2611 HR. (Objects)

*Deventer*

j. Museum voor Tropische Landbouw en Volkenkunde (Museum for Tropical Agriculture and Ethnography). 184 Brinkgreverweg, 7415 CH. (Objects)

*The Hague*

k. Iconografisch Bureau (Iconographical Bureau). 26 Prins Willem Alexanderhof, 2595 AB. (Pictures)

l. Informatiebureau van het Nederlandse Rode Kruis (Information Bureau of the Dutch Red Cross). 27 Prinsessegracht, 2514 AP. (Card index of civilians, prisoners of war and repatriates from Indonesia)

m. Rijksbureau voor Kunsthistorische Documentatie (National Bureau for Art Historical Documentation). 7 Korte Vijverberg, 2513 AB. Department of Topography, 64 Kazernestraat. (Documentation)

*Leeuwarden*

n. Gemeentemeuseum "Het Princessehof" (Municipal Museum "Het Princessehof"). 11 Grote Kerkstraat, 8911 DZ. (Objects, drawings)

*Leyden*

o. Centrum voor de Geschiedenis van de Europese Expansie (Centre for the History of European Expansion). 4 Middelste Gracht, 2312 TW. (Documentation)

p. Rijksmuseum voor Volkenkunde (National Museum for Ethnography). 1 Steenstraat, 2312 BS. (Objects, paintings, drawings, pictures, photographs)

q. Sinologisch Instituut (Institute of Sinology). 33 1e Binnenvestgracht, 2312 BB. (Documentation)

*Rotterdam*

r. Maritiem Museum "Prins Hendrik" (Maritime Museum "Prins Hendrik"). 48 Scheepmakershaven, 3011 VC. (Objects, maps, drawings, pictures)

s. Museum voor Land- en Volkenkunde (Museum of Geography and Ethnography). 25 Willemskade, 3016 DM. (Objects, pictures, photographs)

t. Stichting Atlas van Stolk (Foundation Atlas van Stolk). 12 Aelbrechtskolk, 3024 RE. (Maps)

*Tilburg*

u. Nederlands Textiel Museum (Dutch Textile Museum). 23 Gasthuisring, 5041 DP. (Japanese fabrics, batik, ikat)

v. Nederlands Volkenkundig Museum (Dutch Ethnographical Museum). 24 Kloosterstraat, 5038 VP. (Objects, pictures)

*Utrecht*

w. Centraal Register van Familiearchieven (Central Register of Family Archives). Rijksarchief (State Archives) Utrecht, 201 Alexander Numankade, 3572 KW. (Register of archival documents present with private persons).

NB. Moved in 1980 to the Algemeen Rijksarchief (General State Archives), 20 Prins Willem Alexanderhof, 2595 BE The Hague.

x. Geografisch Instituut der Rijksuniversiteit (Institute of Geography of the State University). Transitorium II, 2 Heidelberglaan, 3584 CS. (Maps, Centrale Catalogus voor Kartografie (Central Catalogue of Cartography))

y. Munt- en Penningkabinet van 's-Rijks Munt (Numismatic Cabinet of the State Mint). 90 Leidseweg, 3531 BG. (Coins)

z. Stichting Film en Wetenschap (Foundation Film and Science). 29 Hengeveldstraat, 3572 KH. (Films, photographs, sound archieves)

# INDEX

# Introduction

This index is based principally on the inventories and lists of the archives and collections mentioned in this guide. The numbers in the index refer to the numbers assigned to these archives and collections in the text. The numbers in italics refer to particular archives or collections as a whole.

In the compilation of the index I was confronted with the following remarkable features:

The great variety in the entries, ranging from very rough lists to extremely detailed inventories. As a result the roughly described archives are much more poorly represented in this index than those with detailed inventories.
Generally speaking, most of the names and subjects listed in this index also occur in the archives of Koloniën (Colonial Affairs, no. 25 and 276), though this is not evident from its inventory. The systematic and alphabetic indexes of the archives themselves will have to provide the decisive answer in this matter. To avoid underrepresentation of the archives of Koloniën, I have also placed the number of Koloniën after several entries in this index, though only there where it is certain that the name or subject concerned occurs in these archives, even if this is not apparent from the inventory.

The changing terminology, nomenclature and orthography in the course of one century and a half. I tried to introduce some uniformity here as appears below, from my observations on the adaptation of the entries to this index.

The defective or even completely missing indexes to the inventories. For this reason the present index has been based on the text of the inventories mentioned in this guide. So it is very well possible that a certain item cannot be found by means of an index to an inventory, while it nevertheless is described in the actual inventory.

## Adaption of the entries

*General.* The entries of the index are based principally, as indicated above, on the contents of the documents as described in the inventories, viz. by names of persons, organizations, geographical areas and subjects. Besides, some entries have been inserted which indicate the nature of the document, such as a diary, map, photograph, picture, oriental manuscript etc. The whole has been arranged as a "dictionary catalogue" in one alphabetic arrangement.

I have tried to avoid the mention of more than 40 items after any entry. Those entries which yielded more than 40 items were provided with a sub-classification according to period and/or geographical area. In the case of entries which referred to practically every archive and collection mentioned in this guide (e.g. "Indonesia"), the term "passim" has been used.

See also the heading "classification according to period".

*Names of persons*. For the listing of Indonesian personal names, my guide has been: J.N.B. Tairas. *Pedoman tajuk nama-nama pengarang Indonesia*. Jakarta, 1973. For non-Indonesian names was used *Names of persons;* national usages for entry in catalogues; 3rd ed. London, 1977. IFLA-edition.

The practical result for this index is that European names are arranged alphabetically according to the first element of the family name, with a cross-reference, where relevant, from the next element of the family name to the first. Chinese names are arranged alphabetically according to the first element of the name as it occurs in the inventories, with a cross-reference from the second element to the first. The remaining names are arranged alphabetically according to the last element of the name, with cross-references from the first element to the last. Where it was uncertain whether a name referred to a person or a subject, it was alphabtically arranged according to the first element, and no cross-references were given.

In the case of European names titels have been left out; but titles have been included with the other names. In the case of these non-European titles (where known) cross-references have been given to the names of persons, e.g.: "Raden. *See* under the names of the persons".

Names after the elements "bin" and "gelar" were considered as not being part of the actual name. So there are no cross-references from this part of the name to the part coming before the elements "bin" and "gelar".

With respect to Indonesian names it has sometimes been necessary to standardize the spelling. Cross-references have been made from the various possible spellings to that chosen for this index. For example, Diponegoro has been given the following cross-references: "Diepo Negoro. *See* Diponegoro"; "Dipanegara. *See* Diponegoro"; and "Negoro, Diepo. *See* Diponegoro".

Finally, there are cross-references from pseudonyms to the official name. The pseudonyms used by a particular person will be found in brackets after the name.

For further comments on the spelling of personal names see the headings "orthography" and "alphabetic arrangements".

*Names of organizations*. Organizations which were found under more than one name have been entered under the name which seemed to occur most frequently. Cross-references have been given from the other names to the name chosen for this index, with the other names which have been found being placed in brackets after the chosen name. For example: "Koninklijk Instituut voor de Tropen (Royal Institute for the Tropics). *See* Koloniaal Instituut". After the entry Koloniaal Instituut (Colonial Institute) one then finds the name Koninklijk Instituut voor de Tropen in brackets.

For the purposes of this index a distinction has been made between ministries and departments, in the sense that "ministries" are government institutions domiciled in the Netherlands, and "departments" government institutions domiciled in Indonesia.

For the spelling of the names of organizations see the heading "orthography".

*Geographical names.* For geographical names I was completely dependent on the inventories. For example, from an inventory describing papers about Padang as papers concerning the West Coast of Sumatra, only the entry "Sumatra's Westkust" (West Coast of Sumatra) could be included, instead of the entry "Padang", in order to avoid confusion in the institutions in charge of the papers.

For the processing of Indonesian geographical names I used: A.J. van der Aa. *Aardrijkskundig woordenboek der Nederlanden.* Gorinchem, 1839–1851. 13 vol.; *Atlas van Tropisch Nederland.* Batavia, 1938; and *Encyclopaedie van Nederlandsch-Indië;* 2nd ed. 's-Gravenhage etc., 1917–1939. 4 vol. and 4 supplements. For the remaining geographical names I used: *Times atlas of the world;* comprehensive ed. London, 1974.

Consequently, the most frequently occurring name has been used in each case in this index. From the other names there are cross-references to the name chosen, with the alternative names being mentioned in brackets after the latter. For example: "Aceh. *See* Atjeh"; "Achin. *See* Atjeh" with as entry "Atjeh (Aceh; Achin)".

For further comments on the spelling of geographical names see the heading "orthography".

*The subjects.* Only after long hesitation did I decide to include a number of subjects in this index after all. For in the processing of the subject entries I was completely dependent on the terminology used in the different inventories, which moreover appeared to have changed considerably over the past 150 years. For example, what was referred to as "financial affairs" during the 19th century, appeared to have become "economics" in the 20th century. So, for the subject entries, quite contrary to common practice, I have opted not for the "narrowest" but for the "broadest" entries. In this way I was able to devise an entry like "Financial and economic affairs". For the working out of the subject entries use was made of *Macrothesaurus for information-processing in the field of economic and social development;* new English ed.; prepared by J.Viet. Paris, 1978. OECD-publication.

Subjects occurring in more than 40 archives and collections have been sub-classified according to period and/or geographical area. I was even so unable to prevent some sub-classifications comprising more than 40 items. See, for example, "Military affairs: Asia: World War II". Subject entries without any indication of region in most cases pertain to Indonesia.

*Classification according to period.* A classification according to period has been applied to the geographical and subject entries. As was mentioned above, this has only been done in those entries with more than 40 items. This classification has been based on the period covered by the relevant archives, if possible in combination with the history of the geographical area or subject concerned. In view of this, the classification according to period is not meant to be an exact subclassification, but is intended rather to give a rough idea of which archives and collections are likely to provide data about a certain region or subject for a specific period.

*Orthography.* Whereas in the textual part different spellings may by found, for the index a uniform spelling system has been applied, namely the one in use around 1950. The consequence of this for the names of the Dutch organizations etc. has been that, for example, the original "ee", "oo" and "aa" have been replaced in the index by "e", "o" and "a". So "vereeniging" in the text has become "vereniging" in the index. The word "den" has become "de", and the ending "ch" has been dropped, so that "Nederlandsch" has become "Nederlands" in the index. All this has been done to avoid confusion in the alphabetic arrangement.

As a result of the adoption of this rule the revision of the Indonesian spelling system of 1972 is not applicable for the Indonesian names. In other words, the "c" of modern Indonesian is "tj" for the index (so "Aceh" can only be found under "Atjeh"), "j" = "dj", "kh" = "ch", and "y" = "j". Only the "oe" in Indonesian names has been changed into "u" in the index.

*Alphabetic arrangement.* As a guide for the alphabetic arrangement I have used *Regels voor de alfabetische rangschikking van catalogustitels;* 8th ed. 's-Gravenhage, 1961. The alphabetic arrangement is according to capital letters, so that words like "de", "van" and "voor" (printed without capitals) have been ignored. The same applies to the English indications following the entries in brackets. The "ij" and "y" have been arranged interchangeably.

*Translation.* Though the textual part has been translated into English, I have decided against a translation of the greater part of the index, for the following reasons. It is impossible to translate every name and term from the various Asian countries into adequate English without mobilizing half a university. Secondly, it is often impossible to translate specific terms correctly. So "adat" has remained "adat" for the index, and has not been translated into "customs" or "customary law". Finally, I feel that the translation of names of organizations etc. is confusing. For example, no-one will look for "Sarekat Islam" under "Union of Islam" or "Islamic Union" as "Sarekat Islam" is a well-known term to everyone.

English translations of the names of institutions have been inserted in brackets after the original names in the index only where this was necessary in my opinion. In these cases cross-references have been given in the index from the English translation to the original name. As regards the other institutions only an indication of what the entry concerned roughly covers has been given in English, and only then if I knew what this was.

The difference between translations and indications is shown formally by means of capitals. For translations capitals have been used, e.g.: "Koninklijk Instituut voor Taal-, Land- en Volkenkunde (Royal Institute of Linguistics and Anthropology)", while indications have no capitals, e.g.: "Sarekat Islam (association)".

## Conclusion

Summarizing, I conclude with the statement that this index is anything but complete. The user is advised always to consult several entries in order to overcome its many gaps. It should be obvious that it is quite impossible to point out (and especially to point out consistently) everything that is relevant for 150 years of Asian and Oceanic history. So this index does not pretend to be more than just a preliminary entry to the relevant archives and collections in the Netherlands. Naturally, the entries on the spot are the researcher's most essential aids.

Bei Fragen zur Produktsicherheit wenden Sie sich bitte an:
If you have any questions regarding product safety,
please contact:

Walter de Gruyter GmbH
Genthiner Straße 13
10785 Berlin
productsafety@degruyterbrill.com